THE GEOLOGY OF BUILDING STONES

John Allen Howe

With an introduction by David Jefferson

DONHEAD

First published in 1910 by Edward Arnold, London

This reprinted edition © Donhead Publishing Ltd 2001
Simultaneously published in the United Kingdom and
Massachusetts, USA by Donhead Publishing Ltd.

The publisher requests the copyright holder of this title to contact
them at the address below.

Donhead Publishing Ltd
Lower Coombe
Donhead St Mary
Shaftesbury
Dorset SP7 9LY
Tel. 01747 828422
www.donhead.com

New introduction to this 2001 edition © David Jefferson

ISBN 1 873394 52 7

A CIP catalogue for this book is available from the British Library

Printed in Great Britain by J. H. Haynes & Co. Ltd. Sparkford,
Somerset

Donhead Publishing would like to acknowledge the help of Mike
Chrimes at the Institute of Civil Engineers, London, in loaning an
original copy of the work for this facsimile reprint.

British Library Cataloguing in Publication Data

> Howe, John Allen
> Geology of building stones
> 1. Building stones 2. Building stones – Identification
> I. Title
> 691.2
> ISBN 1873394527

Library of Congress Cataloging in Publication Data
A catalog record for this book has been requested.

Introduction to the 2001 Edition

John Allen Howe O.B.E. started his career as a field geologist, first at the Royal College of Science in London, and then with the Geological Survey of Great Britain. However, his preference was for applied geology rather than fundamental research. In 1902 he became curator and librarian of the Museum of Practical Geology, a post he held until 1920 when he was appointed assistant to the director of the Geological Survey, retiring in 1931. It was with the practical aspects of geology that he became most well known and he was particularly involved in the work of the Institution of Mining and Metallurgy, of which he was president from 1942 to 1944 and who awarded him honorary membership in 1949. Even though his later interests tended towards minerals exploration and mining, it is his earliest book, *The Geology of Building Stones*, for which he is now probably best known.

Although, it is perhaps the medieval cathedrals and castles, together with the stately country houses of the sixteenth and seventeenth centuries, which first come to mind when considering stone buildings, stone was still a major material for both prestigious and domestic projects at the start of the twentieth century. The nineteenth century had been a period of extensive stone building. Projects from the period ranged from Barry's Palace of Westminster and his Royal Institution in Manchester, both built in the first half of the century, to the great Town Halls of Manchester and other northern cities built in the second half. Even the first example of a 'New Town' in Britain, Saltaire in Yorkshire, had been built of stone. By the time Howe joined the Geological Survey, brick and terracotta were being extensively used in addition to stone. Even so, architects such as Sir Edwin Landseer Lutyens, W. R. Lethaby and R. N. Shaw were still using stone. J. L. Pearson's cathedral at Truro, built of Bath stone and local Mabe granite, was being completed by his son, and work was about to start on the new Anglican Cathedral of Liverpool, designed by Sir Giles Gilbert Scott and to be built in local sandstone. The use of local stone was still relatively common. For example, Lutyens used Bargate stone from the Guildford–Godalming area for many of his country houses in that part of the country.

Despite this extensive use of stone, there was no reference work available which provided information on the tremendous range of building stones which had been used in Britain and, perhaps more importantly, were still being produced. The appendices B and C provided by Howe, which list the larger operating sandstone and limestone quarries, give some idea of the scale of the industry at the start of the twentieth century. The first comprehensive study on British building stones, Hull's *A Treatise on the Building and Ornamental Stones of Great Britain and Foreign Countries* had been published in 1872. However, this earlier work is a much more general description of a range of foreign, as well as British, building stones. Hull puts great emphasis on granites and marbles but, as the science of petrography was in its infancy, H. C. Sorby only having made his first petrographic thin section in 1849, the terminology used is now rather antiquated and the book is of little interest today, except in a historical context. Some practical information on the better known stones available at the start of the twentieth century could be found in books such as William Purchase's *Practical Masonry*, the fifth edition of which, published in 1904, contained a new brief section on the nature of building stones, together with sections detailing some of the stones available and their prices. A work bringing together what was known about British building stones would, therefore, be a valuable aid to those wishing to use this traditional resource.

As the curator and librarian at the Geological Survey, Howe was in a position to bring together all the information then available, not only from the field mapping of the Survey, but also from the building stone collection housed at the museum. Being concerned with the application of geology, he realised that the information on building stone was required by those involved in using the material. As a result, and as he points out in his preface, the book is designed to fulfil the 'requirements of students of architecture'. Containing a wealth of geological and other scientific and technical data, the book is nevertheless still eminently readable by those with little knowledge of geology. Furthermore, it relates the various stones to actual buildings. This is not only valuable in the historical context, but also allows the stone to be seen in use. As all those who have been involved with building stones will attest, there can be major differences between a stone straight from the quarry and the same material after a period of time in a building. In addition, comments on locations where stone has been used, particularly the rarer ones, can be of great help to conservators and architects when attempting to identify stone in a building.

The effect of pollution, frost and organic growth on the breakdown of stone had been recognised in the nineteenth century, and academic research had been carried out on various aspects of this. In 1861 a Select Committee had been established to examine 'The causes of decay in the New Palace at Westminster', some of the stone in the new building having lasted no more than about ten years. This report appears to have had few if any practical results. Geologists and other scientists were aware of the various factors which could destroy stone and eventually produce soil. The publication of Howe's book was, however, the first time the subject had ever been brought in any detail to the attention of the users of stone. For the first time architects and others were shown that stone is subject to all sorts of harmful mechanisms, from lichens, to frost, to heat, to pollution, and even to polishing 'where loafers congregate and sun themselves'. Despite the fact that this study of stone decay was to be superseded twenty two years later by Schaffer's seminal work on *The Weathering of Natural Building Stones*, the chapter is still a good introduction to the subject. The same can be said for the section on stone testing. In fact, his views on testing were many years ahead of their time. He considers microscopic examination 'affords a most convenient and reliable means for the investigation of the character of a building stone'. There is also a plea for 'some scheme of *Standard Tests*, and some organization of authority for carrying out tests regularly and quickly'. Almost a century after these comments were written, Britain is now starting to obtain its first standard tests for building stone, in the form of European Standards, including one for petrographic description!

Although still a leading work of reference, some of the geological nomenclature is a little dated. Advances in geology have resulted in some changes to stratigraphic names. For example the Carboniferous strata originally termed the Millstone Grit by Whitehouse in 1778 and the Coal Measures, a term proposed by Farey in 1811, are now both part of the Silesian Subsystem. The Lower Freestone, a traditional building limestone within the Middle Jurassic, has now become the Cleeve Cloud Member of the Birdlip Limestone Formation. However, since geological maps and memoirs are updated relatively infrequently, the traditional names still occur in much of the more recent literature. Where the new terminologies are used, the traditional names are invariably provided, in order that reference can be made to earlier studies. In the case of Howe's discussion of the decay of building stone, although his analysis is sound, recent advances have improved our understanding of the decay mechanisms. It is also pertinent to note that there appear to be occasional errors in the naming of locations from which stone was obtained. Those which have been reported are listed below:

Page 154 Cullington should read Cubbington

Page 218 Clipsham is in Rutland, not near Charlbury in Oxfordshire

Page 316 Stockholm is believed to be Skokholm (or Skokeholme) Island off the Pembroke coast (information supplied by Mr Terry Hughes)

Page 318 Chacombe is in Northamptonshire, not in Wiltshire. This incorrect location is actually taken from an earlier reference (information supplied by Mr Terry Hughes)

Page 323 Atford in Wiltshire, or Atforde as it is shown on early maps, is now known as Atworth

Page 324 Lumdon is almost certainly Swindon, this being the only place where the Purbeck is known to have been quarried in the area (information supplied by Mr Terry Hughes)

These occasional errors are possibly attributable to the poor handwriting of some field geologists, whose field notes Howe probably used when compiling the book! It is perhaps unfortunate that Howe did not include maps showing the location of the various stone types. However, the British Geological Survey's *Building Stone Resources Map of the United Kingdom*, published in 2001, despite not showing all the locations mentioned by Howe, may well prove to be a useful companion to his book.

Despite having been written nearly a century ago, the book still remains an extremely important source of information on building stones. Conservators, restorers, architects and civil engineers, will welcome this facsimile reprint as the starting point for any investigation involving the identification of stone in old or historic buildings. It will not necessarily provide instant answers to the type and source of a stone used in a specific building or area, but it will certainly point the investigator in the right direction.

David Jefferson
May 2001

David Jefferson B.Sc.(Hons), Ph.D., C.Eng., F.G.S., F.I.Q., M.I.M.M.
David Jefferson spent twenty years working all over the world as a geologist with Blue Circle Industries before starting his own consultancy in 1985. He now specialises in mineral-based materials such as building stones, lime and cement, which are related to construction. He is currently geological consultant to the Building Conservation and Research Team of English Heritage, his work including identifying building stones, locating replacement stone, advising on the possible effects of surface treatments and cleaning. He has also worked on a number of National Trust properties and advises a wide range of architectural clients, as well as owners of building stone quarries.

ARNOLD'S GEOLOGICAL SERIES
General Editor: DR. J. E. MARR, F.R.S.

THE GEOLOGY OF BUILDING STONES

BY

J. ALLEN HOWE, B.Sc., F.G.S.

MEMBER OF THE COMMITTEE ON BUILDING STONES AND MORTAR, INTERNATIONAL ASSOCIATION FOR THE TESTING OF MATERIALS

LONDON
EDWARD ARNOLD
1910

[*All rights reserved*]

EDITOR'S PREFACE

THIS is the fourth volume of a series of works treating of economic geology, the previous books having been devoted to 'The Geology of Coal and Coal-Mining,' 'The Geology of Ore Deposits,' and 'The Geology of Water-Supply.'

A work on 'The Geology of Building Stones,' dealing with the great advance made in their study both at home and abroad of recent years, was much needed, and the author of the present work, under whose direction as Curator of the Museum of Practical Geology in London the collection of building stones in that museum assumed its present shape, is peculiarly qualified for the task of writing the work.

J. E. MARR.

AUTHOR'S PREFACE

In the following pages an attempt has been made to gather together some of the facts about the Geology of Building Stones, mainly with a view to the requirements of students of architecture.

Special attention has been given throughout to the materials found in the British Isles; but it has been thought desirable to add brief references to some of the stones of other countries.

Naturally, in dealing with a subject covering so wide a field, selection and much condensation have been inevitable. Limitations of space have prevented the consideration of Ornamental Stones, and the introduction of references to the very numerous sources of information.

The writer's thanks are due to Professor T. Hudson Beare and Mr. W. R. Baldwin-Wiseman for permission to quote certain tables from their exceedingly valuable papers, and to the Councils of the Institution of Civil Engineers and of the Surveyors' Institution for kindly giving their consent.

Thanks are due also to Professor Buckley for permission to reproduce Fig. 28; to the Controller, His Majesty's Stationery Office, for Figs. 11, 19, 20 and 22; to the United States Geological Survey for Fig. 16, from

AUTHOR'S PREFACE

Dr. Nelson Dale's important Bulletins; to Mr. T. C. Hall for assistance with some of the photographic plates; and especially to Professor J. Hirschwald, of Charlottenburg, for generously allowing the reproduction of several figures and tables from his monumental work, 'Die Prüfung der natürlichen Bausteine.'

In common with all who are interested in the subject, the writer is indebted to Professor McKenny Hughes and Mr. John Watson for the fine collection of building stones now forming at Cambridge.

Finally, he wishes to record his indebtedness to his colleagues, and the many architects, quarry-owners, and merchants who have assisted him with information and advice, and particularly to Dr. J. S. Flett, who has most kindly looked through the proofs, and to Mr. H. B. Woodward and Mr. W. Brindley, whose stores of knowledge have always been freely opened and freely used.

J. A. H.

JERMYN STREET, S.W.,
November, 1910.

CONTENTS

CHAPTER	PAGE
I. INTRODUCTORY	1
TABLE OF STRATA	11
II. MINERALS	13
III. IGNEOUS ROCKS—GRANITE	35
IV. IGNEOUS ROCKS OTHER THAN GRANITE	82
V. SANDSTONES AND GRITS	112
VI. LIMESTONES	173
VII. SLATES AND OTHER FISSILE ROCKS	273
VIII. MISCELLANEOUS BUILDING STONES	327
IX. THE DECAY OF BUILDING STONE	333
X. THE TESTING OF BUILDING STONES	362
MOHS'S SCALE OF HARDNESS	411
APPENDICES:	
A. GRANITE QUARRIES	412
B. CLASSIFIED LIST OF LARGER SANDSTONE QUARRIES	416
C. CLASSIFIED LIST OF LARGER LIMESTONE QUARRIES	424
D. LIST OF CHIEF SLATE QUARRIES	430
E. SOME USEFUL BOOKS	433
INDEX	436

LIST OF PLATES

PLATE		FACING PAGE
I.	MINERALS	14
II.	IGNEOUS ROCKS	48
III.	PHOTOMICROGRAPHS OF SANDSTONES	122
IV.	PHOTOMICROGRAPHS OF LIMESTONE — (A) OOLITE, KETTON STONE; (B) CRINOIDAL AND SHELLY, FROM HOPTON WOOD	192
V.	PHOTOMICROGRAPH OF LIMESTONE, HAM HILL SHELLY	206
VI.	PHOTOMICROGRAPH OF LIMESTONE, BATH OOLITE, BOX GROUND	234
VII.	PHOTOMICROGRAPHS OF VARIOUS LIMESTONES	262
VIII.	WEATHERING OF PORTLAND STONE	350

LIST OF MAPS

MAP		PAGE
I.	IGNEOUS ROCKS OF ENGLAND AND WALES	71
II.	TRIASSIC, DEVONIAN ORDOVICIAN AND CAMBRIAN SYSTEMS OF ENGLAND AND WALES	129
III.	GEOLOGICAL MAP OF IRELAND	134
IV.	GEOLOGICAL MAP OF SCOTLAND	137
V.	CRETACEOUS AND CARBONIFEROUS SYSTEMS OF ENGLAND AND WALES	142
VI.	JURASSIC, PERMIAN, DEVONIAN AND SILURIAN SYSTEMS OF ENGLAND AND WALES	189
VII.	SLATE DISTRICTS OF ENGLAND AND WALES	292

THE GEOLOGY OF BUILDING STONES

CHAPTER I

INTRODUCTORY

THERE is no help: sooner or later, in the course of practice, the architect or engineer will have the need of some geological knowledge forced upon him. Force, indeed, is seldom required; for brought as he is into direct contact with geological problems, and having by the nature of his training developed an inquiring mind, the architect or engineer has often become a willing student of the science.

The influence of the geological structure of a country upon the character of its inhabitants, upon their industries, and upon the location of their towns and villages, hardly needs to be elaborated. But more definitely the influence of geological environment makes itself felt in the architectural ideals of a people and their mode of expression. Here I may quote from the 'Stones of Venice,' not because Ruskin's geology is sound—it is bad —but because of the truth underlying his statement. He says, speaking of monolithic shafts: 'It is clearly necessary that shafts of this kind (we will call them, for convenience, *block shafts*) should be composed of stone not liable to flaws and fissures; and, therefore that we must no longer continue our arrangement as if it were always possible to do what is to be done in the best way; for *the style of a national architecture may evidently depend, in a great measure, upon the nature of the rocks of the country.*

'Our own English rocks, which supply excellent building stone from their thin and easily divisible beds, are for the most part entirely incapable of being worked into shafts of any size, except only in the granites and whinstones, whose hardness renders them intractable for ordinary purposes;—and English architecture therefore supplies no instances of the block shaft applied on an extensive scale; while the facility of obtaining large masses of marble has in Greece and Italy been partly the cause of the adoption of certain noble types of architectural form peculiar to those countries, or, when occurring elsewhere, derived from them. . . . The shaft built of many pieces is probably derived from, and imitative of, the shaft hewn from few or from one.

'If, therefore, you take a good geological map of Europe, and lay your finger upon spots where volcanic influences supply either travertine or marble in accessible and available masses, you will probably mark the points where the types of the first school have originated and developed. If, in the next place, you will mark the district where broken and ragged basalt and whinstone, or slaty sandstone, supply materials on easier terms indeed, but fragmentary and unmanageable, you will probably distinguish some of the birthplaces of the derivative and less graceful school. You will, in the first case, lay your finger on Pæstum, Agrigentum, and Athens; in the second, on Durham and Lindisfarne.'

This, in a sense, is true, but other and more complex influences have usually been at work; yet in spite of many exceptions, abundant evidence of the reality of the interaction between geology and architectural forms may be found without even crossing the Channel.

It is surprising to see how sharply and clearly the older domestic and often the ecclesiastical architecture of this country corresponds with the local geological formations.

Look for example, at the Chalk country of the southeast of England; here is the home of flint decoration, best developed in the eastern and southern counties, where this material is most readily obtained. But even where the Chalk has disappeared, and the residual flint, in the later gravels, is all that remains of it, these remnants are still employed in a similar way, as may be seen on the fringes of the Haldon Hills in Devonshire. In the same county, too, may be seen those charming structures the 'cob-walled' farmsteads, only too rapidly passing away. These are built of *adobe*, or earth moulded with stones and straw or hair. They are found also in Cornwall and Somerset, always on the tracts where the decomposed Devonian or Carboniferous shales and slates give rise to surface deposits of stiff greyish clay; or on the tracts of New Red marl, from which cob walls of bright red hue were fashioned. How closely the material follows the geological formation of the ground is shown in a striking manner at Hatherleigh, where on the narrow strip of New Red marl, the extreme limit of a long outcrop stretching out from the east, we find the bright red houses on a belt of rock only a few hundred yards wide, surrounded on either side by cottages of the grey and buff of the Culm.

In a similar way the picturesque half-timbered houses of Cheshire, Worcestershire, Salop, and Gloucestershire, and some of the adjoining counties, are found on the New Red marls or the Old Red marls, which yield material of the same kind, in earlier times for cob walls, and later for bricks.

The limestone and grit districts of the Carboniferous

are equally well distinguished by their own type of building; so is the slate country of Wales, Cornwall, and the Lakes, and the granite country of the West of England and parts of Scotland; and this is true not only as regards general appearances, but in many of the details of construction.

The same close relationship between the buildings and the geological characters is to be seen with equal clearness during a rapid traverse of Europe by rail, no matter what direction be taken.

There are again many minor points affecting the security of dwellings and other structures upon which the knowledge of the geologist is necessary to the architect, but which space forbids us to consider here.

For example, we find the neglect of 'hill creep' resulting in yawning gaps in boundary walls, leaning gateposts, and cracking houses, when a little more depth to the foundation or the addition of an occasional buttress would have insured safety. This steady, slow, but unceasing, descent of the upper layers of all soft rocks, such as shales and clays, towards the bottom of the slope should always be counteracted by greater depth and solidity in the foundations.

Besides the slow downward creep of the surface layers, there is often a tendency, in shaly rocks and clays or among hard rocks resting on clays or shales and dipping towards the valley, for larger masses to move forward in the form of slips.

In the following chapters an attempt has been made to place before the student some of the salient geological characters of building stones; their mode of occurrence and physical and chemical properties.

Here it is necessary that the student should be warned:

INTRODUCTORY

These pages do not undertake to give any adequate explanation of the elementary principles of geology. Thus, although there is a short section on the minerals in building materials, the student must turn to a treatise on Mineralogy if he is to grasp the meaning and technology of crystalline form and habit, and the significance of the association of one mineral with another. Again, although there is a brief introduction to the igneous rocks, he must study some work on Petrology if he is to follow satisfactorily the descriptions of the rocks in thin slices under the microscope and in hand specimens, and he must supplement this by referring to some book on General Geology in order to understand the various relationships of these rock masses in the field and their mode of formation.

Similarly, in the case of the stratified sedimentary rocks, the student may learn from the general textbooks how they are formed and solidified, how they and the igneous rocks are bent, faulted, or broken; how they may be altered or metamorphosed; how they are worn and carved by the agencies of weathering into hills and valleys, plains and escarpments.

It is not, however, from books that geology is to be learned; nothing but personal acquaintance with the things will put the facts and the arguments founded upon them in their true perspective; and this kind of knowledge it is not so difficult to acquire, though it takes time.

It is easy enough to describe a sandstone so that it will be distinguishable from a limestone, and the two stones could be recognized from the description by one who had never seen them before; but to describe twenty sandstones in a way that would lead to their immediate recognition is a task demanding much time and more skill than the writer possesses. Yet how to recognize a stone at sight from amongst a host of others of the same kind is just one of the things that the architect wants to know. The books will not tell him; but once he has seen the stone long enough to fix its appearance in his mind, and has noted the 'feel' of it, it is unlikely altogether to escape him again.

Having become accustomed to the stone, it may then be profitable to look up the microscopic structure, the chemical composition, the physical properties, etc., as set down in the books. In this way the student's own experience and the observations of others may be welded into a stock of sound information.

'Learn to know about your materials,' said Mr. R. W. Schultz recently in addressing a society of architects; 'go to a quarry to see stone actually quarried, go to a mason's yard to see it being dressed, try to do some yourself if he will let you—see it built into position, look at older buildings where the same stone has been used, see how it weathers under different conditions, whether the mouldings decay rapidly (if so, it should not be moulded), what form of surface wears best, and you will find lots of wrong uses and some right ones; but before you have done you will know, or ought to know, what to avoid doing. Learn something in a similar way about limes, sands, wood, plaster, paint—in short, everything that goes to make a building."

More than one President of the Royal Institute of British Architects, and other well-known authorities in the profession, have repeatedly given similar advice.

By all means let this advice be carried out on every opportunity, but even then the busy student will have small chance of seeing more than a few of the better-known quarries during the period when there are so many other urgent calls on his attention. It is hoped that the brief geological and topographical notes on the building stones which follow may be of assistance to the student in

INTRODUCTORY

supplementing the knowledge he is able to gain at first hand.

Now for a word of warning to the seasoned practitioner who may be tempted to search these pages for some information respecting an unfamiliar stone. It is greatly to be feared he will often be disappointed. He would naturally like to find clear descriptions of all the kinds of stone he is likely to meet in the pursuance of his art; he might like to know where they are to be found, in what sizes they may be obtained, how much they cost, for what purposes they are suitable or unsuitable; he might like to know their chemical and physical properties, and their behaviour when subjected to all kinds of tests; and, finally, he would like to know where examples of the stone may be seen in old buildings or in new, in order that he might observe for himself how it had withstood the supreme test of time and exposure.

In no sense does the present volume attempt to fulfil these ideal requirements; if it makes clearer the geological characters of some of the stones, their relationships and distribution, and exhibits the value of some of their chemical and physical properties in true proportions, that is as much as can be hoped.

The treatment of each group is in the first place geological, and in the second place topographical, and many stones of minor importance have been mentioned, because not infrequently a client insists upon having a stone used from his own estate, and a knowledge of what is likely or unlikely to be found on it may be useful.

In this connection it may not be out of place to mention that geological maps on the scale of 1 inch to the mile are available for practically the whole of Great Britain and Ireland, and that maps on the larger scale of 6 inches to the mile have been prepared to show the geology of many large areas. These maps and their accompanying descriptive memoirs (which are full of local detail) may be consulted by anyone in the library of the Geological Survey and Museum, Jermyn Street, London.

A fairly complete series of British and some foreign building stones is exhibited in the same institution. Some colonial building materials are shown at the Imperial Institute, and a fine general collection is housed in the Geological Museum at Cambridge.

The Testing of Building Stones, a subject of great complexity and difficulty, is dealt with—so far as space permits—in Chapter X.; how far the conclusions there arrived at will meet with acceptance will doubtless depend upon individual experience and inclination. Although many of the tests there indicated are confessedly both reasonable and useful, yet it must be admitted that so far in actual practice—largely, perhaps, because really comparable tests are as yet so few—they are rarely, very rarely, made use of by the architect, the builder, or the engineer. In short, the common sentiments in this connection are akin to those of Omar's Pot. 'Said one, They talk of some strict Testing of us—Pish!'

Take, for example, the simplest and most commonly applied test, the chemical analysis; it reveals, amongst other things, the presence of lime; and how often has it not been reiterated by the learned that 'no stone containing lime should be used in towns.' Even if we limit this prohibition to stones containing carbonate of lime, how does theory square with practice? We find fully three-fourths of the large buildings erected in London during the past few years are either wholly constructed or faced with the deleterious substance, or carry it in the prominent dressings. It will be sufficient to mention the Government buildings in Whitehall and Westminster, the Piccadilly Hotel, the Victoria and Albert Museum, to see how our architects are influenced by this test. Even the national monument to Queen Victoria, precisely the type of structure which all men wish to be permanent, is fashioned out of carbonate of lime.

It is true there are limestones and limestones; but it is not comparative and so-called scientific testing which has hitherto determined the use of one or another; it is a question of custom, appearance, and cost. We know that as towns are to-day, the limestones will certainly, and in some cases rapidly, decay; and it does not take long for all ornament to be obliterated, yet we continue to cut into the limestone elaborate designs which cannot last in perfection for ten years. We know, further, that this will continue to be the fate of limestone buildings until we have learned to consume our own smoke.

The choice of a building stone is often no easy matter. The architect may wish to use a granite for one part of the building, as an emblem of strength, and Portland stone for another, that he may have the beautiful effect of its slow and even decay; but the question of cost may prohibit the use of the first, and a north-east aspect may render unsuccessful the employment of the second.

The limitations imposed by situation must always exert a powerful influence on the choice of stone; much is possible in the country that is unattainable in our smoky towns, both as regards colour and texture. The retention of the stone's natural colour and the mellowing effects of age may be reckoned on and allowed for with far more freedom in country buildings—we may even anticipate with satisfaction

> ' the lichen'd wall
> Whose lizard hues are painted by slow Time;'

but the appearance of these delights on a town building would only call forth the steam-brush or the drag.

Cost is the great arbiter oftener than not; fitness and beauty have perforce to give way before £ s. d.

Fear, the fear of risk, prevents the use of many good stones with which the architect is unfamiliar, and it is here that the absence of reliable authoritative tests is felt, and ready information as to the resources of quarries is needed.

There is no single stone which is perfectly appropriate for all purposes, but there is abundant variety to choose from among the granites, limestones, and sandstones, sufficient to meet every requirement as to strength, colour, and texture: it is the duty of the student to familiarize himself with as many of these as possible, and if he makes his choice by the application of sound principles and common-sense, and with personal care sees that the quality is upheld, there is little danger of subsequent disappointment.

TABLE OF STRATA.

Pleistocene	Old river gravels and loams, estuarine beds, and raised beaches	
	Gravel for ballast, road metal, concrete; building sand, brick earth, silt for cement	
	Glacial deposits yielding building sand, brick earth, gravel, road metal, and occasionally rough walling stone	
Pleiocene	Cromer Forest Beds	Building sand, brick earth, gravel, road metal
	Norwich and Red Crags	—
	Coralline Crag	
(Miocene)	—	—
Oligocene	Hamstead, Bembridge, Osborne, and Headon Beds	Building stone, limet, cement, brick earth
Eocene	Barton, Bracklesham, and Bagshot Beds	Glass sand, road metal, setts, building stone, cement stone
	London Clay	Brick earth
	Oldhaven and Blackheath Beds	Flint gravel
	Woolwich and Reading Beds	Brick and tile clay, building stone, setts
	Thanet Sand	Moulding sand
Cretaceous	Chalk	Lime, cement, whiting, building stone, flint
	Upper Greensand and Gault	Road metal, building stone, hearthstone, brick clay
	Lower Greensand	Building stone, road metal, glass sand, building sand, brick clay
	Wealden Beds	Brick clay, sand, building stone, Sussex marble
Jurassic	Purbeckian	Purbeck marble, Swanage stone, lime, gypsum, road metal, tilestones
	Portlandian	Portland building stone, lime, sand, sandstone
	Kimeridgian	Brick and tile clay, cement stone
	Corallian	Building stone, lime, road metal, brick clay
	Oxfordian	Brick and tile clay

TABLE OF STRATA—*Continued.*

Great Oolite Series	Cornbrash	Lime and local building stone
	Forest Marble and Bradford Clay	Lime and local building stone
	Great Oolite	Building stones, stone tiles, road metal
	Fuller's Earth	—
	Inferior Oolite Series	Building stone, lime, stone tiles, sand
	Lias	Building and paving stone, brick clay, lime, cement
Trias	Rhætic	Building stone
	Keuper	Building stone, brick and tile clay, gypsum
	Bunter	Building stone, sand, road metal, pebble gravel
Permian	Red Marls, Sandstone and Magnesian Limestone	Building stone, brick and tile clay
Carboniferous	Coal - measures (and Culm rocks)	Building and paving stone, stone tiles, brick and tile clay, fire clay
	Millstone Grit	Building and paving stone
	Lower Carboniferous Rocks	Building stone, marble, lime, road metal
Old Red Sandstone	Red Marls and Sandstones	Building and paving stone, brick clay
Devonian	Limestone, Shales, and Sandstone	Building stone, lime, marble, road stone, slate
Silurian	Ludlow Beds	Building stone, lime, slates
	Wenlock Beds	Lime, paving stones
	Llandovery Beds	Building and paving stone, lime, slate
Ordovician	Bala and Caradoc Beds	Lime, jasper
	Llandeilo Beds	Slates
	Arenig Beds	Slates
Cambrian	Tremadoc Beds	Slates
	Lingula Flags	Building stone, paving stone
	Menevian Beds	Slates, paving stone
	Harlech Beds	Building stone, slate, road metal
Pre-Cambrian	—	Road metal and local building stone

Igneous rocks occur on several horizons.

CHAPTER II

MINERALS

THE more we know of the true nature of 'minerals,' the greater our difficulty in defining them in a few plain words. It may be said that they are the materials which constitute the solid crust of the earth; but since some of these solid materials are liquefied occasionally, and even converted to a gaseous condition by local heating, in regions of volcanic activity, or are normally liquid and only solidified by local cooling, as in the case of water in the polar regions, it would be difficult to exclude from the term 'mineral' those portions of the crust which commonly exist in the state of liquids or gases. Even from among the building materials water cannot be altogether excluded, for the Eskimos find in it a convenient and durable 'freestone.' Interesting relations may also subsist between organic bodies and minerals; for instance, the wooden piles which form the foundations of the houses in parts of Nicaragua are found after about fifty years to be converted *in their lower parts* into hard siliceous stone.

When, however, the term 'mineral' is employed in its everyday sense, certain definite kinds of mineral matter are connoted.

The different kinds of mineral substances can be recognized and distinguished by the possession of certain definite characters which are the peculiar property of each kind.

Two distinct properties must be possessed by a mineral substance before it can be recognized as a separate kind; these are its chemical composition and the expression of its peculiar molecular construction. Two mineral substances may have precisely the same chemical composition, but if the form-type differs—due to differences in molecular arrangement—they will not be regarded as one mineral kind. Thus, calcite ($CaCO_3$) and aragonite ($CaCO_3$) have the same composition, but their molecular construction, as expressed by their crystalline form, is different. With the peculiarities of molecular arrangement are associated several properties which serve to distinguish the mineral and give it characteristic optical, electrical, thermal, and elastic features.

One kind of mineral may occur in many minor variations of form, colour, and habit, without losing its essential characters. Minerals may occur with regular external crystalline form as crystals—(Plate I.); or they may have easily recognizable crystalline structure without the external regularity of form (for example, most of the quartz grains in granite)—(Plate II.); or they may be amorphous, without crystalline structure.

Minerals are formed and grow from aqueous solutions, cold or warm, or from molten liquid rock magmas. Some mineral substances have been formed through the mediation of organisms—*e.g.* from plants, coal and diatomaceous earth; from animals, many limestones. A few minerals are deposited by sublimation from a gaseous condition, but they are not of the kind that should appear in building materials.

Through the action of the various agencies in their environment, minerals of one kind are continually being modified or broken up and re-formed as minerals of other kinds.

PLATE I

MINERALS.

Felspar: *a*, *c*, and *d*, crystals (*d* is a twinned form); *b*, a cleavage face, red felspar; *f*, felspar (light) intergrown with quartz (dark) in 'graphic' granite.

Quartz: *e*, crystalline mass; *g*, transparent crystal.   Pyroxene: *h*, Augite.

Amphibole: *i*, hornblende; *j*, actinolite in quartz schist (mica in upper right corner).

Mica: *k*, portion of large crystal.

Calcite: *l*, dogtooth spar; *m*, cleavage rhomb in yellow calcite; *n*, nail-head spar.

MINERALS

Rocks are built up of minerals associated in an infinite variety of ways, and the following brief notes upon some of the more important rock-forming minerals are inserted here for the assistance of the reader, who will find them referred to in some of the subsequent chapters.

To gain any knowledge whatever of the real character of these minerals, it is essential that the student should see and handle the minerals themselves. Their physical and chemical properties, and their mode of occurrence, will be found described in works on mineralogy (see p. 433), but for their appearance and characters when examined in thin slices under the microscope, books on petrology must be consulted.

The number of different kinds of mineral that go to build up the earth's crust is comparatively small, well under one thousand, and of these only a dozen or so are important rock-formers.

Quartz.—Oxide of Silicon (SiO_2). H.= 7. Sp. gr.= 2·65. Crystallizes in the Hexagonal system. Insoluble in acids. Fuses at about 1600° C.; vitreous lustre, conchoidal fracture.

This is the most widely spread and important of the minerals occurring in building stones. It occurs in grains as an essential constituent of all granites and felsites, and as a subordinate mineral in a host of other igneous rocks. It forms the great majority of sands, sandstones, and grits, in the state of grains derived from pre-existing igneous and sedimentary rocks. As a cementing material it is common in many sandstones and shales, and there are few limestones, clays, or slates, in which it does not appear.

In mineral veins which traverse rocks of all kinds, and in hollows and druses, it is frequently found with well-marked crystal faces. (Plate I. and Fig. 1.)

The colour of quartz is subject to a wide range of varia-

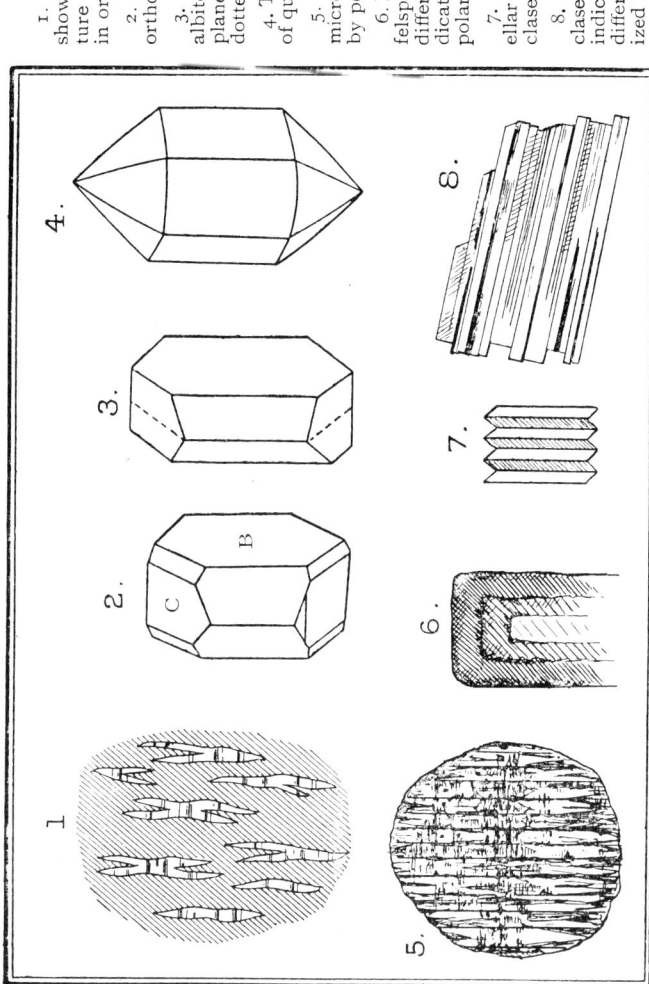

1. Section of felspar, showing 'perthite structure'; plates of albite in orthoclase.

2. Typical form of orthoclase crystal.

3. Typical form of albite crystal; twinning plane indicated by dotted line.

4. Typical crystal form of quartz.

5. Micro-section of microcline felspar seen by polarized light.

6. Diagram of a zoned felspar crystal; zones of different composition indicated by shading — polarized light.

7. Diagram of lamellar twinning in plagioclase felspar.

8. Section of plagioclase felspar; twinning indicated by bands of different tint in polarized light.

FIG. I.

MINERALS

tion. Very commonly the grains are colourless and glasslike; large crystals of this character are known as 'rock - crystal.' It may be 'smoky,' yellow, brown to black; amethystine, violet or purple, reddish (inclusions of hæmatite); rose-coloured (titanium or manganese); milky or opaque, as in vein quartz. Most of these varieties may be observed in the quartz of granites. The opalescent appearance of some quartz is caused by a fibrous structure due to a mineral like asbestos which has been replaced by silica.

Granitic quartz frequently contains many sheets of minute vacuoles, which may contain bubbles of liquid (water or carbon dioxide) and gas (Fig. 2, *d*). Their presence may possibly exert an influence upon the cleavability of the stone and upon its behaviour when subjected to fire.

Although quartz has no tendency to cleave readily in any direction under normal conditions, if it is plunged into cold water suddenly after being heated it tends to fracture with a rough rhombohedral cleavage.

Silica plays an important part in many rocks in other forms than that of quartz.

Chalcedony is one of these varieties of natural silica; it is an apparently amorphous substance, having a waxlike lustre and even fracture. It is translucent; its specific gravity is 2·6, and its hardness is equal to that of quartz. Examined under the microscope, it is seen to be composed of minute crystalline fibres; it is not really amorphous. It may be colourless, grey, white, yellow, brown, dusky blue, or black. It occurs in sedimentary rocks, sandstones, and limestones, as a cement, filling up interstices or appearing in patches or replacing organic fragments. In igneous rocks it is found filling vesicles, lining geodes, and forming small veins. Banded varieties are called agates.

Opal is a form of hydrated silica ($SiO_2 \cdot nH_2O$). H.=6. Sp. gr. = 2·1. Unlike chalcedony, it is amorphous. The ordinary content of water ranges between 5 and 10 per cent.; but some varieties have as much as 30 per cent., while others are almost anhydrous. It occurs in rocks in much the same way as chalcedony. It is found in a dull white condition, or with the lustre and play of colour of the well-known precious opal. Wood opal is a replacement of woody tissue by opal (see p. 13).

Jaspers are impure siliceous rocks composed of quartz in very minute particles mixed with argillaceous matter, and coloured by yellow, red, and dark brown hydroxides of iron. Jasper is opaque and somewhat dark in appearance; its strong coloration and hardness have led to its frequent employment as a decorative stone. Most jaspers are due to the metamorphosis of argillaceous or calcareous rocks by aqueous solutions.

Next to quartz in importance, among the minerals of building materials, are the **Felspars.** These are silicates of alumina, alkalies, and lime; they form a very clearly defined group, the members of which are closely related by their physical and chemical properties. According to their mode of crystallization, they fall into two groups: Orthoclase felspar (monoclinic) and plagioclase felspar (anorthic). An important character in these minerals is the presence of two well-marked directions of cleavage nearly at right angles.

Orthoclase is a potash felspar ($KAl.Si_3O_8$); it crystallizes in the monoclinic system. Its hardness is 6; it melts at 1150° C., and it is insoluble in acids. When it has had space to build perfect crystals, it assumes forms such as those indicated in Fig. 1.

Of course, the perfect forms indicated in the figures are rarely seen in the broken or polished surfaces of the stones

MINERALS

in which orthoclase is found; the crystals appear usually as stout prisms or broad plates (Plates I. and II.). Twinning is usually recognizable by the change in the amount of reflection from the adjacent portion of the crystal as the stone is turned about.

In appearance orthoclase is colourless and glassy in the varieties Sanidine and Adularia, but it is usually somewhat turbid white, grey, yellowish or flesh-coloured, due partly to a cloudy kaolinized alteration of the crystal, stained more or less with oxides of iron. The milky appearance of the felspar in granite, etc., is partly due to the effect of the minute cleavage flakes in reflecting the light.

The pearly, pale bluish opalescence of some orthoclase is due to minute inclusions or decomposition, or to layers of different composition (see below) parallel to the orthopinakoid or orthodome.

Orthoclase is a very characteristic constituent of granites and syenites.

Among the plagioclastic felspars there are two which must receive prior consideration, because the researches of Tschermak and others have shown that many of the plagioclases are neither more nor less than mixtures or solid solutions of these two—**Albite**, soda felspar, and **Anorthite**, lime felspar—in varying proportions. These mixtures may be represented by the general formula

$$m \text{ Albite (Ab)} + n \text{ Anorthite (An)}$$

or

$$m \text{NaAl.Si}_3\text{O}_8 + n\text{CaAl}_2\text{Si}_2\text{O}_8.$$

The close relationship between the composition of the felspars and their physical properties may be seen by reference to Table I.

The chemical composition of the pure felspars is show in Table II.

Between the plagioclase felspars enumerated in the two

tables there are others of intermediate composition which do not receive distinctive names. The names Oligoclase, Andesine, etc., are retained for convenience, to denote felspars of a certain range of composition in a series of mixtures of the albite and anorthite molecules.

TABLE I.

Name of Felspar.	Short Formula.	Specific Gravity.	SiO.	Extinction on Face C.	Extinction on Face B.	Mean Refractive Index.	
Orthoclase	—	2·565	64·7	0°	+ 5°	1·523	↑ Increasing fusibility Increasing solubility in acids ↓
Microcline	—	2·56	64·6	+15° 30'	+ 5°	1·523	
Albite	Ab	2·624	68·7	+ 4° 30'	+19°	1·534	
Oligoclase-Albite	Ab_6An_1	2·645	64·9	+ 2° 45'	+12°	—	
Oligoclase	Ab_3An_1	2·659	67·0	+ 1° 4'	+ 4° 36'	1·542	
Andesine	Ab_2An_1	2·694	60·2	− 5° 10'	−16°	1·558	
Labradorite	Ab_1An_3	2·728	49·3	−17° 40'	−29° 28'	1·570	
Bytownite	Ab_1An_6	2·742	46·6	−27° 33'	−33° 29'	—	
Anorthite	An	2·758	43·2	−37°	−36°	1·582	

TABLE II.

Felspar.	SiO_2.	Al_2O_3.	K_2O.	NaO.	CaO.
Orthoclase	64·7	18·4	16·9	—	—
Albite	68·7	19·5	—	11·8	—
Oligoclase	62·0	24·0	—	8·7	5·3
Labradorite	60·2	25·2	—	7·9	6·7
Andesine	51·7	30·9	—	4·0	13·4
Bytownite	47·4	33·8	—	2·0	16·8
Anorthite	43·12	36·7	—	—	20·1

The fact that some of these plagioclase crystals are mixtures of the kind indicated cannot be observed except indirectly by chemical analysis or examination of the optical or other properties which accompany the changes in composition (see Table I.). In some crystals, however, the mixture is revealed by the microscope, when minute inclusions of one felspar appear intergrown with another (see Perthite); and not infrequently a single crystal is shown, by its behaviour in polarized light, to be composed of a series of felspars passing gradually from a less siliceous variety at the centre to a more siliceous one at the periphery.

It has been pointed out that, since there is usually a little soda in orthoclase, presumably due to an ultra-microscopic intermixture of albite, it is quite possible that its monoclinic character may be apparent and not real —that it may, in fact, be anorthic with a pseudo-symmetry. There is, moreover, a distinctly anorthic potash felspar of the same composition as orthoclase; this is **Microcline**, an important rock-forming mineral. Its outward form cannot be distinguished from that of orthoclase in the rough stone.

Another alkali felspar, **Anorthoclase** — $(NaK)Al.Si_3O_8$ —containing soda (and a little lime), shows a cross-hatching like microcline. It is exemplified in the well-known 'Rhomb-porphyry' of the Christiania district, Norway.

Albite $(Na.Al.Si_3.O_8)$ is a common felspar in some granites, gneisses, schists, and porphyrites; it is usually rather opaque white, sometimes pearly and opalescent.

Albite very often occurs in parallel intergrowth with orthoclase (Fig. 1). This is called the 'perthite structure' (from Perth, Ontario); it may be visible with the naked eye, or only distinguishable with aid of the microscope (micro-perthite); or its presence in an ultra-microscopic condition may be assumed, as in anorthoclase, when it is called 'crypto-perthitic structure.'

Anorthite $(CaAl_2Si_2O_8)$, the lime felspar, is a rare mineral. The plagioclase felspars, intermediate between albite and anorthite, the albite-anorthite mixtures, are important rock-formers. They rarely assume the perfection of form seen in albite, anorthite, or the orthoclastic felspars, but occur rather in irregular grains, stout prisms with ill-defined outlines, or as elongated, lath-shaped bodies. They cannot be distinguished in the hand specimen, but by one or other of their optical characters they may be differentiated without much difficulty under the microscope. Frequently, however, it is more satisfactory to examine crushed fragments of the mineral rather than the thin slice of the rock.

The more basic plagioclases (those with less silica) are naturally found most commonly in the more basic rocks. One of these, labradorite, in the rock found on the coast of Labrador, deserves attention because its beautiful iridescent colours in shades of blue, green, yellow, and purple, have led to the stone being employed in small slabs for ornamental objects. The iridescence of the rock is due to both of the essential minerals composing it—namely, hypersthene (p. 27) and labradorite felspar. The iridescence is caused by minute inclusions of hæmatite, göthite, ilmenite, or diallage, arranged in a lamellar manner within the crystals. This rock is known as 'labradorite' or 'labrador-spar.' The so-called 'labrador' or 'labrador-granite' of Scandinavia is not the same kind of stone (p. 84); it has acquired the name from its superficial resemblance to the original rock.

In some rocks the place of felspars is taken more or less completely by the *felspathoids*, **Leucite**, $KAl(SiO_3)_2$, and **Nepheline**, $K_2Na_6Al_8Si_9O_{34}$. The former mineral is pseudo-cubic, and the latter is hexagonal; the one commonly forms icositetrahedra, the other hexagonal prisms. Nepheline in the older rocks is frequently granular, and is called **Elæolite,** from its greasy appearance.

Mica Group.—This group embraces a number of aluminous silicates of variable composition, but with closely related properties. They crystallize in the monoclinic system, but, except in certain lavas and in pegmatites, they do not often develop their perfect form, which is that of nearly true hexagonal plates or pyramids. They are best characterized by their extremely perfect basal cleavage and by the elastic nature of the thin cleavage flakes. From the chemical point of view there are two distinct classes of micas:

1. Alkali micas—

Potash mica	= *Muscovite*	⎫ For convenience
Soda mica	= *Paragonite*	⎬ often called
Lithia mica	= *Lepidolite*	⎭ 'white mica.'

2. Ferro-magnesian micas—

Iron-magnesian mica	= *Biotite*	⎫ For convenience
With much magnesia	= *Phlogopite*	⎬ often called
With much ferric iron	= *Lepidomelane*	⎭ 'black mica.'

The alkali micas are paler in colour than the ferro-magnesian kinds. The former vary from silvery white to pale yellow, pale green, or pinkish tints; the latter range from pale brown, brownish greens, and reds, to black.

There are many other varieties with more or less well-defined individuality, but for the present purpose it is sufficient to class all the micas, according to their appearance, as 'white' or 'black.'

Most important are the two kinds—muscovite (and its varieties) and biotite (and its varieties).

Muscovite—potash mica—has a composition which may be represented by the formula $H_2K.Al_3(SiO_4)_3$. H. = 2–3. Sp. gr. = 2·76–3·1.

It is not attacked by HCl, and, as regards weathering and other forms of alteration, it is one of the most stable minerals. It is found in many granites and gneisses and mica schists; in quartz porphyries, but not in recent eruptive rocks. In sandstones, grits, and shales, it often plays a prominent rôle, and, by the tendency of the flakes to lie with their flat faces in one plane, it has a marked effect in emphasizing the fissibility of the stone along bedding planes. Besides occurring as an original mineral in granites, it is frequently found as one of the alteration products of felspar and other aluminous minerals. The varieties known as *sericite* and *Liebenerite* often occur in this relationship.

Biotite.—The composition of biotite may be represented approximately by $K_2HMg_6Al_3(SiO_4)_6$; but iron and other substances are present. H. = 2·5–3. Sp. gr. = 2·8–3·2.

This mineral is a common constituent of many granites, gneisses, syenites, diorites, andesites, and trachytes, also of some basalts and many metamorphic schists.

In igneous rocks and metamorphic rocks whose flow structure or gneissose structure is strongly developed, the presence of abundant mica flakes may tend to produce weakness in a stone in a direction parallel to that of the flow or schistosity. Mica has little influence on the strength of unfoliated rocks or those without marked orientation of the grains. It is, however, a source of trouble if it occurs very abundantly or in large flakes in a stone that has to be polished, for the flakes themselves do not take the polish, but tend to drag out in the process and leave hollows.

Dark micas, especially those rich in iron and magnesia, are much less stable than the light kinds. They pass readily through a series of bodies of indefinite composition, ' vermiculites,' into chlorites. For example :

Biotite.	Vermiculite.	Chlorite.
$Al_2Mg_2KH.(SiO_4)_3.$	$Al_2Mg_2H_2(SiO_4)_3 3H_2O.$	$Al_2(MgOH)_4.H_2(SiO_4)_3.$

The chlorites resemble the micas in appearance. They form similar ragged scales or hexagonal plates; their colour is usually dark green, and their lustre is inclined to be pearly. They crystallize like mica in the monoclinic system. Unlike mica, the flakes are flexible but not elastic. The composition is very variable; it resembles that of the micas, but is more strongly hydrated and more basic, and alkalies are often completely absent. They may arise in many kinds of rocks, from secondary changes wrought upon various aluminous minerals besides dark mica—*e.g.*, amphiboles, augite, epidote, garnet, tourmaline.

The green colour of some granites is due to this mineral, and it is the cause of the same tint in many other rocks; it is especially abundant in certain schists.

Very important rock-forming minerals are the **Amphiboles** and **Pyroxenes**, two groups of silicates which are intimately related both by their chemical and physical attributes. The more prominent members are included in the following list:

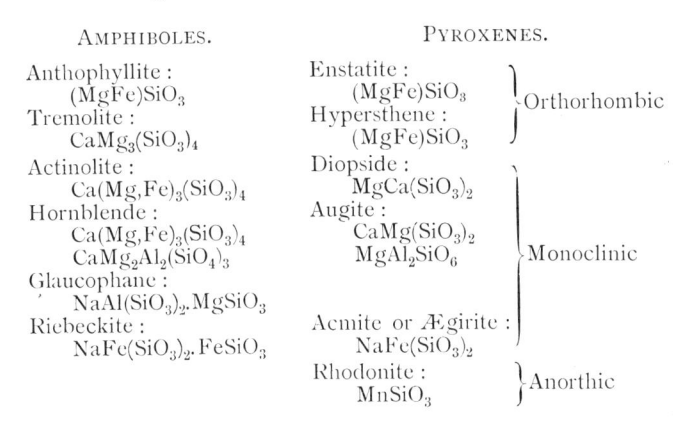

AMPHIBOLES.	PYROXENES.	
Anthophyllite: $(MgFe)SiO_3$	Enstatite: $(MgFe)SiO_3$	Orthorhombic
Tremolite: $CaMg_3(SiO_3)_4$	Hypersthene: $(MgFe)SiO_3$	
Actinolite: $Ca(Mg,Fe)_3(SiO_3)_4$	Diopside: $MgCa(SiO_3)_2$	
Hornblende: $Ca(Mg,Fe)_3(SiO_3)_4$ $CaMg_2Al_2(SiO_4)_3$	Augite: $CaMg(SiO_3)_2$ $MgAl_2SiO_6$	Monoclinic
Glaucophane: $NaAl(SiO_3)_2.MgSiO_3$		
Riebeckite: $NaFe(SiO_3)_2.FeSiO_3$	Acmite or Ægirite: $NaFe(SiO_3)_2$	
	Rhodonite: $MnSiO_3$	Anorthic

Most of the amphiboles are monoclinic, and are characterized by a well-marked cleavage parallel to the prism faces; hence in the sections normal to these faces the cleavage cracks appear crossing at angles of 56° and 124° (Fig. 2). The coloured varieties are strongly pleochroic.

Hornblende (H. = 5–6. Sp. gr. = 2·9–3), the most common form, is an important constituent in many syenites and granites (monzonites), in which it may be brown or green; in gabbros, diorites, phonolites, and trachytes, it is usually

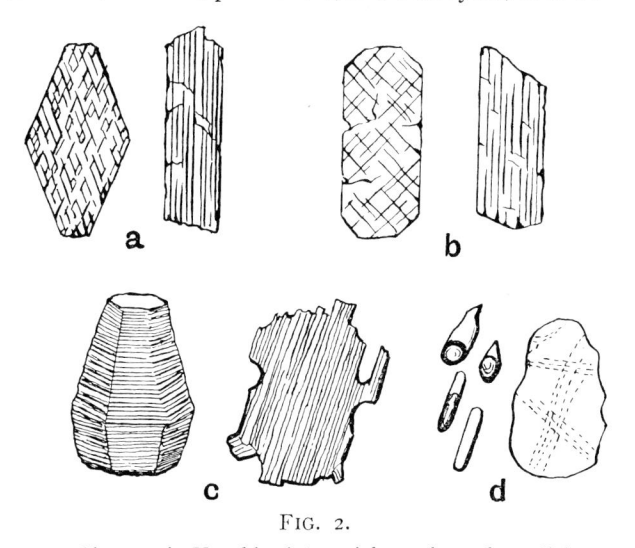

FIG. 2.

a, Cleavage in Hornblende ⎫ at right-angles and parallel
b, Cleavage in Augite ⎭ to the prism faces.
c, Cleavage in Mica: a crystal and an irregular flake.
d, Vacuoles in Quartz: highly magnified (left); in streaks less highly magnified (right).

green; in andesites it is brown. It likewise appears in many gneisses and schists.

Actinolite forms green needles and bladelike crystals in certain schists.

Tremolite is green to white, and occurs in elongated narrow crystals. The tendency to elongation is very marked. The true asbestos is a variety of this mineral in which the *fibres* (crystals) may attain the length of a yard;

they are silky and flexible. Jade or nephrite is a compact form of the same mineral, or of actinolite.

Riebeckite is a blue mineral; it is found in some granites, micro-granites, and trachytes.

Important rock-forming minerals occur in the **Pyroxene** group in both the monoclinic and rhombic systems. The coloured varieties may or may not be strongly pleochroic, and the cleavage parallel to the prism face is not so well marked as in the amphiboles; the traces of the cleavage plane form angles of $87\frac{1}{2}°$ in cross-section, but there is in addition a fairly perfect pinakoidal cleavage (Fig. 2). In the well-formed crystals the prism faces are less well developed than the pinakoids.

Among the rhombic pyroxenes, the minerals **Enstatite**, **Bronzite**, and **Hypersthene**, may be mentioned. They are similar in composition, but differ in the amount of iron they contain.

>Enstatite contains less than 5 per cent. of FeO.
>Bronzite contains 5 to 14 per cent. of FeO.
>Hypersthene contains about 14 per cent. of FeO.

Bronzite and hypersthene are very prone to a type of alteration which produces the appearance known as 'schiller' (schillerization). Its effect is mainly due to interference of the light caused by minute layers of hydroxides or oxides of iron deposited in closely adjacent cleavage planes or in glide planes; or to cavities in the same positions.

These minerals are found forming short prisms or large irregular plates in diorites, some granites (monzonites), dolerites, porphyrites, andesites, and in gabbros and peridotites.

Augite (H. = 5–6. Sp. gr. = 3·2–3·6), the most important monoclinic pyroxene, often forms well-shaped crystals in volcanic rocks (andesites), or large irregular plates in more crystalline rocks and some basaltic rocks (gabbros, dolerites). It is often difficult to distinguish in the fractured rock from dark hornblende; its colour viewed in this way is black, but in the thin section it is readily recognized by its optical characters and cleavage. In thin sections it is pale green to pale brown or violet-brown; pleochroism is feeble or absent.

Ægirite or **Acmite** is green in colour; it occurs in some syenites, nepheline syenites, and phonolites.

In many gabbros augite occurs in an altered condition, resembling that of bronzite and hypersthene; it is then called **Diallage**, and is often laminated (parallel to the ortho-pinakoid), due partly to interlamination with bronzite.

An interesting type of alteration, to which the name **Uralitization** is given, is the transformation of an augite crystal more or less completely to hornblende. The form of the augite is retained, but the cleavage and composition are those of hornblende. This change, known as 'uralitization,' is usually accompanied by the formation of calcite and epidote in the immediate neighbourhood (see Table III.).

It is most essential to observe that these analyses do not stand for anything more than *examples;* they are neither average results nor extremes, and in some of the minerals —the dark micas, for instance—very wide divergencies from the above composition are to be found.

Olivine is a silicate of magnesia and iron, $(MgFe)_2SiO_4$, in which the magnesia is liable to considerable variation. H. = 7. Sp. gr. = 3·3–4. It has imperfect cleavage, conchoidal fracture, and vitreous lustre. Usually it appears as greenish or reddish glassy grains in basic rocks (gabbros, peridotites, dolerites); in some basalts it forms short prisms, which appear hexagonal in section. This mineral alters very readily along internal cracks and round its margins into serpentine, often accompanied by the liberation of iron, which gives it a red, brown, or black stain.

Serpentine ($H_4Mg_3Si_2O_9$. H. = 3. Sp. gr. = 2·6) is a massive or fibrous mineral with greasy lustre; green, red, or yellow, in colour. It is formed by the decomposition of

TABLE III.

	A.	B.	C.	D.	E.	F.
Silica—SiO_2	45·23	47·06	39·48	43·76	43·42	47·95
Titanium oxide—TiO_2	4·28	1·82	0·30	0·78	—	—
Alumina—Al_2O_3	7·73	7·77	12·99	11·62	19·00	30·26
Chromium oxide—CrO_3	—	trace	—	—	—	—
Ferric oxide—Fe_2O_3	2·95	1·30	7·25	6·90	17·64	2·43
Ferrous oxide—FeO	4·07	8·15	10·73	10·47	—	3·10
Manganese oxide—MnO	0·07	0·20	1·00	0·50	—	—
Magnesia—MgO	12·25	13·52	11·47	12·63	0·54	0·94
Lime—CaO	23·37	19·33	12·01	9·84	1·81	0·98
Sodium oxide—Na_2O	0·47	0·33	1·70	3·43	3·66	2·00
Potassium oxide—K_2O	0·12	0·11	2·39	1·28	8·77	10·19
Nickel oxide—NiO	0·05	trace	—	—	—	—
Phosphorus pentoxide—P_2O_5	—	0·06	—	—	—	—
Water—H_2O	0·37	0·20	—	—	4·30	1·13
Fluorine—F	—	—	0·05	1·82	—	—
	100·96	99·85	100·25	100·49	99·14	100·25

A and B = Augite; C and D = Hornblende; E = Dark Mica (Lepidomelane); F = Light Mica (Muscovite).

several kinds of ferro-magnesian minerals. It is an important feature in many greenish marbles (serpentinous marbles), often with associated talc and magnesite. A very markedly fibrous variety is known as **Chrysolite**, a mineral which is largely exploited as 'asbestos' (Canadian asbestos).

Another basic magnesium silicate is **Talc**—$H_2Mg_3(SiO_3)_4$. H. = 1. Sp. gr. = 2·7. Like serpentine, it is one of the final decomposition products of ferro-magnesian silicates. When pure it has a silvery-white appearance, but it is very often stained by the presence of from 1 to 4 per cent. of FeO, and thus becomes green, yellow, or red. It occurs in rocks as small flakes resembling mica, but pliable and non-elastic; also in irregular patches and small veins. It is best known in its massive form **Steatite**, or Soapstone. Its resistance to fusion and to the attack of acids are noteworthy characters.

A few other silicate minerals may be briefly mentioned here, as they will be referred to later; they are not, however, of such importance as the foregoing from the point of view of building material formation.

Epidote ($HCa_2Al_3Si_3O_{13}$. H. = 6. Sp. gr. = 3–4), monoclinic, is a dark green mineral which owes its colour to a variable amount of iron. It has a vitreous lustre, and in thin sections is strongly pleochroic. It has a good cleavage. It usually occurs in grains or short stout prisms.

Zoisite is a mineral of very similar composition, but orthorhombic in its crystallization. A massive red or pink variety called **Thulite**, from Telemark in Norway, is used as an ornamental stone.

Both these minerals may occur as alteration products of felspars, and the green colour of some granitic rocks is due to the minute grains of epidote that have arisen in this manner.

Garnets are minerals of the general composition $R''R'''(SiO_4)_3$, where $R'' = Ca, Mg, Fe, Mn$, and $R''' = Al, Fe, Cr$. They crystallize in the cubic system, the common form being the dodecahedrom. Their lustre is vitreous and the fracture uneven; they are readily fusible and brittle.

MINERALS

There are several varieties, but it will be sufficient to mention three of the more common ones—namely:

Pyrope—$Mg_3Al_2(SiO_4)_3$ ... ... Magnesia garnet.
Almandine—$Fe_3Al_2(SiO_4)_3$ ... ... Iron garnet.
Grossular—$Ca_3Al_2(SiO_4)_3$ ... ... Lime garnet.

Garnets are common in some crystalline schists, gneisses, hornfelses, and other metamorphic rocks, and in some basic rocks. In certain limestones they appear as alteration products.

Zircon ($ZrSiO_4$) is not uncommon in many coarsely crystalline rocks and their derivatives, sandstones, etc., as a subordinate non-essential mineral, occurring in short prisms and grains.

Tourmaline, a mineral crystallizing in the hexagonal system, is a borosilicate of alumina and alkalies, with more or less magnesia and iron. H.=7. Sp. gr.=3·1. It has a subconchonoidal fracture and only imperfect cleavage. Commonly, as seen in rocks, it occurs as black needle-like crystals, short prisms, or irregular grains. In thin slices it is brown or blue, and often strongly pleochroic. Usually it is a subordinate mineral in granitic rocks, and some sandstones, etc.; rarely, as in the case of luxullianite, it becomes an important part of the rock.

Andalusite (Al_2SiO_5) occurs in grains or elongated prisms in argillaceous schists and some slaty rocks. A variety with dark enclosures, which give a crosslike pattern in the transverse section, is called 'chiastolite.'

Kaolinite ($H_4Al_2Si_2O_9$), possibly with a variety of minerals of similar composition, forms a considerable part of most clays, and is one of the most common results of decomposition in felspars. The cloudy opacity of the felspars in many rocks is usually put down to this mineral, although it is rarely possible to actually determine it by its physical properties.

Several important rock-forming carbonates may now be considered. The first of these is **Calcite** ($CaCO_3$), carbonate of lime. This mineral crystallizes in the hexagonal system, and it assumes a great variety of forms; but as a rock-builder it nearly always exists as irregular grains. It possesses a very perfect cleavage in three directions, parallel to the faces of a rhombohedron, and lamellar twinning is prevalent. H. = 3. Sp. gr. = 2·72. The cleavage planes, along which it nearly always splits when broken, have a vitreous lustre. The colour is white unless the mineral has inclusions, and often it is colourless and transparent. Calcite builds up the bulk of all limestones and marbles, even in many of those which appear to consist of amorphous carbonate of lime. It appears in granular patches in many igneous rocks, as the result of the breaking down of lime-bearing minerals. In some sandstones it may act as a cement.

Dolomite is a double salt with a composition represented by $CaMg(CO_3)_2$. H. = 3·5. Sp. gr. = 2·85. Like calcite, it has a perfect rhombohedral cleavage, and it crystallizes in the same system. Its lustre is pearly, and it is usually rather yellower than calcite, although some dolomite marbles are beautifully white. It occurs in many limestones as isolated rhombohedra and irregular grains; also forming massive bedded rocks.

Chalybite ($FeCO_3$. H. = 3·5. Sp. gr. = 3·8) forms crystals much resembling dolomite, but they are usually darker. It occurs as scattered crystals or in groups in many clays and some limestones. Sometimes it is found in the form of small spheroidal nodules in clays (sphœro-siderite). It also occurs in veins and in some sandstones as a cementing material.

Magnesite ($MgCO_3$) is not commonly a rock-forming mineral; it occurs in veins in serpentine and other decomposed magnesia-bearing rocks and in massive beds. It is a white opaque mineral with conchoidal fracture. It is employed in the manufacture of refractory bricks.

Iron-bearing minerals are not present in bulk in building materials, but they are so widely spread throughout all kinds of rocks, and in their oxidized and hydrated state they act so powerfully as colouring agents, that several of them must be noticed here.

Magnitite (Fe Fe_2O_4) is a black opaque mineral with a metallic lustre; it crystallizes in the cubic system, usually forming octahedra or dodecahedra.

Hæmatite (Fe_2O_3) is black, iron-coloured with brilliant metallic lustre, or red, brownish-red, and earthy. It crystallizes in the hexagonal system. H.=6. Sp. gr.=5·2.

Limonite ($2Fe_2O_3.3H_2O$) is an amorphous brown or black mineral. H.=5·5. Sp. gr.=3·8.

Göthite ($Fe_2O_3.H_2O$) is blackish-brown, and crystallizes in the orthorhombic system. H_2.=5. Sp. gr.=3·8-4·4.

Pyrites—Iron Pyrites (FeS_2. H. = 6. Sp. gr. = 5·1) occurs as cubes, octahedra, or pyritohedra, and as irregular grains and masses of a brass-yellow colour and bright metallic lustre.

Marcasite (FeS_2. H. = 6. Sp. gr.=4·8) has the same composition as pyrites, but it crystallizes in the orthorhombic system, and its prevalent habit is that of radiating fibrous nodules. When fresh its colour is pale yellow, but it decomposes with great readiness, giving rise to a white efflorescence (ferrous sulphate). This tendency to rapid decay causes its presence in building material to be very harmful.

Well-crystallized pyrites withstands weather very well, and is much less harmful; but when it does decompose, it forms the injurious sulphuric acid and the yellow or brown stain of the oxides and hydrates mentioned above.

Pyrites is found in many of the coarsely crystalline igneous rocks, in limestones, particularly earthy varieties, and in clays.

Three sulphates are of sufficient interest to be mentioned here:

Gypsum—Sulphate of Lime ($CaSO_4.2H_2O$). Monoclinic. H. = 2. Sp. gr. = 2·3—is a colourless transparent mineral when in the crystalline form, called **Selenite**. These crystals are of common occurrence in many clays, where they are found as isolated individuals or in aggregate groups; to labourers they are often known as 'congealed water.' A brick clay, otherwise quite good, may be rendered unfit for use by their presence. In its massive crystalline form gypsum is called **Alabaster**; it may also be stalagmitic or fibrous (satin spar) in veins.

Plaster of Paris derives its name from the excavations of Montmartre near Paris, where the gypsum beds are worked.

A natural anhydrous sulphate of lime is **Anhydrite** ($CaSO_4$). **Barytes**, the sulphate of barium ($BaSO_4$), or 'heavy spar,' is mentioned here because it occasionally forms the cementing material in sandstones.

Ilmenite ($FeTiO_3$) and **Titanite** ($CaSiTiO_5$) occur as subordinate minerals in many rocks.

Glauconite is a green silicate of iron, with magnesia, potash, and other substances; of variable composition. In the form of grains and casts of fossils it constitutes an important ingredient in certain rocks—*e.g.*, the Greensands, Kentish rag, etc.

CHAPTER III

IGNEOUS ROCKS

ROCKS may be classified broadly into three groups: (1) *Sedimentary*, rocks formed by deposition at the surface of the earth of material lying in beds or layers. (2) *Igneous*, rocks formed by the fusion of rock matter deep within the earth's crust; often found penetrating the sediments. (3) *Metamorphic*, rocks of sedimentary or igneous origin that have suffered certain changes in structure and composition, since their first formation.

IGNEOUS ROCKS.

It is not to be supposed that all igneous rocks have penetrated the others in which they lie, by injection from below, for some, like the lavas of existing active volcanoes, may be seen actually hardening at the surface and forming regular beds there. We do not depend upon this criterion to distinguish igneous from sedimentary rocks, for it is possible to recognize the igneous origin of rocks even in hand specimens and microscopic fragments from their peculiarities of internal structure.

An igneous rock is one that has solidified from a molten state, and the variations in the conditions of solidification lead to variations in character which permit a classification of these rocks into three tolerably well-defined groups:

1. Those that have solidified deep within the earth's crust under what are called Plutonic or Abyssal con-

ditions—that is, under great pressure and at a slow rate of cooling.

2. Those that have solidified in conditions of more rapid cooling—Hypabyssal conditions.

3. Those that have solidified at or quite near the surface under conditions of comparatively rapid cooling, and subject to pressure little above that of the atmosphere.

From this point of view, therefore, we may divide igneous rocks into three broad groups corresponding to the circumstances of their origin:

1. Plutonic or Abyssal rocks—Intrusive (Granite).
2. Hypabyssal rocks—Intrusive (Porphyry).
3. Volcanic rocks—Extrusive (Rhyolite).

Each of these three types of rock has been derived from a molten rock magma, which, as we shall see later, may present many variations in composition from place to place; indeed, variations may frequently have arisen from time to time within the same magma mass during the process of solidification.

We may now briefly notice some of the modes of occurrence and more obvious peculiarities of these three types.

The **Plutonic** rocks, of which granite may be taken as the type, appear at the surface of the earth only as the result of the removal by ordinary agencies of weathering of vast masses of superincumbent strata. Thus, they appear in the cores of mountain ranges or in great masses which have been injected under and into overlying beds, sometimes lifting the latter up, as in the injected masses with the form known as 'laccoliths'; sometimes being squeezed along planes of weakness in horizontal, vertical, or intermediate directions; occasionally absorbing some of the surrounding rock, and almost invariably causing modification (metamorphism) by virtue of their high temperature.

Most of the material so injected into the hardened portion of the crust was probably at one stage of its existence in the condition of a structureless fluid, containing all the elements of granite, for example, in a state of mutual solution.

As cooling set in, some of these substances crystallized out of solution in the form of definite minerals; others followed, according to the regular laws of solutions, until the whole mass had crystallized into a solid rock; the later minerals fitting in between and moulding themselves round those formed at an earlier stage.

In the deeply-seated plutonic masses the cooling must of necessity have been extremely slow, with the result that the whole rock became completely crystalline (holocrystalline, Plate II., *b*), and no non-crystalline, undifferentiated material remained. Moreover, the crystalline structure of such rocks is usually coarse, and each of the principal component minerals is well developed.

The **Hypabyssal** rocks, like those of the plutonic group, are all intrusive in others that have consolidated earlier; but the bodies of molten material from which rocks of this type have consolidated were smaller than those giving rise to plutonic rocks; their rate of cooling was consequently more rapid.

Individual crystals of earlier-formed minerals may be of comparatively large size, but there is usually a distinct ground mass of much finer grain than is associated with the plutonic type (Plate II., *d*). This ground mass may be holocrystalline and granular, like that of granite on a small scale, or it may be formed of a microscopic intergrowth of the essential minerals (granophyric structure); it may be a densely felted mass of crystalline material in which he individual crystals are often too small to be disinguished (felsitic structure), or it may be glassy.

Two or more of these varieties of ground mass may be associated within the same rock.

Rocks of this type occur as offshoots from the great plutonic masses; they may even appear on the outer, rapidly chilled surfaces of such masses. They occur also in the forms of veins or dykes, forced across the bedding of stratified rocks or through joints in earlier-cooled igneous rocks; and they appear as bedlike sills injected between the beds of stratified rocks; or they may form small bosses and laccoliths.

The **Volcanic** rocks differ from those in the other two groups in that they are always extrusive in character; they have been poured out or blown out at the surface of the earth or on the bottom of the sea. They are represented by modern lava flows and the outpourings of so-called volcanic 'ash,' pumice, and similar fragmentary material of various sizes, and by their ancient equivalents formed by the volcanic activity of previous geological periods.

Such rocks usually occur stratified as bed upon bed of lava or of volcanic ash (tuff). Beds of lava may be interstratified with those of tuff, and either form of volcanic deposit may lie between strata of the ordinary sedimentary kind.

Naturally, the rate of cooling in these circumstances has been more rapid and irregular than in the case of the two other groups. A large proportion—sometimes the whole of the rock mass—is found in a glassy or only feebly devitrified condition (Fig. 3, V).

The free movement of the rapidly cooling rock has frequently left its mark in fluidal or flow structures, and the relief from pressure on extrusion has often given

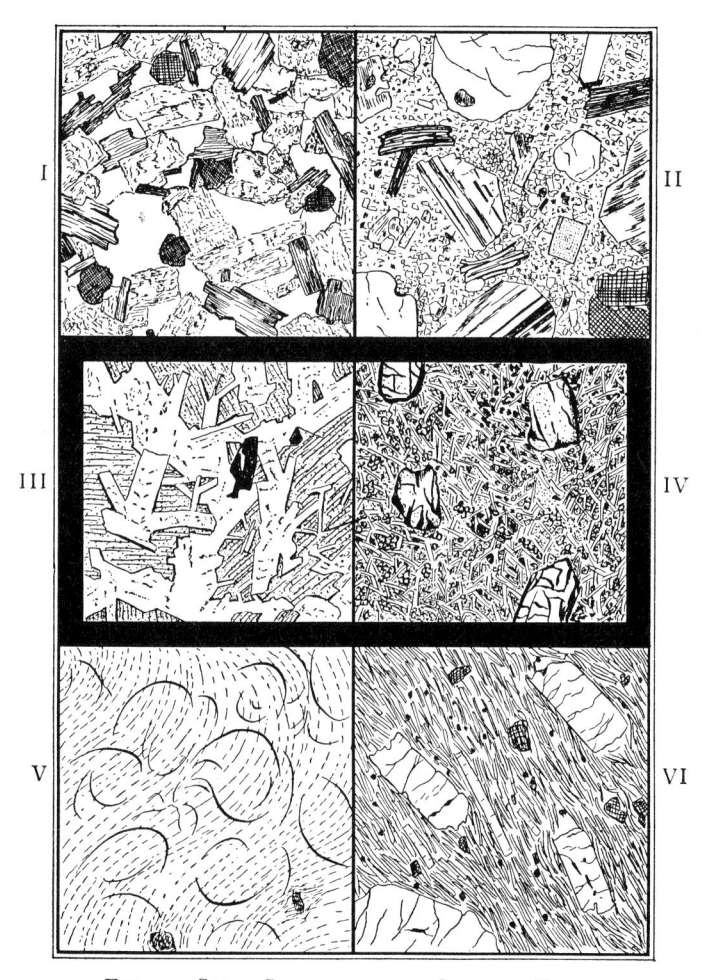

FIG. 3.—SOME STRUCTURES IN IGNEOUS ROCKS.

I. *Granite:* quartz, clear; felspar, clouded; mica, dark. II. *Granite porphyry:* a ground mass of the same minerals as in (I.), with similar structure, but finer grained; porphyritic quartz, plagioclase, orthoclase, and mica (polarized light). III. *Diabase* or *dolerite:* typical ophitic structure; large crystals of augite enclosing crystals of plagioclase. IV. *Basalt:* granular crystals of augite with lath-shaped felspars; porphyritic olivine, cracked and stained; some residual glass and small grains of magnetite. V. *Obsidian:* a glassy volcanic rock; streams of microlites (undeveloped crystals) showing the flow structure; curved 'perlitic' cooling cracks. VI. *Trachyte:* showing flow structure; large crystals of sanidine felspar in a ground mass of smaller crystals of the same.

rise to vesicular structures, so well exemplified in the frothy nature of pumice.

There is a type of texture or internal structure common to the three rock groups, which is of considerable importance, since it materially modifies the appearance of the stone. It is due to the crystallization of one or more of the characteristic minerals of the rock at an early stage, so that it had freedom to assume its proper form with well-developed crystal faces. These well-formed (idiomorphic) crystals or phenocrysts are said to be porphyritic; the rock, too, which contains such crystals is called a 'porphyritic' rock. Several of the minerals in a single rock may be porphyritic, and occasionally a single mineral may appear in two distinct crops of phenocrysts. Porphyritic structure is less prevalent in plutonic rocks than in the two other types.

The classification of igneous rocks may be undertaken from an entirely different standpoint—namely, that of their chemical composition. They all contain silica, alumina, and varying amounts of potash, soda, lime, magnesia, and oxides of iron. The silica exists in the free state as the mineral quartz and in a variety of combinations with the other constituents, and it has been found convenient in practice to recognize a grouping based upon the total silica contents of the rocks. Thus we have an *Acid* group, with 66 per cent. or more of silica; an *Intermediate* group, with silica varying from 52 to 66 per cent.; and a *Basic* group, with the silica content below 52 per cent. This simple relationship is complicated, however, by the fact that in a given magma earlier-solidified portions are frequently more basic than what is left over to solidify later; therefore apophyses formed at this period may be basic, while those found later may be acid.

IGNEOUS ROCKS

The two systems of grouping, one founded on the mode of origin and the other on the chemical composition of the rocks, are quite independent. Examination of any area where igneous rocks of diverse origin are exposed shows that the great masses of slowly cooled plutonic rock have sent out prolongations or apophyses into the surrounding strata, which have cooled more rapidly than the parent mass, and therefore have the structures peculiar to the hypabyssal type; and, further, when these tongues have reached the surface and outpoured there, the still more rapidly cooled material has all the characters of the volcanic type.

Here, then, if the parent magma were of the acid variety, as indicated by its chemical composition, we should find plutonic, hypabyssal, and volcanic rocks all of the same chemical type; conversely, if the parent magma were basic, the associated apophyses and erupted volcanic rocks would likewise partake of the same nature.

Since the specific characters of an igneous rock—those characters by which it is distinguished from other igneous rocks—are determined by (1) its chemical composition and (2) its internal structure or fabric, both of these factors have to be taken into consideration simultaneously in any scheme of classification for practical purposes.

An igneous rock is an aggregate of discrete minerals together with more or less undifferentiated matter. The kind of rock is recognized by the nature of its component minerals, by their mutual relations with one another in the aggregate, and by their relative proportions in the rock.

The nature of the minerals—their specific character and peculiarities of habit—is not fixed solely by the chemical composition of the original liquid magma, but by a very complex and variable set of conditions, which include changes in pressure, in temperature, and in the relative masses of different compounds in solution at any given moment during the cooling process, to which must be added the introduction of fresh magma streams, of the same or slightly different composition, into that already partially crystallized.

The chemical analysis of an igneous rock may tell us whether it should be classed as acid, intermediate, or basic; or in one of the innumerable gradational stages between one or other of these divisions; it will tell us whether the rock has a high or low percentage of potash, soda, magnesia, iron, and so on, but with all this we shall know nothing of the rock as a rock.

It is not sufficient to know the chemical composition; we must know also the minerals of which the rock is composed, and the manner in which they are mutually associated.

Since the composition of the rock-forming minerals is well known, it is often enough for us to recognize in the hand sample, or in thin sections under the microscope, their kinds and relative proportions, in order to be able to form a very fair idea of the composition of the rock. Chemical analysis is necessary to confirm and give exactitude to such observations, and it is, of course, essential for the determination of the nature of the undifferentiated, crypto-crystalline, or glassy matter which cannot be resolved by the microscope.

In Table IV. a small number of the better-known rock types are arranged in such a way as to show roughly their relationship, both genetically and chemically. Petrologists have recognized and described very many other types to which distinctive names have been given; but the student should clearly understand that all these divisions are artificial, and though some names are necessary for convenience, many of the names in the literature are ill-chosen and of trivial signification. The rocks themselves show no such marked individuality of type except locally, for even the most characteristic plutonic rock is found in Nature to pass by insensible gradations into those that would be recognized as hypabyssal or volcanic; and chemically there is no sharp line to be drawn anywhere between the most acid and the ultrabasic igneous rock.

TABLE IV.

Acid. —— *Intermediate.* —— *Basic.*

| | | Orthoclase Felspar. || Plagioclase Felspar. ||
		With Quartz.	Without Quartz.	Hornblende or Augite.	Augite and Olivine.
Completely crystalline; coarsely grained, sometimes porphyritic	Plutonic	Granite	Syenite	Diorite	Dolerite Gabbro
Completely crystalline, or with small proportion of uncrystallized glassy material; fine-grained ground mass, sometimes with porphyritic crystals	Volcanic / Hypabyssal	Rhyolite, Quartz-porphyry	Syenite-porphyry	Andesite	Basalt
Glassy, non-crystalline	Volcanic	Obsidian Pitchstone	Trachyte-glass Pitchstone	Andesite-glass	Tachylyte

GRANITE.

Granite, as its name implies, is a rock made up of granular particles. These grains are crystalline individuals of three or more kinds of minerals; they are united firmly together by their intergrowth to form a rigid mass.

Granites exhibit many variations of colour, texture, and strength, but all true granites are built up of the same limited number of mineral species. The essential minerals forming granite are felspar, quartz, and mica; frequently with hornblende, and sometimes with augite in place of, or in addition to, the mica.

Many totally different kinds of rock have been called 'granites'—sometimes from ignorance of their true nature, sometimes for trade purposes; thus, one hears of 'Mendip Granite,' which is in reality a limestone; 'Petit Granite,' a dark Belgian limestone; 'Ingleton Granite,' a conglomerate; 'Black Granite,' a name applied to many varieties of basalt, diabase, and other basic igneous rocks, and so on. But in the interests of everyone concerned, it is desirable to employ the name only with its proper significance.

The Minerals forming Granite.

Felspar.—This is the most obvious of the granite minerals, and in the majority of granites is the most abundant. Its colour—white, grey, yellow, pink, or red—together with its opacity, renders it easily distinguishable in the freshly fractured or polished surface of any granite. The felspar of granites is very rarely transparent, as it is occasionally in a few varieties of volcanic rock, though it may be translucent.

The felspar of granite is not all of one kind; while the most prevalent form is orthoclase, it is very frequently associated with a plagioclastic variety. In typical granites the potash felspars predominate, in the form of orthoclase, microcline, or perthite; more rarely a soda-felspar, soda-orthoclase, albite, or anorthoclase, may assume predominance. Rocks of the latter character have been called 'alkali-granites.' Again, in some granites the soda-lime felspars are associated with alkali-felspars in about equal proportions (Adamellites). Finally, there are granites in which the soda-lime felspars exceed the alkali-felspars in abundance (Grano-diorites). It is important to remember that these distinctions cannot be recognized by the unaided eye.

The form taken by felspar in the majority of granites is that of irregular grains, either without any indication of crystalline form or with its faces only partially and imperfectly developed. In many granites the felspar grains are grouped in clots or aggregates. It sometimes happens that some of the felspar occurs in the form of tabular crystals: small, $\frac{1}{2}$ inch to $\frac{3}{4}$ inch, as in Penryn; or large, 1 inch in Shap granite, 1 to 5 inches in Lamorna or Colcerrow granite. These well-formed crystals constitute an earlier crop of felspars, which has usually been succeeded by a later crop of ill-developed grains. Quite frequently the earlier porphyritic crystals are orthoclase or perthite, while the later-formed grains are albite or oligoclase or orthoclase. Occasionally there may be two crops of porphyritic crystals, larger and smaller.

The colour of granitic felspar has a fairly wide range of variation, as indicated above. As a rule the orthoclase crystals are more prone to coloration than the plagioclases, the latter being very commonly dead white, pearly white, or light grey. The pink, yellow, buff, and red tints in orthoclase are all due to the inclusion of very minute granules of iron oxides. This coloration may be an original character of the felspar, or it may be, in the case of much-weathered granites, a rusty stain introduced from without. Green tints in felspar, exemplified in the granites of Lamorna and other Cornish examples, is usually due to the presence of minute scales of chlorite; in other cases it is caused by granules of epidote produced during the partial decomposition of the felspar.

The lack of transparency which characterizes the felspar of nearly all granites is due in part to the frequency of cleavage planes, and in part to the presence within the crystals of minute granular decomposition or alteration products; in either case the effect is the same—namely, the repeated reflection of the incident light from the numerous surfaces so formed.

Quite apart from the processes of ordinary weathering, the felspars of granites have usually undergone some form of alteration whereby more or less of the original felspar crystal has been transformed to an aggregate of other minerals.

Quartz in granites almost always takes the form of irregular grains, mostly commonly moulded upon the other minerals; sometimes it occurs in rounded blebs, and more rarely some of the crystals are found with an approach to the perfected crystal outline.

It sometimes happens that the quartz and felspar in portions of a granite mass, more frequently in granite veins, have crystallized simultaneously, either mutually interfering with one another or forming parallel growths, as in 'graphic granite'; this may take place either on a large scale, visible to the naked eye, or on a microscopic scale, when it is known as 'granophyric' structure (p. 51). Most of the quartz of granite is clear and glasslike, and for this reason it usually appears dark grey in the fractured or polished surface, since the light which enters it is absorbed by the surrounding minerals, and little is reflected. Some granitic quartz is tinted dark grey, 'smoky quartz,' or it may be faintly yellow, pink, violet, or bluish.

There is no distinct cleavage in granitic quartz, but its place is taken to some extent by the multitude of extremely small cavities, which tend to lie along certain planes within the quartz grains. Frequently it will be observed that within a block of granite there is a uniformity of direction among the planes containing cavities in the majority of the quartz grains; thus for example, we may find one series of approximately parallel planes with a fairly constant north and south orientation, while these are intersected by another series running east and west, or horizontally. The possible significance of these cavities in the process of quarrying will be noted farther on. The cavities have been briefly described on p. 17.

Mica.—The mineral which has taken the third place in order of importance in granites is mica. It will be sufficiently accurate for our present purpose to recognize two kinds of mica, dark and light. The former is present in the great majority of granites; the latter is comparatively rarely found alone, but is frequently associated with the dark kind; sometimes the two are united in parallel growth. The dark mica is dark brown in colour; it is commonly called Biotite, a term which here embraces more than one variety of the mineral (haughtonite, containing much ferrous oxide; lepidomelane, with much ferric oxide). The light mica is often called Muscovite. Mica is usually present in the form of small ragged crystalline plates, readily distinguished in the fractured or polished rock by its colour, irregular outline and softness.

Rarely, granitic rocks are found quite free from mica, consisting only of felspar and quartz.

The minerals which sometimes more or less completely take the place of mica in granite are hornblende, augite, enstatite, and tourmaline. They are all dark in the hand specimen, and not easily distinguishable without the aid of a lens. Granitic hornblende in thin sections is brownish-green or green, showing the characteristic cleavages; granitic augite is usually pale-coloured in microscopic sections; and enstatite is also pale when examined under the same conditions, and is rarely very abundant. The tourmaline in thin sections is blue or brown; some of the Cornish granites (luxullianite) contain large quantities of the mineral.

The essential mineral constituents of granite are those already mentioned—namely, felspar, quartz, and mica or its representatives. In addition to these there are many others—accessory minerals—quite subordinate in bulk, though of great interest to the petrologist; their influence upon the behaviour of the stone as applied to structural purposes is insignificant, and the mention of the names of a few of them will suffice. The accessory minerals include magnetite, pyrite, hæmatite, apatite, sphene, andalusite, cordierite, and zircon; tourmaline may be included here.

Certain other minerals are met with in granites that have undergone more or less decomposition; these will be noticed in the section on weathering.

Texture.—The coarseness or fineness of grain, the disposition of the several minerals forming the rock, and their mutual relationships, are the characteristics embraced in the term 'texture' ('fabric' or 'structure').

To obtain a complete knowledge of the texture of a granite, it is essential that its appearance should be examined in large polished slabs, and in thin slices in the microscope; and these aspects of the stone should be compared with the characters exhibited upon a freshly fractured surface. If attention is confined to one only of these methods of examination, a very misleading conception of the texture of the stone is likely to result.

PLATE II

IGNEOUS ROCKS.

a, Gneiss, Scotland; *b*, Granite, Cornwall; *c*, Granite, fine-grained, with pegmatite vein showing the same minerals on a large scale; *d*, Porphyritic structure, Porphyrite, Cumberland; *e*, Andesite (both black and white specks are mica scales).

IGNEOUS ROCKS

Size of Grain.—There is possibility of considerable latitude in the size of grain in granites; as a general rule there is a fair amount of concordance among the three important minerals in this respect : if the felspars are large, the quartz and mica tend to be large also, although there are exceptions in which quartz, for instance, is proportionately larger than usual.

At the outset it is necessary to distinguish two types of texture—the non-porphyritic, or typical granitic texture, and the porphyritic type, in which one or more of the felspars stands out prominently from the rest by reason of its greater size, frequently associated with greater perfection of form. Large porphyritic felspars may be developed in a crystalline ground mass which is either coarse- or fine-grained, but it is obvious that a stone of the latter kind will act under stresses more like a coarse-grained than a fine-grained one.

It is sufficient for our purpose to estimate the size of grain of non-porphyritic granite by reference to the felspars. Dale regards as *coarse* a granite containing felspars of 1 centimetre, or $\frac{2}{5}$ inch, and over; in a *medium*-grained granite they will be between 0·5 centimetre ($\frac{1}{5}$ inch) and 1 centimetre ($\frac{2}{5}$ inch); while in a *fine*-grained granite they fall below this limit. Many of the granites in constant use are coarser than the limit indicated above, and granites occur in which the grains are very small, 0·175 centimetre (about 0·007 inch), but such fine-grained granites are not much used.

Porphyritic crystals range in size from $\frac{1}{4}$ inch to 5 inches; examples of this magnitude may be observed occasionally in the Colcerrow stone.

In estimating the size of individual grains, either of quartz or felspar, the fact is very often ignored that the apparently homogeneous grains are themselves formed of an aggregate of distinct crystalline individuals in close juxtaposition. This can always be readily verified in the thin slice under the microscope. Bearing this in mind, it is evident that some granites are much finer-grained than they appear.

Mutual Relations of the Minerals.—It has previously been stated that in a granite all the minerals are crystalline; there is no glassy or non-crystalline material; moreover, with the exception of the porphyritic felspar and some of the small minor accessory minerals, it is exceptional for the individual mineral grains to have perfected crystal outlines. The crystalline grains have formed *in situ*—not simultaneously, but in regular sequence, the more basic minerals first, followed by the others in order of decreasing basicity.

Thus, the normal order in most plutonic rocks is approximately as follows: First appear the small accessory minerals, magnetite, apatite, zircon, sphene; then the ferro-magnesian minerals, augite, hornblende, biotite, muscovite; these are followed by the felspars, from the more basic lime-felspars through the soda-lime, to the soda and potash varieties; finally comes quartz and, if it is present, microcline.

This order holds good in a large proportion of granites, but partial reversals of the normal sequence are not uncommon; for example, orthoclase may be found enclosing grains of quartz—hence later—as in Shap granite and some Cornish examples; muscovite may precede biotite, as in Rubislaw granite; orthoclase may precede albite, as in some Cornish granites; and so on.

The texture of granite is frequently influenced by the simultaneous crystallization of pairs of minerals. What appears to the unaided eye as large simple crystals of orthoclase are in reality intergrowths on a microscopic scale of that mineral with one of the more basic felspars. Similarly, biotite and muscovite are occasionally found intergrown. Again, felspar and quartz may have grown together, either on a coarse scale, as in 'graphic granite' (Plate I.), or on a microscopic scale, as in 'micro-graphic' (granophyric) or micro-pegmatitic structure. Here it is interesting to note, as throwing light on the intimate bonding of granitic minerals, that what appear as small isolated specks of quartz in the midst of a felspar crystal are in fact often the portions of large individuals seen in section. It is as though a sponge-like framework of felspar had been filled and surrounded by a single quartz individual.

In general, among granites suitable for constructional purposes, the distribution of the minerals is very regular and constant throughout large masses. The grains are commonly roughly equidimensional, and where they have unequal dimensions (as may be in the felspars and mica) they are arranged haphazard without definite orientation. On the other hand, there are few granite masses which do not exhibit a tendency to orientation of the elongated or flat grains in some portion or other of the mass. The micas may tend locally to lie with their flat faces all in one direction, and the same may apply to the tabular felspars. Such orientation is due to flow movements in the partially consolidated mass, and is known as 'flow structure.' This tendency, developed among the pale tabular felspars in some Cornish granites, produces in the quarry faces and in large polished slabs the beautiful effect of eddies and drifting in falling snow. A slight tendency in this direction is observable in the flesh-coloured porphyritic crystals in large slabs of Shap granite. Flow structure in non-porphyritic granites is best made evident by the mica flakes.

Rarely, granites exhibit an orbicular or spheroidal arrangement of the minerals similar to that described on p. 86 (napoleonite), and occasionally the minerals are found roughly arranged in bands. Irregularities in the normal texture of granites may arise from the following causes:

1. Sudden changes in the coarseness or fineness of grain may appear either in the form of irregular rounded clots in the midst of the uniform granite, or as veins traversing the mass. These clots or veins (pegmatitic veins) may or may not consist of the same minerals as the surrounding mass.

2. Sometimes cavities occur (geodes, druses, vughs), which are lined with perfectly-formed crystals, quartz, felspar, mica, similar to those in the surrounding rock.

3. Dark inclusions, xenoliths ('heathen' of quarrymen) are sometimes local aggregates of the darker, more basic minerals; sometimes they are included portions of the rock into which the granite has been protruded, much altered, of course, by the heat.

Characteristics of Granite in the Mass.—Certain features which are common to most granite masses in all parts of the world may be noticed here; they are *joints*, *rift* and *grain*, *veins* and *dykes*. These structural characters are best observed in quarries, where their presence exerts a considerable influence upon the method of getting the stone.

Joints.—The joints in granites are of two kinds—vertical and horizontal. The former are divisional planes which run through the rock either vertically or inclined at a very high angle to the horizon; they usually extend with great regularity to great depths. It is no unusual thing to see a single joint-face in a quarry 50 to 100 feet high, and apparently reaching to still greater depths. In any single mass of granite these vertical joints are generally constant in direction, and two sets are commonly found approximately at right angles with one another, with only subordinate deviations from these two directions. These divisional planes in the best part of a quarry are of no breadth, the two adjoining faces of rock being closely adjacent; but there is often a thin film of iron stain on the faces, due to the passage of water and air within the cleft. Occasionally it is obvious, from the smoothed and slickensided faces of the joint, that movement has taken place along the plane, the rock on one side having been shifted slightly relatively to that on the other side, either vertically or horizontally.

For the greatest convenience in quarrying, these joints should neither be so near together as to break up the rock into blocks too small for use, nor so far apart as to render the abstraction of large blocks a difficulty. Sometimes, in place of a single clearly defined joint, the granite is traversed by a series of closely adjacent parallel joints; such tracts are a source of inconvenience in quarrying, since they have to be removed as waste or left standing. The stone in the neighbourhood of these joints is often badly stained or weathered. Highly jointed tracts of this kind are called 'end-grain' or 'grain-end' in the Cornish district; 'headings' in some American quarries.

The second kind of joint is horizontal or nearly so; by these joints the rock is split up into 'sheets' or 'beds.' While the vertical joints are strictly analogous with the like structures in sedimentary rocks, the horizontal ones are of entirely different character; they have no relationship with the horizontal bedding planes of the sediments or stratified rocks, although it is usual and convenient to speak of the slabs of rock so formed as beds. They are certainly the result of stresses set up within the mass subsequent to its solidification, but what the nature of the causative stress may be is a problem about which there is considerable diversity of opinion; it is not even clear whether the cause is the same in all cases.

To speak of this kind of joint as 'horizontal' is not strictly correct. When they are well developed in the middle of large granite masses, they are often as near horizontal as may be, and form extensive level floors in the quarries hundreds of yards square; but more frequently they depart from the horizontal attitude to a greater or less extent, and tend to follow with rough parallelism the external form of the granite mass. In all cases, whether they are developed on a large or small scale, they tend to break up the granite into lens-shaped masses. As a rule they are more numerous near the exposed surface of the granite than they are lower down in the mass. The tendency to follow the contour of the granite surface is well exhibited in the Cornish quarry districts and on the borders of Dartmoor, as indicated in Figs. 4 B and 5, but it may be observed wherever the granite has an undulating surface, and near the borders of granite masses.

Except near the weathered surface, the sheets formed by horizontal joints are seldom thin; they range from 6 inches to 30 or 40 feet.

When joints are running 'tight' or close together, the effect of weathering has been considerable, resulting in a rotten, friable stone. Tracts of this character are sometimes referred to by the quarrymen as 'sand,' 'sandbars,' or 'bars'; 'heads' in America. The term 'sand-streaks' or 'sand-seams' is applied to thin micaceous veins in some of the granite of Concord, New Hampshire, U.S.A.

Rift and Grain.—Many granites, though not all, show a marked tendency to split more freely in one direction than another. When this tendency is present, it is due to what is called the 'rift' of the stone. In some cases the cause of the rift is readily ascertainable; in others it is very obscure. In granites in which a flow struc-

IGNEOUS ROCKS

ture is well marked—that is, when there is a definite orientation of some of the minerals—it is clear that the rift is conditioned by this arrangement. Thus, the mica flakes may all lie parallel with one another, with their broad faces and cleavage planes in the same direction; or the felspars may have a similar disposition. In either case both the form of the crystal and its internal cleavage will assist the rock to split.

On the other hand, rift is quite well developed in rocks in which no such regular orientation of the minerals is discernible; it appears then to be due to the effects of stresses, perhaps similar in kind to those which have produced the joints, but differing in degree, and more diffuse in operation.

It has been demonstrated in the case of certain granites that the rift is parallel in direction to sets of minute, rather irregular cracks, which were observed to traverse several crystals, quartz, or felspar, indifferently, without change of direction. Further, in some granites it seems to be associated with the prevalent direction of the sheets of microscopic bubbles and inclusions in the quartz (see p. 17). These sheets of bubbles are very abundant in some granites, and they are often seen to run in two directions, approximately at right angles, both of which are maintained in an irregular manner as the sheets are traced from one particle of quartz to another.

The 'grain' of granite is the direction in which the stone may be split with a degree of ease second to that of the rift direction. [The term 'grain' used in this sense has nothing whatever to do with the coarseness or fineness of grain (texture).] Grain is due to the same causes which induce the rift, but they have operated in a lesser degree. The direction of the grain is approximately at right angles to that of the rift.

There are many peculiar features in the disposition of

rift and grain within granite masses which have not yet been clearly explained. For example, the rift may be very constant in direction in granite masses extending over a considerable area; on the other hand, it may vary in closely neighbouring quarries, or it may even be different in direction and degree in adjacent sheets within the same quarry. It may run vertically or horizontally, conforming to the direction of one or other of the three main sets of joints; or it may run diagonally across from joint to joint; or in one part of an area it may be horizontal or vertical, and gradually pass over into an intermediate or inclined direction towards another part of the area.

To whatever cause the rift may be due, its influence upon quarrying operations is most marked. Where it is well developed the stone may be split with ease by plug and feather, while in its absence blasting would be imperative. This means, of course, a saving both in material and labour. Certain granites cannot be economically fashioned into setts or Klein-pflaster, simply because this property of ready splitting is lacking.

A curious feature in the behaviour of some granites is that the rift is more perfect if the stone is split from one direction than from another. This is illustrated in the diagram (Fig. 4A): if the plugs are placed in the face b, the split runs true and square; but if they are placed in the face a the split tends to leave the right plane, and turns off diagonally or irregularly, as shown by the dotted line.

Besides being of so much importance in quarrying the granite, the presence of rift has a marked influence upon the behaviour of the stone under pressure, although I have never seen an example of a granite block that had suffered in a building through being laid with the rift vertical; it is, indeed, extremely difficult or impossible to recognize even in old blocks when they are set in position.

The three directions in which blocks are cut in the

IGNEOUS ROCKS

quarry usually receive distinctive names from the quarrymen; thus, they may be designated the 'rift,' the 'grain,'

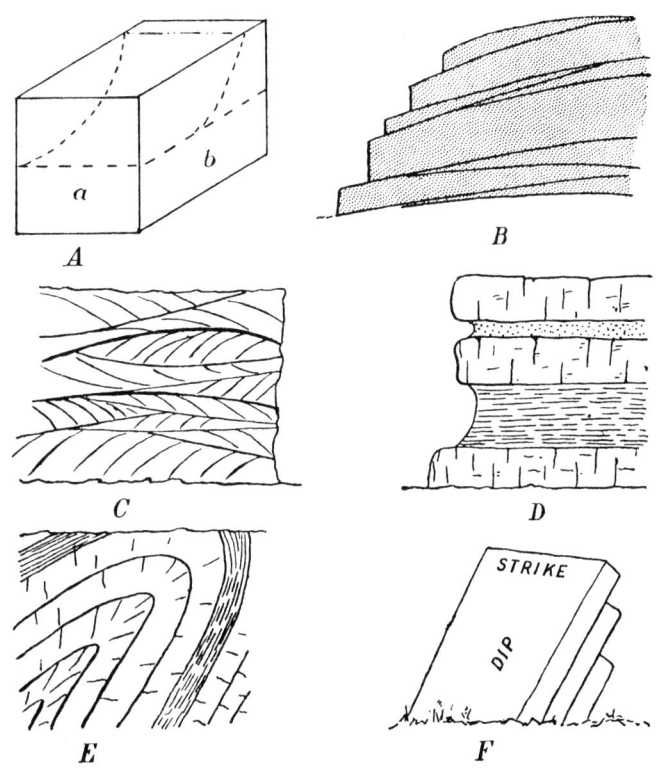

FIG. 4.

    A. Rift and grain in granite.
    B. Pseudo-bedding in granite.
    C. Diagonal or current bedding in sandstone.
    D. Horizontal bedding: rocks of different kinds.
    E. Folded strata, originally horizontal.
    F. Diagram of the relations of dip and strike.

and the 'hard' or 'hard-way'; or the 'cleavage-way,' the 'quartering-way,' and the 'tough-way.' These terms are employed somewhat differently in different granite districts, and care has to be exercised, when visiting the quarries, to avoid confusion due to thinking that the quarryman means what you mean by the same expression. Thus, while the rift is always the direction of most easy cleavage, it may be coincident with what in one quarry is called the 'cleavage-way,' and in another with what they call the 'quartering-way.' Careful observation in the quarry will usually enable one to distinguish the three directions even in detached blocks while in their rough state; but it is sometimes by no means easy, and requires considerable experience.

Dykes, Veins, and Allied Structures.—The regular homogeneous character of the granite in many quarries is occasionally interrupted by portions which differ from the main mass, both in texture and composition; and where they are numerous or developed upon a large scale they may interfere seriously with the processes of quarrying.

The most striking of these variations are the dykes, sheets of igneous rock which traverse the granite in any direction, either as simple wall-like masses and tabular layers or as branching and anastomosing intrusions. In all cases the dykes are younger than the granite in which they lie. As regards composition, dykes may be very similar to the surrounding rock, differing merely in the average size of grain; or they may be more siliceous (acid) than the granite, or less siliceous (basic).

The most common form of siliceous dyke is the rock known as *pegmatite*, in which it is customary to find the same type of felspar and mica as in the surrounding rock; but the individual crystals are much coarser, and often very perfectly formed. To anyone familiar only with the granite of commerce, the size attained by the crystals in some pegmatites would excite no little wonder; for it is not uncommon to find mica in tabular masses from 1 foot to 4 or 5 feet in diameter, or felspars many feet in length and of proportionate girth. One crystal of felspar in the

University Collection at Christiania is about 7 feet long. In some pegmatite veins or dykes a parallel intergrowth of the felspar and quartz (pegmatitic structure) is a marked feature. In addition to the ordinary minerals of the surrounding granite, other and rarer species are frequently developed, and may assume commercial importance.

Veins or dykes of granitic rock in which paucity of mica is associated with comparative fineness of grain are known as *aplite*. The majority of veins or dykes of pegmatite and aplite are narrow, from a few inches to a foot or two; but they may occasionally be as much as 20 to 30 feet in width, and where the pegmatite is as a marginal feature of the granite, it may assume great dimensions, reaching to hundreds of yards. Similar variations of texture and composition are found in the midst of some granites in the form of spheroidal or irregular masses.

It has been indicated above that dykes and veins interrupt the regular working of the quarry, and the material therefrom is usually waste; but in some districts they are put to use in the form of copings for garden walls, and in the formation of rockeries and stone borders. Blocks of pegmatite have been brought to the London district from Scandinavia for rockeries, in which, by reason of their bold 'figure,' they make an attractive substitute for the familiar and sordid-looking clinker or burnt brick.

It should be pointed out that the characteristics of pegmatite—namely, coarseness of crystallization and a tendency to greater acidity than the parent igneous rock— are not limited to the granites, but may be observed in relation to other more basic rocks—syenite, gabbro, etc.

In all cases—whether their mode of occurrence be as broad dykes, narrow ramifying veins, druses or geodes, coarse isolated patches, or in the form of marginal tracts— the pegmatites may be regarded, according to Harker, as representing the residual 'mother liquor' at the end of the process of crystallization—a view which readily accounts for the peculiarities of structure and composition of this rock variation.

The more basic dark veins and dykes that are sometimes found penetrating granite masses have usually no genetic relationship with the rock they traverse, but are of a newer and later phase of activity. The small 'knots' or rounded masses, rich in dark minerals, mica, hornblende, and the like, are usually early basic segregations or secretions formed from the liquid crystallizing magma. If the knots are small they are disregarded in quarrying and dressing the stone, but when they are large they are treated as waste. They are to be seen in blocks of some granites

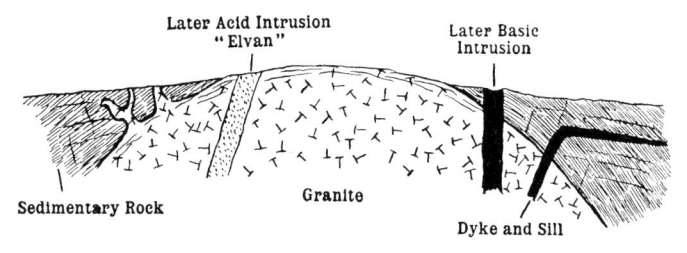

Fig. 5.—Granite and Elvan Dykes and Veins of Basic Rock.

when set up in buildings, but they should not be permitted to appear in stones that have to take any moulding or carving. Dark included masses, often similar in appearance to the basic segregations, are portions of the surrounding rock caught up by the granite during its intrusion.

Dykes and veins receive various names from the quarrymen, which differ in different districts. Practically all the veins in the Cornish area are called 'elvans' (see p. 94) whatever their petrological characters; some are diabases and similar rocks, but the majority are micro-granites and quartz-felsites or quartz-porphyries. Sometimes the veins are called 'horses,' white or black as the case may be.

In the immediate neighbourhood of dykes the granite is frequently unfit for use, and has to be left standing or removed as waste.

Colour.—Except in the case of engineering works, the colour, or rather the appearance, of granite is the quality which perhaps more often than any other determines the final selection of the stone. Notwithstanding this fact, the variable nature of the causes which produce the colour effect and the subtle nuances of tint and texture in the rock make it wellnigh impossible to convey in words the innumerable different impressions that the mind readily assimilates through the eye.

It is easy to speak of dark red, red, pink, flesh-colour, greenish, and grey granites, and to convey thereby a general idea of the colour; but when it comes to distinguishing between the numerous pinks and the infinite variety of greys, language fails, and nothing can take the place of a personal acquaintance with the stones themselves.

The dominant colour of granite is nearly always determined by that of the felspars (*q.v.*), but this is modified by the colour and disposition of the other minerals, the quartz, mica, or hornblende, and still further by the degree and kind of alteration that has taken place in these minerals by the production of secondary minerals.

The *appearance* of granite is influenced by more than the mere colour of the fresh stone; it is compounded of the several effects produced by the state of aggregation of the minerals—felspars of one or more kinds, porphyritic or non-porphyritic, large or small; quartz in small individuals evenly distributed, or in large clots of grains—the presence or absence of flow structure or of basic knots at frequent intervals; the presence or absence of bright cleavage faces in the felspar, which may glitter and shine by the light they reflect; and finally by the nature of the surface given to the stones by polishing, axing, picking, etc. Polishing and rubbing invariably produce a darker colour effect than fine-axing.

Chemical Composition ; Mineral Composition.—The chemical analysis of a granite informs us of the composition of the aggregate of minerals which compose the rock; if, therefore, we are familiar with the average composition of the essential minerals, we may foretell the rock analysis with sufficient accuracy for most practical purposes, if the minerals can be recognized and their proportionate representation in the stone can be estimated. The necessary determinations can be made by observation and simple measurements on thin micro-sections of the stone, and on a polished face or clean fracture. For this reason the two characteristics—mineral and chemical composition—will be considered together.

Given a stone with granitic structure, be it coarse or fine in texture, it may be regarded as a granite if its silica content does not fall below a minimum of 66 per cent. This is the normal minimum for the silica content; it is true that some abnormal rocks with a slightly lower percentage would be classed with granites by some petrologists, but they would not be so recognized by others.

A rather curious error is not at all infrequently made by many who read analyses prepared for trade purposes; it is to regard the silica expressed in the analyst's report as representing the crystallized silica, quartz (this applies to all analyses of stone). This, of course, is by no means true. The total silica content embraces not only the free quartz, but that which is combined with other elements in the other minerals.

The analyses of granitic minerals in Table V. will make this clear:

IGNEOUS ROCKS

TABLE V.

	*1.	2.	3.	4.
SiO_2 (Silica)	67·99	47·95	35·55	54·89
Al_2O_3	19·27	30·26	17·08	1·50
Fe_2O_3	0·28	2·43	23·70	5·06
FeO	—	3·10	3·55	7·46
MnO	—	—	1·95	—
MgO	0·02	0·94	3·07	16·01
CaO	0·75	0·98	0·61	12·08
Na_2O	6·23	2·00	0·35	0·37
K_2O	3·05	10·25	9·45	0·38
H_2O	0·90	2·85	4·30	2·72
Total	99·03	100·76	99·61	100·47

* 1, Felspar, Anorthoclase (or AbAn); Minnesota, U.S.A. 2, A White Mica, Muscovite; Fichtelgebirge. 3, A Black Mica, Lepidomelane; Ireland. 4, A Hornblende (Green); Schwarzwald.

These analyses will serve to illustrate how widely, not only the silica, but almost all the other simple molecules, are distributed among the main constituent minerals.

There is a sufficiently wide range among the granitic felspars to permit of a simple classification of these rocks into *alkali*-granites and *lime-alkali*-granite. In the former group, depending upon the relative predominance of potash (or soda-bearing) felspars, we have *potash-granites* (with orthoclase, microcline, or perthite) or *soda-granites* (with albite, anorthoclase, or soda-orthoclase); in the latter group a certain amount of lime-bearing felspar (oligoclase) is present along with, but not in excess of, the alkali felspars. These varieties of granite pass by insensible gradations one into another, and by the gradual increase in the proportion of lime-bearing felspars they shade off into rocks which are near to diorites (p. 85);

the intermediate stages are the grano-diorites or quartz-monzonites of some authors. In order to have some agreement as to the limits of these arbitrary subdivisions, it has been proposed to class them as follows:

Most alkaline	...	Alkali-granites, in which more than two-thirds of the felspars are alkali-bearing.
Medium alkaline	...	Intermediate, 'adamellite' type, in which less than two-thirds and more than one-third of the felspars are alkali-bearing.
Least alkaline	...	Grano-diorites, in which less than one-third of the felspars are alkali-bearing.

If it could be clearly proved that the granites with lime-bearing felspars were actually the less durable as employed in constructional work, it would be worth while to emphasize the importance of this classification. In the present state of knowledge, however, it does not appear to possess much practical significance.

Another mode of distinguishing the varieties of granite is to group them according to the dominant coloured or ferro-magnesian mineral. Thus, if dark mica alone is present the rock is a *biotite-granite*—this is a very common type; a *muscovite-biotite-granite* contains both light and dark mica; a *muscovite-granite* is almost devoid of dark mica; if hornblende is associated with dark mica the rock is a *hornblende-biotite-granite*; with hornblende alone it is a *hornblende-granite*. Then there are the less common varieties *tourmaline-granite, augite-granite, riebeckite-granite,* and so on.

The student should bear in mind that this nomenclature has been employed in a very rough-and-ready way by writers.

For the purpose of giving a general idea of the composition of granites from different parts of the world, a few illustrative examples will now be considered:

1. *Red Granite: Percentage of Minerals estimated by Measurements.*

	1	2.	3.	4.
Felspars	65.30	55.91	49.92 to 70.83	68
Quartz	28.65	35.66	23.04 to 41.08	27
Mica (biotite)	5.55	8.43	4.72 to 11.29	5

2. *Grey Granite.*

	5.		6.	Average.		7.
Felspars	58.86	Felspars	55.80 to 69.51	60.02	Felspars	50.05
Quartz	33.88	Quartz	22.06 to 33.71	30.60	Quartz	37.55
Hornblende	7.26	Riebeckite and Ægirite	7.47 to 11.10	9.37	Mica	12.40

A more detailed estimate of the minerals in Delank granite obtained by Rutley with Delesse's method yielded the following result:

Felspar, orthoclase	30
Felspar, plagioclase	6
Mica, biotite	7
Mica, muscovite	11
Quartz	46
	100

Compare with this the result of measurements on a fine-grained grey granite from Milford, New Hampshire:

Felspar, soda-lime (oligoclase)	34.03
Felspar, potash (microcline, 14.15; orthoclase, 15.57)	29.72
Mica (biotite)	8.58
Quartz	27.09
Magnetite	0.25
Minor accessories	0.33
	100.00

It may be of interest to compare the chemical analyses with the mineral percentages in the foregoing table; the numbers at the head of the columns in Table VI. correspond with those used on p. 65.

TABLE VI.

Chemical Analyses of Some Red Granites

	1.	2 and 3.			4.
SiO_2	71·44	72·02	76·07	77·08	68·55
Al_2O_3	14·72	14·43	12·67	12·54	16·21
Fe_2O_3	2·39	1·25	2·00	—	2·26
FeO	0·46	0·89	—	0·95	—
CaO	—	1·8	0·85	0·75	2·40
MgO	0·96	trace	0·10	0·01	1·04
Na_2O	7·66	5·85	3·37	3·54	4·08
K_2O	0·89	5·41	4·71	4·93	4·14
TiO_2, ZrO_2, etc.	0·78	—	—	—	—
H_2O (ignition)	0·61	0·35	—	—	—
	99·91	0·71	99·80	99·96	99·13

Chemical Analyses of Some Grey Granites.

	5.	6.	7.	8.
SiO_2	77·61	73·93	72·84	72·05
TiO_2	0·25	0·18	—	—
Al_2O_3	11·94	12·29	16·25	15·83
Fe_2O_3	0·55	2·91	0·14	0·39
FeO	0·87	1·55	1·49	1·50
CaO	0·31	0·31	1·10	1·14
MgO	trace	0·04	0·55	0·51
Na_2O	3·80	4·66	2·25	2·65
K_2O	4·98	4·63	5·19	4·79
MnO	trace	trace	—	—
H_2O (ignition)	0·23	0·41	0·63	0·64
	100·54	100·91	100·44	99·50

Brief descriptions of these granites are appended:

1. *Redstone Granite, Conway, New Hampshire.*—Coarse, pinkish, mottled, with amethystine grey quartz and spotted with black. A biotite granite passing into a biotite-hornblende granite. The pinkish felspar is orthoclase minutely intergrown with soda-lime felspar (oligoclase-albite); there is also a little oligoclase-albite separately crystallized.

Compressive strength, 22,370 pounds per square inch.

2. *Milford Granite, Massachusetts.*—A biotite granite of medium to coarse texture, not porphyritic, even-grained. The felspars are pink to cream-coloured. Orthoclase and microcline minutely intergrown with soda-lime felspar, a subordinate yellowish to white albite to oligoclase-albite, slightly kaolinized; faint blue quartz somewhat fractured. No. 3, p. 65, shows the range of estimated mineral percentages in the more gneissose stone from the same locality. Nos. 2 and 3, Table VI., show the analyses from three neighbouring quarries in the district.

Compressive strength, 20,000 to 29,200 pounds per square inch.

4. *Shap, Cumberland* (p. 70).

5. *Rockport Grey Granite, Pigeon Hill Quarry, Massachusetts.*—A hornblende granite, of medium to coarse texture, even-grained. Colour, medium grey, sometimes with greenish or bluish tinge, with black spots. Felspars, grey orthoclase with occasional microcline, with minute intergrowths of albite to oligoclase-albite; very little of the latter kind is separately crystallized. The quartz is smoky, and contains many cavities and bubbles. The hornblende is dark, and a very little black mica is present.

Compressive strength (three tests), 20,716, 20,522, 17,772 pounds per square inch.

6. *Quincy, Massachusetts.*—A riebeckite-ægirite-granite. Colour, a medium grey with greenish, bluish, or purplish

tinge, passing to a very dark bluish-grey, with black spots. The trade names for these shades are, 'medium,' 'dark,' and 'extra dark.' Texture even, medium to coarse. Felspars, orthoclase, with minute intergrowths of albite to oligoclase-albite. The several colour shades in the grey felspar are produced by minute inclusions of epidote (green), hornblende (dark brown), and riebeckite (bluish). Some of the lime-soda felspar is crystallized separately. Riebeckite, looking blue black, and ægirite, greenish-black, take the place of the usual mica, etc. The quartz is smoky and bluish. The stone takes a high polish.

7. *Penryn, Cornwall* (see Appendix A).—Compressive strength, 19,450 pounds per square inch, 1,250 tons per square foot.

8. *Carnsew, Cornwall* (see Appendix A).—Compressive strength, 22,336 pounds per square inch, 1,436 tons per square foot.

Physical Characters of Granites.—In a discussion of the physical properties of any rock group, it must be constantly borne in mind that we are dealing with materials which show a remarkable amount of variability. Take the granites, for instance; even if we confine our attention strictly to those rocks which are granites in the petrological sense, how diverse is the material in its obvious structural conformation alone, to say nothing of those obscure differences of rift and fracture which make one granite easy, another difficult, and another impossible, to work with profit.

In this and subsequent chapters dealing with other stones, the remarks upon the physical characters are therefore to be interpreted as generalizations only.

Specific gravity	2·6 to 2·8.
Weight per cubic foot	160 to 200 pounds.
Resistance to pressure	Usually from 1,000 to 2,200 tons per square foot, but samples may fall as low as 800, or rise as high as 2,740.

Distribution of Granites.

England.—All the granites now worked in the West of England are grey. Reddish granite does occur on a small scale in the St. Austell district, but it is quite unimportant. The red granite of Trowlesworthy Tor, which has a pleasing but rather peculiar colour—not pink—is not now quarried.

The constituents of *Cornish granite* are orthoclase and micro-perthite, often large and porphyritic, quartz, and a fair amount of plagioclase which is largely albite. Both light and dark mica are present, and tourmaline is a common accessory mineral. The large porphyritic crystals are not present in all the stones. They often exhibit zonally-arranged inclusions of dark mica, and not infrequently twinned crystals appear in section as white crosses. The large crystals sometimes lie with a fairly regular orientation, but this is not strongly marked as a rule. On the borders of the Penryn mass, in the Bodmin mass, and elsewhere, a granite of much finer grain appears; it is rarely quarried, however, beyond the upper stained portions, though it appears to be a good stone.

One of the standard textbooks on building materials refers to Cornish granite in rather disparaging terms; it is hardly necessary to-day to point out how ridiculous it is to compare the locally kaolinized granite—here the work of deep-seated agencies—with the granite in its normal condition. Nor is it any longer correct to refer to the Cornish granite as 'moorstone,' a name which was quite applicable in the early days of the industry prior to the formation of deep quarries. The earlier stones were obtained from the large loose blocks which in those days were, and are still in many places, strewn over the moors. (For further details as to granite quarries see Appendix A.)

In the *Midland Counties* true granite is found only in

Worcestershire and Leicestershire, and in neither district is the granite quite of the normal type. In Worcestershire a gneissose granite occurs at North Hill, and a hornblende granite appears north of Wych, Great Malvern. The Mount Sorrel granite in Leicestershire is richer in plagioclase than ordinary granite, orthoclase is subordinate, and hornblende occurs in addition to biotite. The rock has in parts the character of a grano-diorite. The stone varies a good deal in colour; a light kind is rather dark grey, with paler roundish felspar crystals scattered rather sparingly in a smudgy, indefinite-looking ground mass containing grains of bluish-grey quartz and patches of dark hornblende. Another darker variety has the same structure, but the ground is dark reddish-brown, and the porphyritic felspars are paler red. Yet another variety is very fine-grained, without any obvious structure, and of a rather dead brown appearance. Some varieties are greener than others; it is not much used for buildings, although it takes a good polish, but it is largely quarried for setts and macadam.

Passing northward, no more granite appears until *Shap Fell* is reached in Westmorland, where the rock crops out as a compact mass some two and a half miles by two miles. This rock has a very striking appearance on account of its distinctive colour, a rather brownish-red, with large porphyritic crystals of flesh-coloured orthoclase; the finer-grained part of the stone occurs in several shades from grey to warm brownish-red (Light and Dark Shap). The porphyritic crystals frequently exhibit a roughly parallel and meandering flow arrangement. The rock is a biotite granite with plagioclase and orthoclase; it is of interest to note that, contrary to the usual rule, the quartz is enveloped by the orthoclase.

Shap granite is obtainable in large blocks, and has been much employed as a decoration stone; it polishes very

IGNEOUS ROCKS

MAP I.—IGNEOUS ROCKS OF ENGLAND AND WALES.

well. It is also used in engineering work, dressings, macadam, and for concrete paving slabs and artificial stone. It has been used in all parts of England, in America, and other places abroad. It may be seen in the columns at St. Pancras Station, the posts round the enclosure at the western entrance to St. Paul's Cathedral, Temple Bar Memorial, and very many shop-fronts; also in the graving-docks, Southampton, the harbour works, Heysham, North-Eastern Railway, and the viaduct over the Severn at Shrewsbury.

A large mass of granite occurs at Eskdale, in Cumberland; it is irregular in shape, about thirty-five square miles in area, and extends from Wastdale and Boot, near Eskdale, southwards by Muncaster and Dayockwater nearly to Bootle. There are several varieties, but most of it is coarse-grained, grey, and porphyritic; the felspar is mainly perthite with orthoclase and oligoclase, abundant quartz, and both light and dark mica. Some of this rock is granophyric, and the quartz in some cases has crystallized early, as in the Shap granite. It has not been much worked.

Scotland.—The principal occurrences of granite in Scotland are those of Kirkcudbrightshire, Aberdeenshire, and Kincardineshire; to these must be added the granites of the Highlands and the Western Isles, most of which are too inaccessible to be of value for commercial purposes.

The Kincardineshire-Aberdeenshire granites, embracing the well-known rocks of Peterhead and Aberdeen, cover a large area, and present many points of great interest to the petrologist; but over a great part of the region much of the stone is unsuited for constructional purposes. Most of these worked rocks are biotite granites, with oligoclase in varying proportions along with orthoclase or microcline; on the whole they are less like grano-diorites, and more siliceous, than those in the Galloway district.

IGNEOUS ROCKS

In the *Aberdeen* district, where granite is essentially a material for construction, preference is given to certain kinds of the stone for specific purposes. For domestic architecture, ashlar stone is generally specified to be from Rubislaw, Kemnay, Sclattie, Oldtown, Toms Forest, and Tillyfourie; while Kemnay is specified for lintels, sills, and strong courses, or wherever finely dressed parts are required for emphasis. This stone has a very bright appearance when finely tooled. The darker grey granites, such as Rubislaw, are very effective in rough ashlar. Of the coloured granites, Corrennie is much used. Kemnay, Tillyfourie, and Toms Forest, are used for engineering work, and when dressed are only distinguished, if at all, with difficulty.

In the *Peterhead* district the characteristic stone is a biotite granite, with microcline and some plagioclase, and smoky or clear quartz. The texture is moderately coarse and not porphyritic; the common colour is a dark flesh tint; the grains of the constituent minerals have ill-defined boundaries, but stand out clearly. The stone from the Cairngall quarry is of a very beautiful cool grey colour, with small irregular-shaped white felspars sprinkled in the grey ground; good examples of the polished stone may be seen in the wall decoration of the hall of the Museum of Practical Geology, London. Large blocks were used in the Prince Consort's sarcophagus; eight columns in St. George's Hall, Liverpool (shafts 18 feet high); round the fountain at Trafalgar Square; the lintel over the door of the Duke of York's Monument; and the pedestal of the Duke of Wellington's statue at the Royal Exchange, Glasgow.

The red Peterhead stone is almost exclusively used for polished work. It has been employed in the Duke of York's Column (1830); the columns of St. George's Hall, Liverpool; in the Fishmongers' Hall; and many monuments and shop-fronts.

The granites quarried in *Kincardineshire* are similar to those in the Aberdeen district; the Cove quarries are the most active. The stone is dark grey, and of medium-sized grain; the felspars are small and white, and the quartz is in rather larger grain; a roughly-marked foliation is sometimes apparent. The stone is mostly used for roads. Nigg Quarry, a little north of Cove, yields a finer-grained, darker grey stone. From the Hill of Fare quarries, north of Banchory, a fine-grained dark red stone, finer-grained than Peterhead, Birsemore, or Corrennie, was obtained and employed in a memorial to Queen Victoria, Windsor Castle, and in the polished part of the Byker Bank, Newcastle. In an axed state it may be seen in the single-span bridge at Kelvin Park, Glasgow.

The granite of *Kirkcudbrightshire* and the adjoining counties appears protruding through the sedimentary rocks in three large and a number of small quarries. The large protrusions are (1) Criffel and Dalbeattie; (2) the Cairnsmore of Fleet; and (3) the Loch Dee mass. All the principal quarries are situated in the first of these. The stone is usually a bright clean grey; in mineralogical composition it is subject to some variation, but generally it is much richer in lime-bearing felspar than the Aberdeen rocks. The Criffel and Dalbeattie stone is moderately fine-grained, and consists of a white oligoclase felspar in a finer ground mass of quartz and orthoclase, together with dark mica or hornblende, or both—sometimes with augite. The rock is evidently in parts much nearer a quartz-diorite than a normal granite.

For further details as to quarries of Aberdeen, Peterhead, and Kirkcudbrightshire granite, see Appendix A.

Granites of the same period as those worked about Aberdeen occur in Nairn, Elgin, and Inverness, Ben Rinnes, Grantown, and farther north at Ben Loyal, where there are considerable quarries; in Strath Halladale, Lairg,

IGNEOUS ROCKS

and large masses on the borders of Sutherland and Caithness, and in the Ord of Caithness.

The granites of the South-West Highlands are richer in plagioclase than those of Aberdeen, and more closely resemble those of Kirkcudbrightshire and Wigtonshire. The larger masses are those of Rannoch Moor, Ben Cruachan, and Loch Etive; here are the large Bonawe quarries at Taynuilt, conveniently situated near the loch. This is a strong, moderately fine-grained stone, greyish-blue or greenish-grey in colour, a shade darker than Kemnay, with much plagioclase, little quartz, biotite, a little green hornblende, and some augite. It is mainly used for setts and macadam. In the same county are the quarries at Craigmore, near Taynuilt, Ardshiel, and Blackwater Dam at Ballachulish. This is moderately fine-grained, rough when axed, but with a lively appearance when polished; it is darker than Bonawe stone; the felspars are pale salmon or buff, evenly mixed with a white variety (not porphyritic), and a considerable amount of mica.

Some of the granites of Loch Etive are pale brown or brown in tone, of coarse to medium-coarse texture; that from Bars has both white and pinkish felspars developed porphyritically; the High Rock stone is less regular in appearance.

A red granite is found in Argyllshire, with bright red felspars in clear white quartz; both minerals form irregular grains.

An extremely coarse rock (pegmatite) from Portsoy, Banff, consists of irregular masses of pink felspar along with great masses of quartz full of strong flakes of mica.

In the Western Isles there are several granites, many of which are much younger than those of the mainland. Most of the granites of Arran, that of St. Kilda and Beinn an Dubhaich in Skye, are biotite granites. Those of the

Red Hills in Skye, and in Mull and Rum, have a lower silica percentage, while hornblende and augite take the place of mica. The Ross of Mull granite is warm red in colour, with red orthoclase and some white plagioclase. It is used in the piers of Blackfriars Bridge.

The fine-grained drab-coloured rock of Ailsa Crag may be mentioned here as an example of a riebeckite-microgranite.

The rock largely quarried at Furnace, Lochfyneside, though known as a granite, is a quartz-porphyry; it is used for setts, curbs, and crushed stone. In appearance it is much like some of the white, speckled-grey Cornish elvans.

Ireland.—Times without number it has been pointed out—quite truly—that there is abundance of good granite in Ireland. The principal quarrying district embraces the irregular granite mass near Newry, Co. Down, the smaller mass between Carlingford Lough and Dundale, and that forming the Mourne Mountains.

The Newry granite is much older than the other two; it consists of quartz, orthoclase, and an alkali-lime felspar, with biotite and hornblende. The colour is generally greenish-grey or grey-blue. Setts constitute the main product from this district, but large blocks may be obtained, and the stone has been widely employed as a building and polished monumental stone.

The next important granite district is in the South-East of Ireland, where a large elongated mass stretches from the coast at Kingstown in a south-westerly direction for about fifty-six miles, mainly in the counties of Carlow and Wicklow. The bulk of this rock consists of quartz, potash felspar, often microcline, subordinate plagioclase (albite and oligoclase), and both light and dark mica. Except in local variations, its texture is moderately coarse. The principal quarries in this granite are Parnells (Arklow), Glencree, Ballyknocken, and Ballybrew. In Dublin County there are numerous small quarries producing granites of medium texture and various shades of grey: Glencullen, Barnacullia, Ballyedmanduff, and Balally. Some of the local granite used in Dublin City has been very badly selected. Kingstown Harbour was built with stone from Dalkey quarries, and it is used in the churches of St. Paul and St. Werburgh, Dublin. Quarries on Killiney Hill supplied stone for the Thames Embankment, the pier and harbour, Kingstown, and many large buildings in Dublin. (See Appendix A for Irish Granite Quarries.)

Granites (silvery grey and red) have been quarried in Donegal, at Fanad, Mulroy, and Fairy Castle quarries. Good red granite is found in Galway; some of it is rather like a fine-grained Peterhead in general tone, but with a tendency to have large pink porphyritic felspars, in a ground of pink and cream-coloured felspars, abundant quartz, and a moderate amount of mica; other varieties are rather coarser, and some blue-grey stone also occurs. The principal quarries are at Shantallow.

In the **Isle of Man** granite is worked in the Dhoon quarries near Ramsay, and in the Foxdale quarries; these stones are muscovite-biotite granites with much oligoclase, some microcline, and abundant quartz; it is of medium texture and light grey colour. The Dhoon stone is more porphyritic than the other.

A biotite granite is quarried at St. Brelades, Jersey.

In **Wales**, although there is a great development of igneous rock, there is little granite; the largest outcrop is in the Lleyn Peninsula, Carnarvonshire. This is five miles long from north to south, and about two miles in breadth. The rock is a biotite granite with altered orthoclase and some oligoclase.

Some Foreign Granites.—Of the foreign granites, those which exercise the greatest influence on the British trade come from Scandinavia. Large quantities of stone are shipped for curbs, setts, ornamental, engineering and architectural work.

For ornamental purposes a good deal comes to Scotland to be polished.

Sweden.—On the west coast the Bohuslän district yields many grades of stone, from light grey to pale red in colour, and fine to medium-coarse in grain. There are quarries at Lysekil (coarse reddish), Idefjord, Malmön, and many other places.

In the Halland district much of the stone is gneissose, and variable in colour; it is largely used for curbs. The 'Varberg granite' is a grey, red, or dark green rock, used for polished work; it does not weather well. From the western side come the red granites of Vånevik and Virbo in Småland; these have a medium to coarse grain with the quartz; they are known as 'Red Swedish' in Britain and America. The brilliant crimson Uthammar granite is called 'Bon Accord Red' (some of the Swedish Bon Accord is an olivine-gabbro). From Jungfrun Island a red stone known as 'Virgo granite' is obtained. Red and grey granites are worked near Norrtälje, for use in the Stockholm district. The 'Stockholm granite' much resembles the Aberdeen grey. From the Graversfors district the 'Swedish Rose' granite is obtained; this has dark red felspar with deep blue and purple quartz. Other varieties are coarse-grained with brown felspar and blue quartz. The granites and gneiss of Blekinje—fine-grained, red, and grey—are mainly used for curbs and setts, and sent to Germany.

Norway.—Great quantities of curbs and setts are obtained from the Norwegian granite, but the principal architectural stone comes from the Liholt quarries in the Idefjord. This is an excellent fresh grey stone of medium grain; it has been much employed in large buildings in

Christiania—*e.g.*, the New Theatre, Freemasons' Building, Sparebank; and in the Royal Liver Building, Liverpool, and many others in this country; it is known as 'Grey Royal.' A similar stone from this district is the 'Imperial Grey.'

The United States of America.—Granites are worked in many of the Eastern and Western States; the stone is often gneissose in structure. In Maine the proximity to deep water, and the freedom from 'stripping' or overburden, produce favourable conditions, similar to those which obtain in Scandinavia. The most extensive quarries are those of Vinalhaven or Fox Island, a coarse stone, grey to pink in colour. The Augusta and Hallowell quarries yield a light grey stone used for the large figures in the Pilgrim Monument, Plymouth, Mass. The Mount Desert and Crotch Island granites are light to dark grey and dull pink. Red Beach is pink to red; Otter Creek, medium-grained, dull red; Mount Waldo, light grey; East Blue Hill, grey to pinkish, and sometimes porphyritic.

The Quincy, Rockport, and Gloucester granites of Massachusetts are rather coarse granites and gneisses, often hornblendic, and dark blue-grey to pink in colour.

In Wisconsin the principal granites are Montello, bright red and grey-blue; Waushara, pale pink and red; Amberly, fine grey to coarse red; Warsaw, grey, brown, and brilliant red; Granite City, reddish-grey and red; Waupaca, coarse, with brown, red, or pink felspar in a greenish to black ground.

According to Merrill, a very beautiful coarse red granite is found near Lyme in Connecticut; the red felspars have the unusual property of being clear and transparent; it is used in Newport and Rhode Island.

Granite is found in California, Georgia, Maryland, Minnesota, Utah Territory, Virginia (fine grey), Washington, New Hampshire, and several other States.

In *Canada* there is plenty of good granite, but it is not

yet exploited to any great extent. In Eastern Quebec, near Stanstead and Staynerville, the pale grey granite is quarried for roads and for building. Red granite has been worked at Kingston, Ontario, and at Barrow Island, Chatham, Wentreath, Granville. The so-called 'Bay of Fundy' granite is a medium-grained red hornblendic rock from St. George, New Brunswick. In Halifax County, Nova Scotia, a grey granite is quarried; a grey stone is found at Burard Inlet, British Columbia.

France.—In France granite occurs in Normandy, Brittany, the Vosges, and in the Central Plateau.

The following are a few selected types: Laber granite, near Brest, occurs in large blocks; its colour is grey, pink, and blue, fine-grained, with large porphyritic felspars; used in sea-walls at Bordeaux, Cherbourg, Havre, and Thames Embankment, and in the pedestal of the Luxor Obelisk, Paris. Alençon granite, fine to medium grain; pale yellow to grey-blue; used in the principal buildings of Alençon. Iles Chausey granite, near Granville; fine-grained, hard, bluish-grey. Combourg, medium grain, bluish. Pontivy, moderately hard, yellowish-grey. Vosges granite, Geradmer; porphyritic, variable colour. Remiremont, very hard, bluish-grey. The granites of Chamonix and Epierre are talcose or protogine granites, hard and white, spotted with green. The so-called Kersanton granite, from near Brest, is a dyke rock consisting mainly of plagioclase felspar, dark mica, and quartz; it is not a true granite, but a variety of dyke rock called Kersantite. It is largely quarried and very durable; it is fine-grained, and light bluish or greenish-grey, which darkens on exposure; the sixteenth-century church of Vannes is built of it, and it is used for architectural and engineering work.

Austria.—Most of the granites are grey or bluish-grey, fine and coarse; they are quarried in the south at Meiszau and Gmünd in the Tyrol; in Steiermark; Budweis,

Carlsbad, Przibram, Petersburg - Jechnitz (Pilsener granite), and elsewhere in Bohemia; Setzdorf and Friedeburg in Silesia; in the north at Mauthausen (much used in Vienna), Dornach, Neuhaus, etc.

Germany.—The granites of Saxony are worked in the districts of Meissen, Oberlausitz, and in the Erzgebirge and Fichtelgebirge. Those of the first-named district are reddish and rather coarse; those of the second are pale blue and of medium grain; both grey and red granites of fine to medium grain are found in the third, and there are many quarries.

The Silesian granites are mostly pale grey and of medium grain; there are quarries at Striegau, Strehlen, Görlitz, and Oberstreit. In the Odenwald a dull red medium-grained granite is worked at Fahrenboch and Felsberg. In the Bavarian Mountains the Blauberg granite is quarried at several places. Granite is worked in the Black Forest at Oberkirch, Triberg, Waldshut, Gernsbach, etc.; and in Thuringia a red stone comes from Brotterode, Ruhla, Mehlis, etc.

Among the granites used in ancient times that from between Assuan and the first Nile cataract—the red hornblende-biotite-granite of Syene—is one of the best known.

In India there are many occurrences of granite and gneiss, some of which were used in ancient temples, and carved with extraordinary patience; sometimes the whole edifice was cut out of a single rock mass. More recently they have been employed for engineering, as in the case of the Raichur granite for railway bridges and works of the Bombay Port Trust.

CHAPTER IV

OTHER IGNEOUS ROCKS

As a material for construction, whether in architecture or in engineering work, granite far exceeds in importance all other igneous rocks. This is due in part to the inherent good qualities of the stone, its strength, its adaptability to various styles of dressing, its satisfactory colour and appearance, and its susceptibility to polish; but its popularity is also accounted for by the fact that its occurrence is not limited to a few isolated tracts—it appears in tolerable abundance in every quarter of the globe.

Many of the other igneous rocks are less widely distributed, and though they may be familiar building stones in certain localities, their qualities are not generally so well known to architects.

In the following brief descriptions no attempt will be made to include all the igneous rocks that at one spot or another have been employed in building; a selection has been made to embrace some of the more prominent types, and these are presented according to no strictly petrological order, but in groups according to their mode of origin —namely, (1) Plutonic, (2) Hypabyssal, (3) Volcanic.

PLUTONIC—HOLOCRYSTALLINE ROCKS.

Syenite.—The syenites are rocks of granitic texture, more often coarse to medium-grained than fine-grained. The prevalent minerals are alkali felspars (orthoclase), etc., or the felspathoids may more or less completely take their place, and in either case there may be in addition a little lime-soda felspar; micro-perthite is common. The typical ferro-magnesian mineral is hornblende (hornblende syenite). *Quartz* is *absent*, or, if present, is usually in small bulk in the condition of micrographic intergrowth with the felspar. The hornblende may be replaced to a greater or lesser extent by other minerals, giving rise to what are known respectively as mica-syenite (granite-syenite) and augite-syenite. The most common felspathoid-bearing syenite is nephiline- or elæolite-syenite; quartz is never found in this variety. When subordinate quartz does occur in syenites, they may be distinguished as quartz-syenite, quartz-mica-syenite, etc. As in the granites, the syenites may be subdivided into a group rich in potash and another rich in soda.

Chemically syenites resemble granites, but they contain less silica. The specific gravity of these rocks ranges between 2·5 and 3·6. Less variable than granite in its resistance to pressure, the average strength is 1,170 to 1,280 tons per square foot, with a minimum of 730 and a maximum of 2,200.

The absence of quartz renders syenites softer to work than granites, but they are on the whole quite as tough, if not more so. It is only their comparative rarity that prevents their more extended employment for building; they are excellent in every way, and for polished work they are often superior to granites, on account of the sparsity of mica-flakes. The porosity coefficient is about 1·3. Hirschwald notes that many of the syenite garden-walls in Dresden are upwards of a hundred years old; the surface of the stone has peeled only slightly here and there, otherwise it is in good condition. It is equal, in fact, in point of wear to the better grade granites, and is superior to those of medium quality.

Although true syenites are limited in occurrence, the name has been used in the stone trade as an appellation for rocks that have no relationship with those described in this section. This is not the fault of the trade alone, for there has been some ambiguity displayed by petrologists in defining the term in earlier years; indeed, the original 'Syenite,' from the ancient Syene, in the neighbourhood of Assuan, Egypt, is now reckoned a hornblende-biotite-granite, on account of its quartz content. The so-called 'Swedish Syenite,' from Southern Sweden, a coarse-grained green stone, speckled with white, is a diabase; so also is the 'Lausitz Syenite' of Spremberg, Prussia; 'Odenwald Syenite' is a diorite. All these stones resemble one another in appearance.

DISTRIBUTION.—No true syenite is worked in Great Britain for building stone, though some of the Leicestershire plutonic rocks approach it in character. The rock occurs in the Channel Islands (St. Helier), in the Highlands of Scotland, and at Llanfaglen and Glan-y-mor in Carnarvonshire. One of the best-known syenites used in Britain is the so-called 'Norwegian *Labrador*,' the now familiar 'Dark Pearl,' 'Light Pearl,' and 'Imperial Pearl' of our market; it is an augite syenite (Laurvigite) from Laurvig and Fredericksvaern. This beautiful stone varies from a rich velvety blue-black to pale bluish-grey; the appearance of sparkling patches is caused by the schiller structure (p. 27) in both felspar and augite crystals. The bulk of this rock consists of felspar (cryptoperthite, orthoclase, microcline, anorthoclase, and some albite), and with the augite several varieties of amphibole frequently occur along with biotite. Some parts of the rock contain, in addition, olivine and nepheline.

Another Norwegian syenite, *Nordmarkite*, which occupies a large area north of Christiania, consists principally of red micro-perthite, with albite veins, and a little oligoclase; there is only a small representation of ferro-magnesian minerals. This stone is used a good deal in Christiania and the neighbourhood in house construction. Syenites are worked on a small scale in Saxony; they occur at Monzoni in the Tyrol (monzonite, an augite-syenite), in Southern Portugal, Transylvania, the Urals, and in North America. In the last-named country it is worked to some extent in Arkansas, near Little Rock, where the stone is called *Fourche Mountain Granite;* and at Allis Mountain and Magnet Cove an elæolite syenite (Diamond Jo Granite) has been wrought. The Allis Mountain stone is blue-grey when fresh, but according to Merrill it fades badly; it has been employed in the cathedral at Little Rock.

Diorite.—This is one of the rocks embraced by the old ambiguous term 'greenstone.' Dark green, indeed, and dark greenish-grey are its prevalent tints; very compact varieties are nearly black. Diorites are composed of soda-lime felspars, usually oligoclase, andesine, or labradorite, with green or brown hornblende. Quartz is absent in the typical rock, but it is frequently present in some varieties, either in fine granophyric intergrowth with felspar or in small grains. Biotite is not uncommon along with hornblende and pale green augite and enstatite, or other rhombic pyroxenes occur occasionally. Grains of magnetite, pyrites or ilmenete are usually present, and as secondary minerals resulting from alteration we find calcite, chlorite, and epidote. Structurally the rock is granitic, moderately coarse to fine in grain, and even in texture; porphyritic structure is comparatively rare. In some diorites the hornblende has well-formed crystals, and in others the felspar has crystallized early and has good faces, but more often the crystals are irregular in outline.

The diorites are not very well defined rocks, and in nature they pass on the one hand by increase of silica into the grano-diorites and granites, and by decrease of the same material into gabbros. With the introduction of orthoclase they approach the monzonites.

The typical diorite is *hornblende-diorite*. Other recognized varieties are *mica-diorite* and *augite-diorite*, those with obvious quartz are *quartz-diorites* (tonalite).

The specific gravity of diorite is 2·8 to 3; the porosity is small, 0·25 per cent.; the hardness is moderate, and the strength fairly good. Pressure tests give 1,640 to 1,830 tons per square foot. Its weather-resisting quality is fairly high, but the presence of much pyrites tends to lower that of some examples.

It is a difficult rock to work, mainly on account of the absence of a good rift, such as that possessed by granite, and it has not been very extensively employed for architectural work on account of the dull colours of the dressed surface. When polished, however, some of the coarser diorites are extremely handsome rocks, the white felspars showing up strongly in contrast with the dark hornblende, which often forms the greater bulk of the rock. The polish is brilliant and durable, but it is difficult to produce. As a mural decoration and for small columns and monument bases it may be used with very good effect.

A very striking stone is orbicular diorite, *Corsite* or *Napoleonite*, from Corsica. In this rock the hornblende and felspar have arranged themselves in large globular masses, 1 to 2 inches in diameter, with radial structure and concentric rings of alternating dark hornblende and white felspar. The stone is only employed on a small scale for special decorative effects or in small polished articles.

Diorites are limited in occurrence: in the Channel

Islands they are found in Jersey at St. Aubyn and St. Clement's Bay, at Fort Tourville in Alderney, and forming much of the northern part of Guernsey, where they are extensively quarried, mainly for road material. In England they appear at Hestercombe, near Taunton (tonalite), and at Brazil Wood, near Mount Sorrel in Leicestershire. Diorites occur in South-Eastern Ireland, and they are extensively developed in the Southern Uplands and Highlands of Scotland, but they are nowhere worked for building stones except for purely local purposes. There are numerous occurrences in the European Continent and America, but here, too, they are very little wrought.

Gabbro is a rock with granitic structure and a coarse to medium texture. There are many varieties, but the type form is composed essentially of lime-soda felspar, labradorite, or anorthite, with augite. Occasionally the more acid felspars are present in small amount, and in what are called the alkali-gabbros the augite may be replaced to a greater or lesser extent by the rhombic pyroxenes, hypersthene, and bronzite; olivine, hornblende, and biotite occur in some gabbros as original minerals. Very characteristic of gabbros is the assumption of diallage structure by the augite, and schiller structure is of common occurrence, giving rise to play of colour in patches on the broken or polished surface. The pyroxenes, both monoclinic and rhombic, are prone to uralitization or conversion to green or brown hornblende. Gabbros with much hornblende tend to shade into diorites. Iron minerals and apatite are common accessories.

Classed according to the prevalent mineral, we have *gabbro proper*, with augite or diallage; *norite* or hypersthene-gabbro, *olivine-gabbro*, and *olivine-norite*; and when quartz is present in microscopic intergrowth, *quartz-gabbro* or *quartz-norite*.

One of the peculiarities of gabbro masses is that they tend to be decidedly patchy in mineral composition, due to local variations; thus in one part the rock may be wholly built up of felspar (felspar rock, labrador rock, anorthosite) or olivine, diallage or hypersthene, giving rise to olivine-rock, hypersthene-rock (hyperite), etc.

The specific gravity of gabbros ranges between 2·7 and 3; they are fairly hard, and their porosity is low; their silica content is below 52 per cent.; and their average resistance to crushing is 1,830, and fluctuates between 640 and 2,200.

They do not weather well, and it is said that those with most felspar (labradorite, etc.) behave the worst, because these crystals decompose readily and produce a rough surface, and thus facilitate the attack of destructive agencies. The colour of the rock is against its employment for architectural building, since it is very sombre when dressed or with rock face. The medium-grained varieties are frequently employed in Europe for road metal. It is put to the most satisfactory use in polished slabs for wall-decoration, memorial tablets, small columns, and table-tops. Good large slabs are not easily obtained free from numerous cracks. The presence of schiller, which causes spots and splashes of iridescent purple and blue colour, is a valuable relief to the otherwise gloomy appearance. The effect of a dark dado of this kind is well shown in the stone from Kiev in South Russia, in the Church of the Saviour in Moscow.

Gabbro has been very little used in this country or in Europe for buildings; the so-called 'Norwegian Gabbro' is an augite syenite (p. 84). It may be mentioned here that the original gabbro was a serpentinous stone, which received that name locally in the neighbourhood of

Florence; what is now understood as gabbro is the *granitone*, or *Pietra di maschine*, of the same district.

DISTRIBUTION.—Gabbros occur in Great Britain at the Lizard in Cornwall; at Hanter Hill in Herefordshire; St. David's Head in Pembrokeshire; in the Lleyn Peninsula, North Wales; and at Carrock Fell, near Keswick. In Scotland they are found forming laccolites in the Cuellin Hills, Skye, Ardnamurchan, and at many points in the Highlands, and again in Ayrshire. In Ireland they are known about Carlingford; and in the Channel Isles in Guernsey.

The 'black granite' of Kentallen Quarry, near Loch Linnhe, is an alkali-gabbro (*Kentallenite*); it contains orthoclase in addition to plagioclase, along with augite, olivine, and biotite. It is used principally for setts, but it takes a moderate black polish, which exhibits conspicuous mica plates; when picked or otherwise dressed, it presents a dark dull grey appearance.

In America, as elsewhere, gabbros are not much quarried except for quite local purposes. The 'black granites' of Maine from Addison are black, or black with white specks, or dark grey, and of medium texture; they are obtained in large blocks. The walls surrounding the grounds of the Capitol in Washington are built of this stone; it also appears in buildings in New York, Boston, Quebec, and Montreal. A finer-grained stone, harder, and of similar colour is quarried at Vinalhaven and Tenant's Harbour. In Minnesota a dark blue-grey gabbro is worked near Duluth (*Duluth Granite*), together with a highly felspathic variant of a lavender-blue, grey, or green colour, which is wrought into polished columns. The '*Au Sable Granite*,' from near Keesville, Essex Co., New York, and the '*Labrador Granite*,' from Vergennes, Vermont, are norites, coarse in texture, with bluish schiller; they are used for polished work. A greenish-black norite is quarried in North Carolina.

Several other basic rocks, often coarsely crystalline, may be mentioned here—*e.g.*, the *pyroxenites*, mainly pyroxene with subordinate soda-lime felspar; *peridotites* and *picrites*, rocks rich in olivine—which is almost the only constituent in *Dunite*—associated with diallage, rhombic pyroxene, or hornblende. *Eclogite* is a hornblende-pyroxene rock containing bright green hornblende and augite, with red garnets. These rocks are usually dark green, and are little employed, except in small polished slabs for decorative work.

TABLE VII.—PLUTONIC ROCKS.

	Predominant Alkali Felspar.		Alkali and Lime-bearing Felspars about equal.	Lime-bearing Felspar in excess.	Similar Rocks with Felspathoids.
	Potash Felspar.	Soda Felspar.			
Acid (not less than 66 per cent. silica)	Granites rich in potash	Granites rich in soda	Adamellite granite	Granodiorite	—
Intermediate (not less than 52 per cent. silica)	Syenite rich in potash	Syenites rich in soda	Monzonite	Diorite	Syenite with Nepheline
Basic (less than 52 per cent. silica)	(Shonkinite, etc.)	(Essexite, etc.)	Kentallenite	Gabbro	Gabbro with Nepheline

HYPABYSSAL ROCKS.

Quartz Porphyry, some felsites, granophyres, and pitchstones correspond to the granites.

Porphyry, orthoclase-porphyry, syenite-porphyry, nepheline-syenite porphyry, etc., correspond to the syenites.

Porphyrites correspond to the diorites.

Dolerites and most **Diabases** correspond to the gabbros.

No little difficulty is experienced in framing a brief comprehensive explanation of the rocks in this class for readers unversed in the methods and outlook of petrologists. Perhaps their position may be indicated most briefly by saying that they are *intrusive igneous rocks which have cooled more rapidly than their plutonic equivalents.*

Now, just as there have been all degrees in the rate of cooling in magmas of every grade of chemical composition, from acid, through intermediate to basic, so we find in Nature gradual passages from each kind of hypabyssal rock into its plutonic and volcanic analogue, and the same kind of gradation between the most acid and the most basic forms. Since architects and engineers are less concerned with the mode of origin than with the structure, texture, and mineral contents, and the correlated physical and chemical properties, the question will naturally be asked, What are the characteristic structures and other properties by which stones quarried from hypabyssal rock masses can be recognized? It may as well be said at once that there is no satisfactory direct answer to this question.

Structurally they range from holocrystalline granular to glassy non-crystalline rocks; in texture they vary from medium coarseness to very fine or vitreous; but, on the whole, their mineral ingredients are crystallized on a smaller scale than in the plutonic representatives; they frequently, but not essentially, contain one or more porphyritic minerals, and spherulitic structures are frequently developed. Although they bulk less largely in the earth's crust than the plutonic or volcanic rocks, they appear in many more recognizable variations. A few types will now be described.

Porphyry.—Under this head we have to consider two distinct sub-classes, the Quartz-porphyries related chemically to granites, and Porphyries proper, similarly associated with the syenites. The characteristic felspar of both porphyry and quartz-porphyry is orthoclase.

Quartz-porphyry is a rock consisting of a fine-grained micro-crystalline or cryptocrystalline (felsitic) ground mass composed of quartz and felspar, forming a compact-locking base. In micro-crystalline structure the individual grains can be distinguished by the microscope, but in the cryptocrystalline structure the individuals are indistinguishable; all that can be recognized is that the material is crystalline. In this ground mass there are embedded large crystals of felspar and quartz in well-formed crystals, or in rounded lumps the size of a pin-head to that of a bean. With these, but subordinate in number, there may also be crystals of mica, most often biotite, sometimes pale green augite, green hornblende, or blue riebeckite. If the larger porphyritic crystals are absent, the rock is sometimes called a *felsite*, the felspathic constituent being orthoclase, albite, or anorthoclase. Varieties of felsite rich in soda are *Keratophyre* and *Quartz-keratophyre*.

Quartz-porphyries in which the micro-graphic and cryptographic structures are particularly well developed are called *Granophyres*; they may or may not have porphyritic felspar and quartz.

When the structure of the quartz-porphyries is micro-granitic—like granite on a small scale—the rock is a *granite-porphyry* or *micro-granite*; many of the *elvans* of Cornwall are of this character.

The glassy variant of quartz-porphyry is *Pitchstone*, a rock consisting almost entirely of glassy matter, in which tiny incipient crystals (crystallites, microlites, etc.) are revealed by the aid of a microscope. In this glass there may be porphyritic crystals of orthoclase, quartz, mica augite, and hornblende. Pitchstones have a resinous lustre, and conchoidal fracture and flow-structure is commonly visible, either with the naked eye or with a lens.

Minute perlitic (subconcentric) cracks are nearly always present. The colour of the rock is dark green, grey-black, occasionally red.

Porphyry is similar in structure to quartz-porphyry, but it belongs to the intermediate group of rocks on chemical grounds; as in the syenites, quartz is typically absent or feebly represented. Varieties are syenite-porphyry, nepheline-porphyry, nepheline-syenite-porphyry, etc.

The several varieties of quartz-porphyry and the porphyries are usually of a reddish colour, sometimes pale grey, yellowish, or light brown. They are mostly compact stones, rarely porous, or containing cavities. Their specific gravity is 2·4 to 2·8. The pressure tests of fresh material show high values for the resistance to this kind of stress; the average is 1,645 tons per square foot, and they go as high as 2,740. The average porosity is 0·65 per cent. (0·3 to 1·0). Absorption under normal pressure is between 0·9 and 3·5 per cent. of the weight. Their weather-resisting properties are generally very good. Their hardness and weather-resisting power depend largely on the amount of silicification of the ground mass; much of this silicification may be of secondary origin. The stones may be quite good, even if the felspathic part shows indication of alteration, but in the quartz-free porphyry any indication of earthy appearance on the freshly-fractured surface should cause some distrust of the stone.

Although they are hard rocks and tough, they are not really so difficult to quarry as is sometimes supposed, but the absence of good rift frequently makes them more troublesome to obtain than granites. Most of them take a polish readily.

DISTRIBUTION.—In Cornwall and Devon there are many micro-granites and quartz-porphyries associated with the granite masses; they penetrate both the granite and the surrounding rocks. Locally they are called *elvans*, but the name is applied also to the more basic dykes and to some of the finer-grained granites.

The *Pentuan* (Pentewan) elvan was formerly a favourite stone for ecclesiastical work in the West of England, both for exteriors and interiors, and was employed in other buildings near the quarry in Pentuan Bay, Cornwall. The Roman inscribed stone built into the wall in Tregony Church, is a rough block of this material. The stone has a pale buff tint, and is fine-grained and rather porous. It probably owes its good qualities as a weather-resisting stone to the conversion of the ground mass to an aggregate of quartz and white mica (greisenization). The porphyritic felspars are a good deal kaolinized and silicified.

An altered quartz-porphyry of a warm yellow colour is quarried near St. Columb Minor, and used in the parish church and houses in the village, and in Newquay for lintels, quoins, and facings. The well-selected stone wears very well, turning to a warm grey.

A coarsely porphyritic micro-granite is quarried at Retyn in large blocks; it would look well polished. At Tremore Quarries, near Bodmin, a broad dyke is largely quarried; it varies much in structure and texture in different parts: that from the centre is more coarsely crystalline, and is usually employed for building and polished work. The appearance varies from dark pinkish-red speckled with black to a pale pinkish-grey speckled black and white. Some varieties of this stone are now wrongly sold as 'Luxullianite.' The pale greenish-grey elvan from Newham has been extensively used in Truro.

Many other acid elvans have been quarried from time to time for local buildings, but mainly for road metal—*e.g.*, at Mayon; Land's End; Dowglas, near St. Austell; Helland; St. Neots; Lanivet and Withiel, near Bodmin; Trevailes and Rosecraw, near Penrhyn, etc. They are usually too much jointed to be obtainable in large blocks;

many of them show much secondary alteration, kaolinization, and silicification, and in the latter case they weather well. A good collection of these stones is in the Jermyn Street Museum.

In South Wales micro-granites and granophyres occur at St. David's. Micro-granite, granophyre, and granite-porphyry are plentiful in North Wales in Carnarvonshire, Merionethshire, and Anglesey. A large mass of granophyre lies between Wastwater, Ennerdale, and Buttermere, in the Lake District; the dark mineral in this rock is mainly biotite. A granite-porphyry, with porphyritic orthoclase and small garnets, is quarried at Threlkeld. In the same district there are many other minor occurrences of similar rocks, mostly in the form of dykes and sills.

In Scotland bosses and dykes of quartz-porphyry are very abundant, associated in many cases with the granite masses. The rock quarried at Ailsa Craig contains riebeckite.

Quartz-porphyries, micro-granites, and felsites occur in Ireland, round the granite of Leinster, and again in North Galway; pitchstones and felsites in the Barnsmore range, Donegal.

On the Continent these rocks occur at numerous places —*e.g.*, in Silesia, Saxony, the Odenwald, Thuringia, the west border of the Black Forest, Vosges, and the Rhine Provinces. In the Bozen district in the Tyrol a greenish-brown and reddish rock is quarried. From Sweden may be mentioned the red felsite of Elfdalen, which polishes well; and from Norway the brown rock, with red porphyritic orthoclase, from the Christiania district. In Egypt there are many quarries in the Nile Valley.

In America there are numerous occurrences of these rocks in Massachusetts (bright red, pink, and grey), Missouri, Minnesota, Maine, New Hampshire, Nevada, and Wisconsin; but they are regarded as being very difficult to work, and are little used except for local rough structures.

The fine-grained granitic texture of the micro-granites, the close interlocking of the minerals in the granophyres, and the feltwork of the small crystals in the quartz-porphyries and felsites, all tend in the direction of imparting toughness to these stones; hence they form capital material for foundations and heavy work in embankments, bridges, etc. When polished they have been used for internal wall covering, mantelpieces, columns, landings, and bases for monuments.

Greenish-grey and red granite-porphyries have been used in the basin of the Springbrunnen in the Schlachthaus Hof and the Universitätsfrauenklinik in Leipzig; in the Reichspatentamt and Reichsversicherungsamt buildings in Berlin.

Porphyrite.—The essential difference between these rocks and the porphyries is that in the former the characteristic porphyritic felspar is a lime-bearing plagioclase.

In all other respects, as regards their prevalent structure and physical properties, they closely resemble the porphyries, and behave similarly in buildings. Less bright in colour than most of the porphyries, they may, nevertheless, be employed in a like manner in polished work with good effect.

DISTRIBUTION. — Porphyrites occur in Somerset at Moons Hill Quarry and Beacon Hill, near Shepton Mallet. In Leicestershire the well-known roadstones of Markfield, Groby, Cliff Hill, etc., belong to this class. They have a granophyric ground mass, with porphyritic plagioclase and green hornblende, part of which is an original constituent, and part is an alteration product of original augite. These stones are extremely tough, and are much used for road material. They approach the diorites in some of their characters, and may be compared with the similar stones from the Cleveland dyke and Carrock Fell rocks, and with the Penmaenmawr stone from the coast of Carnarvonshire. Similar rocks, with hypersthene, come from Carn Boduan, near Nevin, and, with mica in addition to hypersthene, from Yr Eifl, in the Lleyn Peninsula; this is a well-known roadstone. In South Wales rocks of this kind, with augite and hornblende, occur near St. David's.

In Scotland there are many occurrences of porphyritic dykes; they are little used for building, but are worked here and there for road metal.

In Ireland the best-known porphyrite comes from Lambay Island, off the Dublin coast; it is usually styled 'Lambay porphyry,' and is an ideal example of a porphyritic rock. The ground mass is dark green, composed of small lath-shaped felspars, with grains of augite; it is an augite-porphyrite. The rock has undergone a good deal of alteration, with the consequent formation of chlorite, calcite, epidote, and magnetite. Scattered throughout the ground mass are porphyritic crystals of pale green labradorite felspar. This is a handsome stone, and takes a fair polish; it closely resembles the rock known as *porfido verde antico*. It is difficult to obtain.

Porphyrites occur at several places in Saxony, Hartz Mountains, Thuringia, Riesengebirge, Black Forest, and Vosges.

The Belgian rock from the Quenast quarries and Lessines is a familiar road metal in England; it is a quartz-porphyrite or quartz-diorite-porphyrite. It is very tough, and the Quenast rock has a crushing strength of 2,520 tons per square foot.

Dolerites and Diabases.—For present purposes these two rocks may well be regarded as one type, since the main difference between them is that diabases usually show more alteration of the minerals than is the case in the dolerites. Moreover, the term 'diabase' has been employed very loosely in this country, while in Germany and in America it is used with a somewhat different connotation.

Dolerites are composed essentially of plagioclase felspar, most commonly labradorite, but sometimes oligoclase and anorthite, together with augite. To these may be added the occasional appearance of other minerals, which give rise to the varieties olivine-dolerite, mica-dolerite, hypersthene-dolerite, and, if a little quartz is present, quartz-dolerite. Chlorite, epidote, calcite, and sometimes serpentine, appear as alteration products; pyrites, and more rarely magnetite, are usually present.

The structure is fairly constant in character and somewhat peculiar (Fig. 3). The felspars in lath-shaped crystals have usually crystallized before the augite, which forms large irregular crystals, enclosing a number of the felspars; this structure is known as *ophitic*. Instead of having this ophitic structure, the augite may appear in the form of roundish grains between the felspars. The rock is nearly always completely crystalline, but occasionally small residual patches of glass remain. Porphyritic structure is rare. The texture is medium-grained to fine-grained. Sometimes amygdaloidal cavities occur.

When freshly quarried the rock is very dark, greenish or bluish black; but it soon becomes greener on exposure, owing in some cases to the drying of the serpentine, which occurs as an alteration product of olivine. The dolerites and diabases have often been called *greenstones*. When weathered the surface tends to assume a dull brown tint, on account of the liberation of hydroxides of iron. The felspars are often stained a greenish tint, but sometimes they retain their pale colour, in which case the trade name 'syenite' has been applied to the stone.

The specific gravity of these stones is about 2·8 to 3. The crushing strength is about that of the mean for granites, and is somewhat lower than that of diorites.

The weathering qualities are good on the whole, but less so in the coarser varieties. The stone is rather more difficult to work than granite. It is frequently well jointed, but there is seldom any definite rift. A high polish is difficult to attain, but where a subdued colour and smooth surface is required it may be very usefully employed.

Dolerites and diabases are more often used for road metal than for building, principally because their dull colour is not appreciated by architects, although when relieved by juxtaposition with brick or lighter coloured stone, they may have a good effect. In the polished state they have been used for walls and monuments.

DISTRIBUTION.—Many of the greenstone dykes, sills, and small bosses of Cornwall and Devon are dolerites and diabases; they are quarried in many places for roadstone, but in building are mainly employed in rough walling. We may mention here the large diabase quarries at Stepper, near Padstow, which produce a compact dark green rock of fairly coarse grain, and the *Catacluse* stone from near Trevose Head, Cornwall, a dark, compact bluish-grey stone. Good examples of the use of this stone may be seen in many Cornish churches, *e.g.*, the old font in St. Merryn's Church.

Polyphant stone, although not a dolerite, may be included here; it is a peculiarly altered peridotite, obtained near Launceston. It has been extensively used for churches in the past for dressings and for carved work, and for fonts inside. It is a dark green stone, with pale green spots of felspar, which take on a warm brown tinge in the weathered samples. The stone is tough and wears well in dry situations, but it is extremely soft and readily carved. At the present day it is being extensively used for ornamental carved and polished work. The copy of an old Cornish cross, forming the War Memorial at Haverfordwest, may be taken as an example.

'Trusham granite' is a dolerite from near Trusham in Devon, and is worked for roadstone and setts; but it might be polished in blocks of moderate size, as the colour in this condition is a quiet dark green.

In the Midland counties they are well developed about Nuneaton; at Atherstone (hornblende-dolerite); Hartshill; the Lickey Hills; Bayston Hill, near Shrewsbury; Lower Wych Quarry, in the Malvern Hills; Ponk Hill, near Walsall; Rowley Regis, near Birmingham (Rowley Rag); the Clee Hills; and other places in the neighbourhood; and in the Welsh border counties. In Derbyshire dolerites form sills in the Carboniferous limestone at Ible, Peak Forest, Tideswell Dale, and elsewhere (locally = 'toadstone').

In South Wales dolerites are abundant in Pembrokeshire about St. David's, Fishguard, and Newport; and in North Wales they are prevalent in the counties of Merioneth and Carnarvon. The Gimlet rock at Pwllheli is one of these, quarried for macadam and setts; in its coarser varieties it is also a very handsome stone when polished, having pale felspars sprinkled through a dark green ground. Carth Head, near Keswick, is a diabase rock, and other examples occur in the Lake District.

Very well known dolerite dykes of great length occur in the North of England—for example, the Cleveland dyke, in Yorkshire, traceable for ninety miles; the Great Whin Sill, stretching from Dunstanburgh Head in Northumberland to Cross Fell, seventy miles; the Acklington dyke, from the Northumberland coast to the Cheviots. There are also many smaller ones. These stones have been used considerably in local building, and bushes for waterwheels have been made of this material.

In Scotland dolerites and diabases are widely scattered; they are worked mainly for road metal at many localities.

In Ireland they appear in Wicklow, Waterford, and Antrim; and there are several small bosses in the Isle of Man.

In America these rocks are quarried in Connecticut, Massachusetts, New Jersey, Virginia, and Pennsylvania, principally for road material. In the last-named State there are important quarries at Collins Station, where the stone is worked for monuments and foundations, and, in a lesser degree, for other building purposes; also at Yorkhaven, York County, whence much stone has been taken for the North Central Railway for bridges, etc. The 'Palisades,' or dolerite cliffs of the Hudson River, New Jersey, have been much quarried for road metal and setts ('specification' blocks, 4 inches by 8 inches by 10 inches, and 7 to 8 inches deep, and square blocks 5 to 6 inches square, and 6 to 7 inches deep). St. Patrick's Cathedral, Hudson County House, and other buildings in New Jersey, have employed this stone.

Dolerites are obtained near Eisenach, Lauterbach in Meiszen, in Oberhesse, Eifel, Kaiserstuhl, Lowenberg in the Siebengebirge, Oberhausen in Amstal, Lausitz, and a fine-grained variety at Brockenham, near Frankfort-on-Main. Saxon diabases have been frequently employed in monumental bases and family vaults. The obelisk Kriegerdenkmal in Bautzen, Saxony, and Kaiserdenkmal in Pasewalk, Pomerania, are of diabase.

VOLCANIC ROCKS.

The volcanic rocks now to be briefly described include lava flows of all ages, from the most ancient to quite recent. Their composition, as the table on p. 102 indicates, shows a wide range of variation; and in texture they show similar diversity, from compact dense basalts to glass-like obsidian and frothy pumice. When fresh they range in colour from white, through neutral greys, to black; but with subsequent alterations they change, in the more basic kinds, to dull browns and greens.

TABLE VIII.—VOLCANIC ROCKS.

	Potash-bearing.	Soda-bearing.		Lime-bearing.	With Felspathoids.
Acid (not less than 66 per cent. SiO_2)	Potash Rhyolite	Soda Rhyolite	Series with alkali- and lime-bearing felspars in almost equal proportions	Dacite	—
Intermediate (not less than 52 per cent. SiO_2)	Potash Trachyte	Soda Trachyte		Andesite	Phonolite
Basic (less than 52 per cent. SiO_2)	—	—		Basalt	Leucite, Nepheline, and Mellilite Basalts

Rhyolites are mostly pale grey rocks with a ground mass that may consist entirely of microlites of felspar with quartz, or there may be, in addition, more or less unresolved glassy matter; if the ground mass is mainly glass the rock is an *Obsidian*. Many of the rocks called felsites belong to this class, for their characteristic felt-work of felspar needles and microlites has resulted from the devitrification of glassy rock, much in the same way that old Roman glass has altered, or modern bottles will devitrify if placed in suitable environment. Flow structure and banded structure, spherulitic structure and small subconchoidal (perlitic) cracks are very characteristic. Porphyritic structure is common; the porphyritic crystals

OTHER IGNEOUS ROCKS

are quartz and sanidine (sometimes albite, anorthoclase, and oligoclase), with plates of dark mica and prisms of augite or hornblende in smaller numbers (Fig. 3, V).

Trachytes resemble rhyolites very closely in external features; the chief difference lies in their chemical composition, which places them in the intermediate class. Like the rhyolites they are grey in colour, except when tinted green or pale brown in older occurrences; they often show marked flow structure; on the other hand, they tend to be less vesicular, although they often have drusy cavities. The porphyritic crystals are sanidine, anorthoclase, oligoclase or andesine, and brown mica, green or blue amphibole, or one of the pale pyroxenes. Quartz is much scarcer in the trachytes than in the rhyolites. The ground mass is of the felsitic type, and is usually more completely crystalline than in the rhyolites. According to the predominant ferro-magnesian mineral, we have amphibole-trachytes, pyroxene-trachytes, and mica-trachytes.

Phonolites are comparatively rare rocks; they resemble trachytes, but differ in containing a higher alkali content, and in the presence of felspathoids in addition to sanidine. They are grey or greenish-grey in colour, and compact. When thin plates of the rock are struck sharply they give out a metallic or bell-like sound, which has given rise to the name 'clinkstone.'

Andesites, unlike the preceding rocks, are characterized by porphyritic plagioclase of the soda-lime series, andesine, labradorite, sometimes oligoclase; with the felspar there may be biotite, light green augite, hypersthene or hornblende, the last-named being the most common. Hence there are hornblende-andesites, mica-andesites, augite- or hypersthene-andesites; and if a little quartz is present, quartz-andesites. These pass into *dacites*, if more than a small amount of quartz occurs. Porphyritic structure is usually well marked in andesites; these crystals lie in a ground mass composed most often of a felted aggregation of felspar needles and microlites, in which a certain amount of glassy matter is enclosed. The glassy matter may be wanting, or it may assume large proportions, as in 'andesite glass.' Flow structures are occasionally evident, and vesicular or amygdaloidal structure is common. When fresh the rock is dark in colour, and the porphyritic felspars are clear and glass-like, resembling sanidine; but both ground mass and porphyritic felspars are commonly stained by alteration products; chlorite produces greens, and hydroxides of iron give rise to browns.

From a technical point of view these rocks—rhyolites, trachytes, and andesites—may for brevity be considered together. They are widely employed as general building materials on the Continent, but, owing to the great range of porosity and the variability in texture, it is not easy on the small amount of evidence to indicate any guiding principles for their employment.

Rhyolites rich in quartz and compact in texture seem to weather better than the more porous or glassy varieties. None of them take a good polish, though obsidian is treated in this way for small ornamental objects. In the younger rhyolites the sanidine crystals, having very often smooth, clean faces, tend to fall out from the weathered surfaces of dressed stone; in the older ones, however, this is not the case, because they become intergrown with the ground mass; and in many cases the more ancient rhyolites are stronger and more weather-resisting on account of the formation of secondary quartz within the rock. The same feature may be noticed in older trachytes and andesites.

Vesicular structure makes for lightness; it reaches its highest expression in the rhyolites, in the rock *pumice*. The specific gravity of this stone in the mass is 0·3 to 0·9; for this reason it is not bad material for vaults and arches, and it has been employed for this purpose by the Romans. The cupola of Agia Sophia, Constantinople, is made of pumice, and in the Lipari Islands, where it is readily obtained, it is employed in many buildings in roughly-carved blocks set in mortar. All the porous or rough varieties of rhyolite, trachyte, and andesites form a good bond with mortar. Mixed with milk of lime crushed pumice is used for the artificial stone 'Niedermendig Schwammstein,' which is employed as a heat-insulator. Crushed pumice alone is used in New Zealand and Southern Europe for steam-packing, heat-insulating, also as a floor-packing.

The behaviour of rhyolites, etc., in buildings is variable: some harden considerably on exposure, others do not.

The castle of Pfalz, Kaiserwerth, 1,100 years old, is built of a compact, well-crystallized quartzose rhyolite, and, probably on account of the quartz, it is still in good preservation. On the other hand, a rather porous rhyolite, with ill-defined sanidine crystals in a fine matrix, employed in the renovation of part of Cologne Cathedral, has peeled badly after only fifteen years' exposure.

As a useful stone phonolite is of only local significance; it weathers pretty satisfactorily, and its crushing strength is about equal to that of the andesites and trachytes; it sometimes reaches 1,830 tons per square foot; its specific gravity is 2·5 to 2·7. Both in weathering and quarrying it tends to break into thin slabs, a feature which makes it handy for walling and paving stones; in the Auvergne it has even been employed for roofing.

Trachytes and andesites give only moderate results under crushing tests—455 to 730 tons per square foot; their hardness is variable, and usually only moderate; the specific gravity is about 2·2 to 2·7, and the porosity of compact varieties about 3 per cent. to 6 per cent. in more porous kinds. The compacter, less porphyritic varieties seem to weather best, for the large felspars give way rather easily. Andesites that are very amygdaloidal are of doubtful character.

Both trachytes and andesites are used in building a good deal in Europe, but not in England or America. They are not difficult to work in moderate-sized blocks, and they key well with mortar. They are used for ashlar, steps, and landings, small columns, balusters, and architraves. As macadam and for local building they are quarried in Scotland, and they are rather favoured on the Continent for cobbles, or *kleinpflaster*, in the steeper streets with light traffic, on account of the good foothold they afford.

Rhyolites, trachytes, and andesites are quarried in England and Scotland, Wales and Ireland, at many places for roadstone.

Trachytes and andesites occur in the Lipari Islands, in Central Italy, the Siebengebirge, Steirmark, and Iceland. Occasionally they are used as millstones. They are quarried in the Siebengebirge district in Germany; in the Rhone Valley (Altberg); in the Eifel district (Kalberg); in the Westerwald, Vosges, and Auvergne. They are quarried in a small way locally in America. In Servia phonolites are worked at Banjvea, andesites at Dobra, trachytes at Kopaonik.

Basalts.—These well-known rocks are typically composed of basic plagioclase felspar (mainly labradorite in

OTHER IGNEOUS ROCKS

the bulk of the rock, with anorthite or bytownite occurring as porphyritic crystals) and augite. To these must be added olivine, which is a very common porphyritic constituent, and occasionally hornblende and mica; grains of quartz are rare. The iron minerals, magnetite, ilmenite, and iron pyrites, are invariably present, and sometimes hæmatite, and the rock is often much altered, with the accompaniment of calcite, serpentine, and chlorite.

Basalts occur as lava flows of all ages, and in structure they may be occasionally quite glassy (tachylyte), or have a glassy base full of small lath-shaped felspars and small grains of augite, or there may be no glassy matter left. In this holocrystalline condition they tend to pass into dolerites—*e.g.*, the dolerite of Linz-on-Rhine. Porphyritic structure may be present or absent, and the same applies to vesicular and amygdaloidal structure. Basalts with the last-named characters are not suitable for employment.

The colour of basalts when fresh is black to dark greenish-grey; when weathered or altered, they take on warm brown tints or dull dark greens. The fracture of fresh varieties is uneven or conchoidal; they are tolerably hard and very tough. Their specific gravity is 2·8 to 3·3, and the crushing strength is very high—2,010 to 3,200 tons per square foot, and it may reach 4,500 tons per square foot. The absorption is about 1 per cent., and the conductivity to heat is considerable, with the natural result that basalt buildings are cold in winter and hot in summer, and this peculiarity is intensified by their dark colour.

Basalts are not particularly pleasing when used architecturally on account of their sombre hues, and except in a few places they have not been much used in this type of work.

The felspar basalts, free from glass and rich in granular augite, give the best results; in buildings 700 years old the decomposed surface in such cases is only 1 to 2 millimetres thick, with bright red spots where the augite has broken down. Even porous varieties show very satisfactory weather results; thus a 600-year-old wall of Niedermendig ashlar exhibited only an insignificant weathered crust, and the stone was quite sound internally. Again, an altered basalt of light grey colour, due to abundant secondary alteration products, from the Roman castle of Saalburg, near Homburg, 1,700 years old, is proved to be still quite strong.

In damp situations and under water basalt does not behave so well; decomposition penetrates farther into the stone, and forms a brown clayey crust. Hirschwald states that out of a number of basalt structures in air 85 per cent. were found very good, and the remaining 15 per cent. in fairly good condition; whereas of basalt structures subjected to the continued influence of water, 50 per cent. had turned out badly, 15 per cent. were moderately sound, and 35 per cent. were well preserved.

Both compact and other varieties on copings and cornices show a tendency to crack.

The felspathoid basalts wear well enough if the minerals are fresh to begin with; should they show signs of alteration in the quarry, they are not likely to stand in the buildings.

DISTRIBUTION.—In Great Britain basalts are rarely used for anything better than rough boundary walls and roadstone. They are found in two or three localities in Cornwall; in Devon, about Ashprington, Totnes, Brent Tor (near Tavistock), Ide (near Exeter); and at numerous places in the palæozoic rocks about St. David's in Pembrokeshire. They appear at a few places in Somersetshire and Shropshire; at Eycott Hill, near Keswick, and at other points in the North of England. In Scotland prominent examples are Arthur's Seat, the Lion's Haunch and Calton Hill, Edinburgh; also there are many other occurrences in Haddingtonshire, Fifeshire, Linlithgowshire, Ayrshire, and in the islands Skye, Rum, Mull, Arran, Raasey, and Eigg. In the Isle of Man they are found at the Stack of Scarlet, and in Ireland they form the familiar Giant's Causeway of Antrim, and occur in Limerick and a few other localities.

Abroad, basalts are widely dispersed, but rarely used extensively for building. They are found in Bohemia, Silesia, Lausitz, the Erzgebirge and Fichtelgebirge, Thuringewald, the Rhone Valley, Hesse (Kamel, Meizner, Habichtswald), Vogelsberg, Odenwald, Black Forest, Siebengebirge; also in Central France, North Italy, Switzerland, and Hungary. In America there are numerous occurrences, but the stone is hardly ever employed for building. Basalts have been employed on the Continent for external walls, foundations, steps, columns, and socles in architecture, also for retaining walls, bridge-piers, dams, etc. They have been much used for macadam.

Very common in basalts is columnar structure; the columns range from the thickness of a lead-pencil up to 3 or 4 feet in diameter, and as much as 40 feet in length. This character is well exhibited in Fingal's Cave, Staffa, at the Giant's Causeway, and in quarries at Lausitz, Meiszen, Habichtswald, and many in the Rhine district. This structure facilitates quarrying and the columnar blocks are frequently employed transversely in walls and in embankments (Rhine), without dressing, also for paving, and as direction and boundary posts.

Volcanic Tuffs.—In recent times the expulsion of fine dust-like material, together with small lumps and large blocks of rock, from active volcanoes is a familiar phenomenon. We may cite, for example, the great outburst of

Krakatau in 1883; and, going farther back, we may point to the material which effaced for so long the town of Pompeii. These materials, whether they fell on land or water, settled down after the manner of sedimentary deposits. The younger tuffs generally form light porous stones, but those of earlier geological periods are frequently—indeed, usually in a much more compact and solid condition through silicification and other mineral alterations that have taken place in them. Among these fragmental volcanic rocks there are coarse varieties—volcanic agglomerates and breccias—rocks of medium texture, and others made of the finest of dust.

They may be classified simply according to the nature of the magma from which they had their origin; thus, there are rhyolite-tuffs, trachyte-, andesite-, phonolite-, and basalt-tuffs; a peculiar basic variety is called palagonite-tuff.

In the older rocks, when the original fragmentary material has undergone change, due to mineralization and earth pressures, very compact rocks result; some of the stones known as 'porcellanite' and 'halleflinta' are of this nature; others that have become markedly sheared become slates (p. 274). The fragments forming tuffs are very varied; they are very often glassy, and, in addition to the volcanic matter, they may contain chips and dust of any kind of sedimentary rock; even fossils may occur in them. They are often highly calcareous. The tuff of Brohltal is used for making hydraulic lime.

As building stones the tuffs naturally vary much in quality: those with the simplest composition, compact and uniformly glassy, are probably the best; phonolitic and other felspathoid tuffs are poor. It is a common occurrence to find a distinct hardening on the exposed surface of dressed stones, but this is not necessarily an indication of good quality.

OTHER IGNEOUS ROCKS

Unaltered tuffs are very porous; in some the pores are nearly half the volume of the stone. It is an interesting commentary on the loose generalization that the porosity of a building stone is an index of its frost-resisting power to find that many of the porous tuffs are less affected by frost than more compact kinds. Most unaltered tuffs are very easy to work and of low specific gravity, and they are less affected by lichens than limestones.

The crushing strength is variable, usually low: that of the Wieburn tuff is 130 tons per square foot, the flesh-coloured porphyry-tuff of Rochlitz, in Saxony, a very good weather stone, has a mean value of 180 tons per square foot; a dirty green fine-grained diabase tuff from Hof, in Upper Franconia, has a specific gravity of 2·8 to 3, and a crushing strength of 1,370 tons per square foot: this is also a good stone.

The trachyte tuff of the Eifel and Siebengebirge is only fit for building in parts—mostly it is too friable; it is often called *Trass* (Andernach and Nettetal), and is largely employed in the manufacture of cement. Tuffs are represented in Great Britain in most of the districts where the older volcanic rocks occur; they are almost always much altered, hardened and sheared, and occasionally used for roadstone (and slates).

CHAPTER V

SANDSTONES AND GRITS

THE degradation and decomposition of rocks like the granites and gneisses gives rise to the formation of quantities of fragmentary material, which in course of time, by the continuation of the disruptive and solvent *weathering* agencies, and by the transporting action of water in its several capacities, assisted in places by the wind, becomes sifted and sorted and laid down in deposits of different grades of fineness and of diverse composition, in *beds* or *strata*.

If for the sake of simplicity, we confine our attention to the behaviour of a typical granite and its degradation products in a moist temperate climate, we shall be able to observe all the essential stages in the production of stratified sedimentary rocks—namely, the sandstones and their allies, the clays and their derivatives, the limestones, and the multitude of deposits of intermediate composition that link these three outstanding classes one with another.

In a granite quarry where there is a considerable amount of over-burden, waste, or 'head' visible, the solid, close-grained rock, some depth from the surface, may be observed to pass upwards into a more openly-jointed stage, when the irregular joints are seen to be more and more occupied by rotten granite material as the surface is approached. The effect of the rotting of the granite along the joints, together with the multiplication of the joints near the surface, is to break up the higher layers of the granite mass into a collection of isolated cores or lumps of undecomposed rock, surrounded by incoherent material. As the subsoil is approached, the size of the loose blocks tends to decrease, and finally, in the higher subsoil, the bulk of the material is incoherent granitic rubbish, capped by soil.

When these places are examined in more detail, it will be found that the low-lying friable granite differs little in appearance from the compact parent rock. The quartz, felspar, and mica are easily distinguishable ; preliminary chemical changes have led to mechanical disruption, but as the material is traced towards the overlying soil, it becomes increasingly pasty and clayey in character, fewer felspar fragments are recognizable, and the colour is usually some shade of brown or yellow. The changes produced in the lower subsoil through the agency of surface water and oxygen are continued with increased rapidity in the region of the soil where the larger plants and microscopic bacteria, aided by the earthworms and other animals, continually operate destructively upon the rock particles.

When such comminuted and decomposed products of he original granite come, as sooner or later they must, under the influence of moving water, they are subjected to prolonged and oft-repeated processes of transportation, with continual degradation and sorting of the particles, until conditions are favourable for their final deposition, in beds of uniform character, on the floor of some sea, lake, or estuary, or the flood-banks of a stream.

Should conditions favourable to deposition arise near the site of the parent rock, the transported material may have suffered so little sorting and modification that the deposited fragments may represent the quartz, felspar, mica, and subordinate ingredients of the granite almost in their original proportion. Such a deposit after consolidation is known as an *arkose;* it will differ scarcely at all from the original granite in chemical composition, though the mode of aggregation of the particles is different. Many of the grits of Carboniferous and Old Red Sandstone age and older palæozoic rocks are of this nature. At an early stage in the process of transportation the finer, clayey, pasty particles—largely the product of the breaking down of the felspars—are washed free from the coarser grains of quartz and carried farther; so that it is usual to find mud deposits in one place, sandy quartz débris, with some undecomposed felspar, in another.

The coarser quartz grains, when aggregated and consolidated, form beds of *Grit;* the finer-grained quartz waste goes to form beds of *Sandstone.* If by any chance a permanent deposit is formed near the original granite source containing *angular* blocks of still undisintegrated granite, we should have a Granite *Breccia;* should the deposit contain *rounded* blocks, pebbles, or boulders, it would be called a *conglomerate.* The fine-grained material first deposited as mud of varying degrees of fineness may go to form beds of *clay,* or *shale,* or eventually *slate.*

During the earlier history of the granite it has often occurred that chemical changes have been wrought in some of the felspars, mica, or hornblende; the alkalies or lime have been taken from the felspars, and the lime and magnesia may have been taken from the other minerals, and these have united again in different combinations to form new minerals. This kind of rearrangement, with less opportunity for recomposition, becomes very active in the upper parts of the rock when exposed, and while the intractable quartz is left intact the alumina becomes hydrated and removed as clayey substance, and the alkalies, lime, magnesia, and iron are slowly and steadily removed in solution.

The processes of destruction thus briefly summarized take place, not only in granites, but in all other rocks, and the materials carried off in solution and suspension are sooner or later deposited elsewhere. Thus the ultimate source of the beds of limestone in their many manifestations is traceable to the minerals of igneous rocks.

In the following chapters we shall deal first with the sandstones and rocks of their kind, and later with the limestones and the derivatives of clay.

SANDSTONES, GRITS, CONGLOMERATES AND BRECCIAS.

It has already been indicated that these rocks have been formed from the degradation of pre-existing rock masses, and it is mainly due to the sorting action of moving water—rain, streams, and marine currents—that we now find in one place coarse breccia or conglomerates, in another grits or fine sandstones.

Since the agencies which effect the breaking up and transport of rock material have been active from the earliest epochs, we naturally find representatives of all these kinds of derivative bedded deposits in every geological system from the earliest to the present day; and then, as now, we find that at any particular instant of time coarse sands were being laid down over one area, fine sands on another, conglomerates in a third, and so on, precisely as at this moment we have a great bank of coarse shingle—a potential conglomerate—in the Chesil Bank of the Dorset coast; broad sandy stretches—potential sandstones—off the eastern counties; or the coarse boulders of some of our rocky shores contrasting with the muddy flats of the Wash or the Thames. Moreover, since the degree of coarseness of any of these deposits is always dependent upon the velocity of the moving water, and as from time to time this velocity has varied, we find the coarse sands giving place to fine muds, clays, and shales, over any one area, and these changes were often repeated many times.

Bedding of Sandstones.—In common with all stratified deposits, the sandstones occur in lenticular beds—that is to say, sooner or later the beds thin out, and finally disappear in every direction. Occasionally great masses of sandstone, hundreds of feet thick, are split up into beds by mere partings or planes of separation, the individual strata ranging in thickness from a few inches to 30 or 40 feet; in other cases the beds of sandstone are separated by layers of shale or by clay seams. The finer-grained sandstones—those deposited in gently moving waters—are usually more regularly and evenly bedded than the coarse varieties; and when mica is abundant in such rocks, the flakes almost invariably tend to lie with their flat faces in the bedding plane; and since there has been a sort of periodicity in the change of velocity of the water movement, the finer flakes have tended to settle down in the quieter phases, in this way producing the planes of ready fissility seen in certain flagstones and in a more marked degree in the more readily splitting tilestones.

Upon the bedding planes of the medium-grained and finer sandstones it is a common occurrence to find ripple marks preserved with great fidelity, along with rain-pits caused by primæval showers, footprints of birds and beasts, and the impressions of plants. Very characteristic of many sandstones is *diagonal* bedding (current bedding or false bedding), due to constant change of direction in the stream or current (or wind) which deposited the sand (Fig. 4). Sandstones and grits are on the whole more prone to exhibit diagonal or current bedding than limestones. This structure may be produced on a small scale, as may be seen beautifully exemplified in some of the fine-grained Coal Measure sandstones, but frequently it is present upon so large a scale in the coarser rocks that building blocks of reasonably large dimensions may be cut without showing the surface of the bed.

SANDSTONES AND GRITS

The coarser deposits, such as conglomerates and breccias, are naturally subject to still greater irregularities of bedding, and often contain large irregular 'pockets' or lenticular masses of shale, clay, and the like.

Among the more notable imperfections to be avoided in selecting stones in the quarry are sporadically scattered clay galls, irregular concretionary iron stains, and incipient joint planes running at high angles to the bedding.

The Mineral Constituents of Sandstones, Grits, etc.—From a petrological point of view sandstones and grits vary considerably in the minerals they contain and in the relative proportions in which these minerals occur. Some, for instance, are locally rich in gold; some contain much titaniferous iron; some galena or tin; some are impregnated with copper, others are comparatively full of apatite, zircon, rutile, magnetite, hæmatite, and so on; some contain hornblende, black mica, augite. Calcareous shell fragments, coral fragments, silicified wood (in Eigg), foraminiferal shells of calcite, or replaced by glauconite, are common in certain sandstones (many 'greensands' and the siliceous variety of Kentish rag); oolitic grains of iron-ore are prominent in some sandstones (Carstone of Hunstanton, and elsewhere). In parts the Thanet sand of the London Basin consists largely of grains of flint, and there are numerous examples of rock resembling sandstones in the Cretaceous system which consist almost wholly of non-crystalline silica (chalcedony)—for example, the malmstone of the South of England and many of the stones known as *gaize* in France.

Iron pyrites and marcasite should be absent from all good sandstones, though they are common enough in impure earthy varieties.

From the point of view of the architect and builder, however, the minerals constituting these rocks may be ranged very simply into two groups: (1) those existing as grains—the original sand grains—and (2) those forming the cement which binds the grains together.

Of the minerals found among the grains only three require notice: these are firstly quartz, secondly mica, and thirdly felspar. Quartz alone may be considered *essential*; in many sandstones this mineral is the sole granular constituent. The quartz grains may be large or small, angular or rounded, simple or complex (that is, a single grain may consist of two or more distinct crystallized particles of quartz, adhering just as they were formed in the parent granite or other rock. This character is clearly shown when the grains are examined in polarized light).

When all the grains of quartz are thoroughly well rounded—as they are in some of the Triassic sandstones—it may usually be assumed that at some stage in their history they have passed through desert conditions, and have been carried about and corroded by wind action. Quartz grains derived from granitic rocks are usually irregular in form; those from gneisses are more flaky and elongated; those from pre-existing sandstones and grits are usually rounded, whatever their primitive source may have been. The grains in marine sands and sandstones are more rounded than those formed by rivers and lakes; only in wind-worn sands are the smallest grains perfectly rounded. Quartz grains, either angular or rounded, not infrequently exhibit the phenomenon of regrowth, whereby they assume, or tend to assume, so far as space permits, the regular form of perfect crystals. A beautiful illustration of this regrowth is to be seen in the Lazonby Stone (see Fig. 6), but it is quite common in many sandstones and grits in a lesser degree of perfection, and it accounts for the minute glistening points visible on the surface of such stones. Quartz is not the only original mineral constituent of sandstones to show this later outgrowth; it has been observed in felspar and in hornblende grains.

The *Mica* in sandstones is not an essential constituent, although more or less is present in the majority of these rocks. Several varieties of the mineral may occur, but their appearance in small flakes is similar in all, and in the grits great majority it is white mica. In some of the coarser the mica flakes may reach $\frac{1}{4}$ inch square; usually the flakes are much smaller, and they tend, as indicated above, to concentrate along certain planes in the rock, especially in those of finer grain.

The *Felspar* grains in sandstones and grits occur in all stages of freshness and decomposition; they are naturally most prevalent in the coarser rocks, though some fine-grained Carboniferous sandstones exhibit a fair amount of this mineral. All the principal varieties of felspar have been observed in sandstones, but it is comparatively rare to find them sufficiently unaltered to be readily recognizable. Some of the freshest felspar grains are the microcline fragments in many of the Carboniferous grits, and some of the Old Red Sandstones and Torridonian beds.

The Cement or Bond of Sandstones and Grits.—The essential grains of sand are held together in a variety of ways; the principal types of cement being siliceous, calcareous, argillaceous or ferruginous, and two or more of these bonding materials may be in association.

The siliceous cement may be introduced into the sandstones by water charged with silica from some other adjoining stratum, or it may be provided by the breaking up of the felspar grains within the rock itself, or by the partial solution of the quartz grains. This secondarily deposited silica, so called to distinguish it from the primary quartz grains, may crystallize in an irregular way between the grains without being in crystallographic continuity with them; or, as in the case cited above, it may so arrange itself around the grains as to build up each grain into as perfect a crystal as the surrounding space permits. By this process of rejuvenescence the nearly contiguous grains are brought into closer apposition one with another, the voids tend to be filled up, and the rock gains in cohesion and density. The secondary silica is not always found in the condition of quartz; it may be deposited in the form of chalcedony or opal.

Calcareous cementing matter rarely occurs alone in sandstones employed as building stones; more often, except in rocks like Kentish Rag or Gatton stone considered farther on, it acts as an accessory to one of the other building materials. Thus it is usually present in crystalline patches in the red Old Red Sandstones and in the more felspathic kinds of Carboniferous grit. In these cases it may have been derived partially from the decomposition of the felspars. An instructive illustration of the way carbonate of lime tends to crystallize as calcite in large crystals enclosing many grains of sand is to be found in the well-known sandstone of Fontainebleau, in France, where small masses of calcareous *sandstone* appear sporadically in the midst of incoherent sand rock, with the characteristic form of calcite crystals. The formation of large crystals of calcite, enclosing many grains of sand, is exhibited in sandstone from Spilsby, in Lincolnshire. In some sandstones the carbonate of lime occurs as a calcareous paste, only crystallized here and there and on a minute scale; the cement in these cases has often been derived from the breaking down of fossil shells. The calcite in this kind of rock occasionally takes the form of a stalagmitic deposit between and around the grains. Parts of the blown sand near Newquay, in Cornwall, are cemented in this way.

In rocks like the Kentish Rag, there is every gradation from a sandstone with abundant calcareous cement— mainly of organic origin—to a shelly limestone with scattered sand grains.

SANDSTONES AND GRITS

In the place of calcite we occasionally find the magnesian carbonate, dolomite, acting as the cementing material, as in the Permian Mansfield stone.

Argillaceous cement in good building stones rarely occurs dissociated from siliceous cement. In the felspar-bearing stones a certain amount of this type of cement is common, but in good stones there is very little clayey matter. In no case does an argillaceous cement by itself form a strong, lasting stone, unless it has undergone considerable silicification, as in many of the Carboniferous, Devonian, and older palæozoic grits and sandstones. A small amount of the white or slightly stained argillaceous paste does not exert any serious influence; in some of the rocks it is not difficult to detect the mineral Kaolinite, and much of the paste may be regarded as kaolin derived from the felspar grains.

Certain sandy rocks contain detrital shale or slate fragments, and others pass by increase of argillaceous muddy matter into sandy shales and mudstones. Sandstones with an argillo-calcareous paste are generally soft and easily weathered, but some of the Molasse sandstones of France and Southern Europe, which are of this nature, are used as building stones; they are rich in very fine mica.

There is a variety of sand-formed rock, known as *Quartzite*, in which the component grains are so firmly united that a stone of great compactness results. A micro-section of such a rock usually shows that the individual grains have been welded closely one to another the angular projections of one fitting into corresponding depressions in adjoining grains, without sign of any intervening cement. This effect is most often produced by pressure, which is always acting in a lesser degree on strata lying beneath a load of overlying rocks, and in a greater degree in regions subjected to the tangential stresses resulting from crustal warping. Hence quartzites are most prevalent in disturbed regions and mountain tracts. A similar effect is frequently noticeable where sandstones have been heated by masses of hot igneous rocks, and in the former case the heat generated by pressure is doubtless functional in bringing about the resultant change in texture.

It is doubtful whether this close welding of the quartz grains could be effected in a perfectly dry mass, and it must be borne in mind that all rock beds naturally contain a great deal of water, even when, comparatively speaking, they are dry; and this water—quarry water—is actively instrumental in assisting the welding operation, for with the increased pressure and heat solution is more rapid than usual. The recrystallization of dissolved silica in any vacuities that remain is a constant concomitant of quartzite formation.

The effect of pressure in welding the quartz grains is visible in many sandstones and grits that do not approach the compact condition of quartzites, and examples may be seen in many of the palæozoic rocks in which some of the cohesion of the stone is due to this cause. In these cases the angular corners of some grains are seen to penetrate into depressions in the neighbouring grains.

The quartzite structure may be produced, however, without pressure wherever there has been a tolerably clean sand permeated by silica-bearing waters; the tertiary 'Sarsen Stones' (p. 169) provide an excellent illustration of this type. In the Folkestone beds of the Weald at Ightham, in Kent, a curious quartzite rock occurs, in which the quartz grains have grown together by a radial arrangement of fibrous quartz round each grain.

In some sandstones the quartz grains are held together solely by iron oxides, but these are rarely used—at any rate, in Great Britain—for building purposes. One of the best-

PLATE III

PHOTOMICROGRAPHS OF SANDSTONES. (Magnification, 30 diameters.)

a, Quartzite, 'Firestone,' Winford ; *b*, Gatton (Reigate Stone) ; *c*, Bramley Fall ; *d*, Quarella ; *e*, Craigleith ; *f*, Pennant.

SANDSTONES AND GRITS

known examples is the Northampton sand formation of the Inferior Oolite, which in places is highly ferruginous; the Lower Greensand, the Kellaways rock, and many others are iron-cemented locally. In practically all the sandstones that exhibit any colour—ranging from a warm cream to rich purple-red—iron plays some part as a cementing medium. It usually occurs either in the form of the hydrated sesquioxides, limonite, göthite, or turgite, or as the anhydrous sesquioxide hæmatite; occasionally the carbonate chalybite is present in small quantity.

It will be gathered from the foregoing remarks that the cement, or bond, uniting the sand grains to form a sandstone or grit is usually complex in character. We do indeed find stones in which the cement is wholly calcareous, siliceous, ferruginous, or argillaceous; but in the great majority of cases the coherence of the rock is due to the combination of these agencies, and not infrequently a single rock will show all these modes of cementation, but in different degrees of importance.

Much more rarely and locally we find other minerals acting as cementing materials: for example, barytes in Nottinghamshire, Cheshire, Oppenheim, in Hesse, etc.; gypsum in the Thanet Sands, Kent, and in Arabia; fluorite in Bavaria; bitumen in Pachelbronn, Alsace; and occasionally also celestite and phosphorite.

According to the predominating cementing material, the sandstones may be classified as (1) siliceous; (2) calcareous, or dolomitic; (3) ferruginous; (4) argillaceous; but the character of the bond presents many variations.

In sandstones in which the cement is siliceous the grains may be united in the following ways:

1. By complete filling of all the pores by quartz, which may be in crystallographic continuity with the grains or in the form of independent grains.

2. By complete filling of the pores, due to the squeezing

together of the grains, so that they are welded together with no *visible* cement.

3. By deposition of a little quartz at the points of contact of the grains alone, leaving more or less unfilled space in the stone.

4. By the formation of a chalcedonic or opaline infilling of the pores and spaces.

5. By the silicification of an argillaceous ground mass, which usually fills most of the space between the grains.

Again, in the case of calcareous and argillaceous and some of the ferruginous sandstones, the cementing material may be small in bulk, lying between the grains, and filling the pores more or less completely, but of no more than sufficient quantity to hold the quartz grains together; on the other hand, there may be so much of this kind of cement that the quartz grains are completely separated one from another, and lie within it like plums in a pudding.

An iron cement may occur in bulk as above, or it may exist merely as a thin film round each individual grain. In this case the grains are only held together by the cement at their mutual points of contact, and such stones are usually rather friable and very porous, as, for example, many of the red Triassic sandstones. The difference between active cementing matter and inert pore-filling material should be noted (Fig. 6[7]).

The Colour of Sandstones, etc.—Perfectly clear quartz grains, united by siliceous cement, produce white stones of great beauty, but they are comparatively rare. Many of the pale grey and pale cream-tinted stones—often called white in the trade—owe their tone to the combination of clear quartz with white argillaceous, and in some cases calcareous, material free from iron stain. Most of the warm cream, buff, or yellow sandstones are coloured by the presence of very small specks of

SANDSTONES AND GRITS

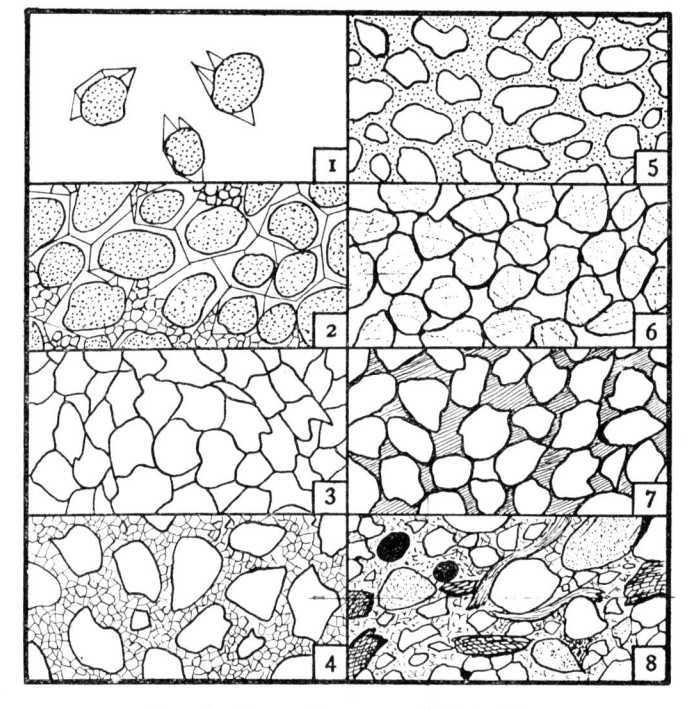

FIG. 6.—SOME TYPES OF SANDSTONE.

1. Individual grains of iron-stained quartz sand, with subsequent crystalline growth of quartz.
2. A sandstone showing the binding of the grains by crystalline development (Penrith).
3. Quartzite: quartz grains pressed together; united by silica.
4. Quartzite with grains of two sizes; silica cement.
5. Sandstone with grains lying in an abundant matrix, which may be argillaceous, calcareous, or of mixed constitution, with or without the addition of silica.
6. Sandstone with quartz grains united by siliceous cement at their points of contact; abundant pore-spaces.
7. A sandstone with the grains united as in 6, but with the intergranular spaces filled with matter taking little or no part in the binding of the grains.
8. A sandstone with mixed grains: quartz, felspar mica, glauconite (black), and fragments of shale or slate.

argillaceous material stained with limonite; the more numerous the specks, the warmer the tint will be. The warm and strong yellow stones—*e.g.*, some of those in the Triassic beds—are tinted by a thin pellicle of the same oxide surrounding each quartz grain. Pink stones are coloured by an admixture of hæmatite with the limonite, or in some felspathic sandstones of the Carboniferous and Old Red Sandstones the tint is derived mainly from the iron stain contained within the felspar grains themselves. The most strongly coloured red and purple stones found in the Old Red Sandstone owe their tint to the last-named cause, and partly to finely-divided hæmatite matter disseminated through the argillaceous paste. The dark grey and greenish stones from the Pennant Series in the Coal Measures are coloured by the shale fragments and carbonaceous matter they contain, often with a certain amount of chlorite; carbonaceous material also colours the very dark Silurian and Ordovician flagstones of South Wales and elsewhere.

The iron stain in some sandstones and grits is unevenly distributed, sometimes being concentrated more strongly in the neighbourhood of joints and bedding planes, or in irregular concentric shells, frequently of great size, around a central nucleus. Such portions of the rock are not employed for the best building blocks unless the stains are quite small. Pale patches in the red sandstones of various ages, but particularly common in the Trias and Permian of this country, are often due to the decolorization of the iron stain by some process of reduction, whereby a colourless iron salt replaces the coloured form; occasionally this action is traceable to the presence of organic matter. Occasionally the process has been reversed.

Chemical Composition of Sandstones.—Among the building stones this presents few features of interest. The silica content is uniformly high in all good stones; alumina

TABLE IX.

	Red Wilderness, Micheldean (T).	Pennant Stone, Craig-yr-Hesg (Q).	Shamrock Stone, Doonagore (T).	Stancliffe, Darley Dale (C).	Giffnock (T).	Craigleith (C).	Quarella (T).	Corsehill (Wallace).	Corncockle (T).	Grinshill (T).
Silica ...	88·70	83·15	84·90	96·40	86·96	98·3	91·20	95·24	92·04	95·46
Alumina ...	3·25	8·10	6·60	} 1·30	} 0·43	} 0·6	{ 3·35	0·56	3·97	1·17
Iron oxides ...	2·10	4·54	3·60		0·16		{ 0·58	1·28	0·71	0·87
Iron carbonate	—	—	—	—	2·53	—	—	—	—	—
Manganese ...	0·10	—	0·65	—	—	—	—	—	—	—
Lime	2·90	0·38	0·90	—	—	1·1	1·23	—	0·39	0·61
Carbonate of lime	—	—	—	0·36	6·00	—	—	1·40	—	—
Magnesia ...	0·11	0·68	1·26	—	—	—	0·25	—	0·33	0·69
Carbonate of magnesia ...	—	—	—	—	2·98	—	—	1·23	—	—
Carbonic acid	1·94	—	none	—	—	—	—	—	—	—
Alkalies ...	0·31	0·78	0·39	—	—	—	traces	—	1·87	—
Water and loss	0·59	2·37	1·70	1·94	0·76	—	3·49	0·56	0·69	1·68
	100·00	100·00	100·00	100·00	99·82	100·00	100·00	100·27	100·00	100·48

T = Trade analysis.   Q = 'The Quarry.'   C = Royal Commission, 1839.

is most abundant in those containing much felspar or clayey matter, while the lime, magnesia, and alkalies are usually present in small amount.

The analyses of Table IX. will serve as examples.

The specific gravity of Sandstones ranges between 1·9 and 2·8; the porosity may lie between 5 per cent. and 28 per cent.

The crushing strength ranges between 130 and 1,470 tons per square foot, the average being 400 to 700 tons.

DISTRIBUTION OF ROCKS OF THE SANDSTONE GROUP.

For a detailed list of the large or more important sandstone quarries, the reader is throughout the following also referred to Appendix B.

Pre-Cambrian.—These ancient rocks contain abundant arenaceous representatives; many are hard quartzites; some are tough greywackes, grits, and conglomerates. Rarely are these employed for building, and then only for the roughest minor local work. This is accounted for partly by the intractable nature of most of the stone; but it must be added that only a small area is occupied by these rocks in England and Wales, and the great tracts in the North of Scotland are inaccessible and almost unoccupied. In the last-named region the variable purple and chocolate-brown Torridonian Sandstone, rich in felspar, is noteworthy; quartzites, grits, and conglomerates occur also in the Dalradian or Younger Highland Schists. Flaggy grits and quartzites of this age are found in the Longmynd district of Shropshire and in Charnwood Forest and Anglesey.

Cambrian.—No building stone of general importance comes from this system; as in the case of the older rocks most of the Cambrian arenaceous stones are either too

SANDSTONES AND GRITS

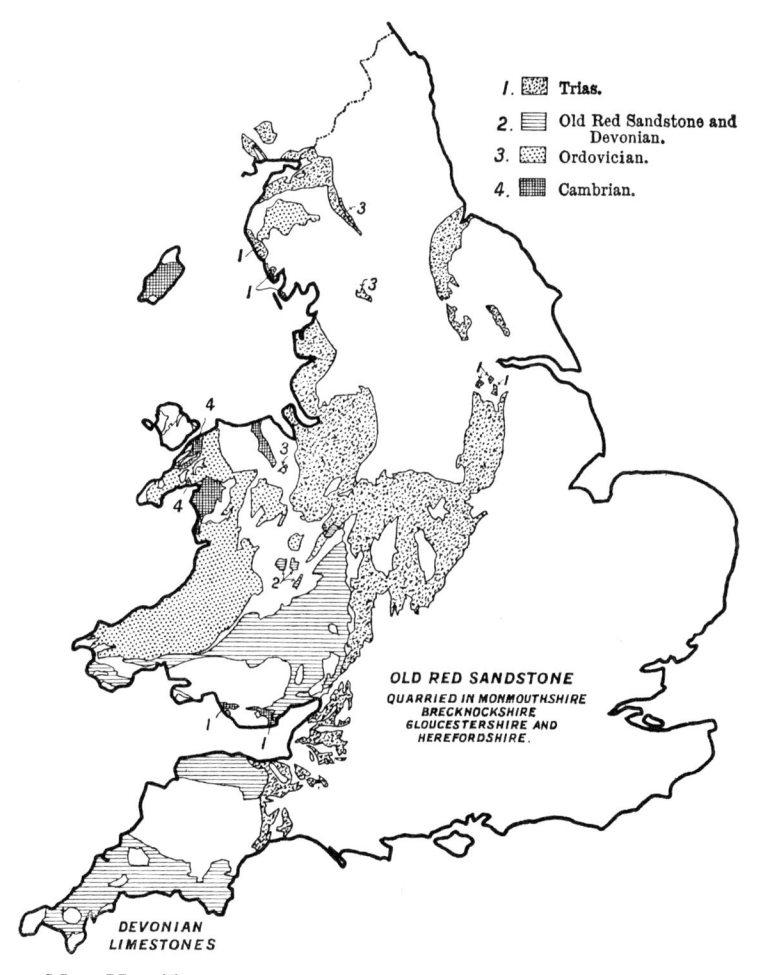

1. Trias.
2. Old Red Sandstone and Devonian.
3. Ordovician.
4. Cambrian.

OLD RED SANDSTONE
QUARRIED IN MONMOUTHSHIRE
BRECKNOCKSHIRE
GLOUCESTERSHIRE AND
HEREFORDSHIRE.

DEVONIAN LIMESTONES

MAP II.—TRIASSIC, DEVONIAN, ORDOVICIAN AND CAMBRIAN SYSTEMS OF ENGLAND AND WALES.

(Older Palæozoic rocks occur in South Cornwall.)

hard or too dull in appearance to be selected for this purpose; moreover, much the greater part of the area they occupy is sparsely inhabited and remote from the centre of population. In South Wales the lower portion of the Cambrian (Caerfai and Solva groups) is mainly composed of grits and flags. Parts of the Cathedral of St. David's are built of this material. In North Wales the Harlech Grits are about the same horizon; they attain a thickness of several thousand feet, and are prominently developed at Harlech and in the country to the north and east of the town. The stone is a very hard, coarse-grained, compact quartzite-grit of a pale greenish colour, weathering grey. Harlech Castle is built mainly of this stone, and it is still employed locally for rough walling and random ashlar, and to some extent for sills, etc. It is a very durable stone and has a pleasant appearance, but it is somewhat harsh under the tool.

To the older part of the Cambrian belong the Hollybush Sandstone, the quartzites of the Malvern Hills, the Wrekin Quartzite, and Comley Sandstones of Shropshire, the Hartshill Quartzites of Warwickshire, Lickey Hill Quartzites of Worcestershire, and the Eriboll Quartzites, and others in the north-west Highlands of Scotland.

In Wales the younger Cambrian strata contain the Lingula Flags, a series of shales, slates, flaggy grits, and sandstones, some 5,000 feet thick, in North Wales, where they are divisible into the Dolgelly beds, Festiniog beds, and Maenturog beds, in descending order. The Lingula Flags are much thinner—2,000 feet—in South Wales.

The Maenturog beds comprise the following strata:

	Feet.
Dark dull blue Slates	1,200
Fine-grained grey Flags	600
Bluish-grey and black Slates, alternating with grey and yellow Flags	300
Fine-grained grey and pyritic Flags	400

The Festiniog beds consist of tough bluish-grey flags, 50 feet, upon brown, grey, and yellow flags, 2,000 feet. From this strata good sound stone is obtained for local buildings in addition to the flagstones. Most of these stones, as already indicated, are dark in colour, and the same remark applies to the flags which occur in the Menevian of Wales.

Ordovician.—Few first-class building stones come from this system, although beds of flaggy sandstones and grits are of frequent occurrence. In the lower, or Arenig, series of North Wales, the Garth Grits lie among slaty beds; dark flagstones appear on the same horizon in South Wales; in Shropshire they are represented in part by the Mytton Flags of the Shelve district, and here, too, are the quartzites which stand up as a massive prominent ridge from near Bishop's Castle to Pontesbury. This is a hard stone, largely quarried at the northern end of the outcrop for road metal at Nills Hill. The Llandeilo, or Middle Ordovician, contains many beds of dark flaggy sandstones in Wales, some of which are worked in a small way in numerous quarries in South Wales. Flags and grits occur also in Shropshire, and in the Skiddaw Slates and Borrodale Series of the Lake District, where they are known as *Calliard*, and in the Girvan district in South Scotland.

The Upper Ordovician, Bala, or Caradoc series contain flaggy beds, sandstones, and grits in Wales; and in Shropshire the sandstones (Soudley sandstone), Hoar Edge, and Spy Wood Grits are quarried to a small extent about Craven Arms. Some of these Shropshire sandstones are highly calcareous, and in places pass into arenaceous limestones. In South Scotland the Barren Flagstone Grits of the Girvan area lie on this horizon.

Silurian.—The rocks of the Silurian system include flaggy sandstones and grits in most of the localities where they are exposed; none, however, is of prime importance as building material, and these rocks are frequently highly calcareous.

In the Llandovery division we find green grits in Pembrokeshire; coarse sandstone and grits about Builth; flags and grits in Radnorshire; the Corwen Grit in North Wales at Corwen and Peny-glog, Denbighshire; and grits, sandstones, and flags in South Scotland.

Flags and grits occur of Tarannon age above the Llandovery beds in Wales, and are worked at Strata Florida, near Aberystwith. The stone has a rather muddy cement, but is very tough, and is suitable for heavy work; it is used in sea-walls. These rocks occur also in South Scotland.

In the Wenlock series calcareous flagstones abound in South Wales; the sandstone, Rhymny Grit, is worked at Rhymny Quarry, near Cardiff. These beds are represented in North Wales by the Denbighshire Grits and Flags (Peny-glog). In the Lake District the great deposit known as the Coniston Flags belongs partly to this series and partly to the overlying Ludlow series. They are exposed in the Brathay and Coldwell Quarries, near Ambleside. The Brathay Flags in the lower portion of the Coniston Flags consist of fine-grained, well-laminated blue flags; the overlying Coldwell beds are coarser, gritty flags, some of which are calcareous. These Coniston Flags have been worked in the Vale of Troutbeck, and at Horton, in Ribblesdale; the 'Horton Flags' come from this horizon. Similar beds have been worked at Studfield, Dryrigg, and Moughton Fell.

The Ludlow Series yields micaceous and rather soft flagstone in the Ludlow district; some of the thicker beds are known as 'pendle' by the quarrymen. Similar, but rather coarser, beds occur in the higher portion of the Denbighshire Flags.

In the Lake District the Coniston Grits attain a thickness of 4,000 feet, and the Kirkby Moor Flags above them, but separated by a great thickness of slaty beds, are 2,000 feet thick; they are marly calcareous flagstones, with beds of grits and coarse slates.

At the top of the Silurian system the Downton Sandstone, micaceous grey, red, and brown sandstone, 80 to 100 feet, is quarried near Downton Castle, Herefordshire, and at Dymock, near Ledbury. The higher beds at Ledbury contain thin sandy tilestones. Coarse red sandstone, with conglomerate and mudstones, represent the beds in South Scotland.

A Silurian grit, composed wholly of grains of oligoclase in close contact, is found south of Parys Mountain at Drys-lwyn-isaf.

The older palæozoic rocks of Ireland contain sandstone and grits in abundance, and in some districts a considerable amount of quartzite; but none of these rocks are utilized except quite locally. A greenish Ordovician grit with some pebbles of quartz suitable only for rough work, occurs in Co. Clare, where it is known as 'Porphyry.' The Rathmichael Round Tower in Co. Dublin is built of a quartz rock and slate; and Reginald's Tower, Co. Waterford, is built of a strong, hard, slaty grit, which has worn very well.

Quartzites occur in Galway, Leitrim, Wexford.

Calcareous and tuffaceous sandstones are used in the old buildings—for example, in the sculptured doorway in the small church of Clone, also in the new church there; it is a light and porous rock and could be sawn. In Wicklow similar stones, greenish in colour, occur in the west of the county, and have been used in ancient structures, as may be seen in St. Kevin's Kitchen, and in the carving on the doorway of the 'Library' at the Seven Churches, Glendalough. From Glencormick, in the same

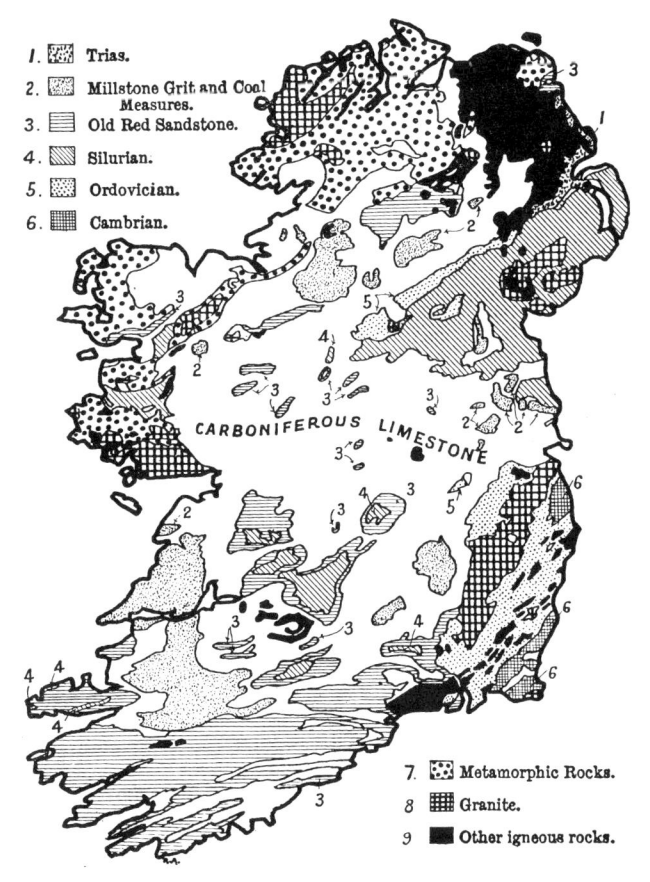

MAP III.—GEOLOGICAL MAP OF IRELAND.

county, Cambrian sandstones, of a warm creamy tint and fine grain, are found in these beds, and have been much used in Bray and the neighbourhood for dressed and coursed work.

From the Cushenden district, in Antrim, massive conglomerates of Silurian age may be obtained, and sandstones from the same place were formerly used in Belfast for ashlar.

A calcareous, ferruginous, brown Ordovician sandstone has been worked at Scrabby, eight miles from Granard, Co. Cavan; it tended to fade on exposure.

In Donegal large blocks of quartzite are obtainable for rough, heavy work; roadstone and flags, also of Ordovician age, occur at Oughterlinn. The quartzites and flags are quarried at several places. A red porous quartzite is obtained at Carrick, eight miles from Milford; it is known locally as 'red granite,' and has been used in Mulroy House. The Ordovician rocks of Down are little used. In Co. Londonderry finely laminated sandstones have been used for general work in Derry, and several of the larger buildings have been built of a bluish slaty stone from Prehan, but with only moderate results. Ordovician flags have been worked in Mayo, between Foxford and Swinford, and in the north-west of the county; massive grits and sandstones also occur.

Devonian and Old Red Sandstone.—In the Devonian rocks of Devon and Cornwall gritty and sandy beds occur of a considerable aggregate thickness, but of little significance as building material, although they are used locally for this purpose and for roads. It will be sufficient to mention the Meadfoot Sandstone and Lincombe and Warberry Grits of South Devon, and the Hangman and Foreland Grits of North Devon, all in the lower division of the system; while in the Upper Devonian of North Devon are the flaggy and micaceous sandstones and purple and green grits of the Pickwell Down beds; the similar beds of the Baggy and Marwood group, quarried at Sloley, east of Marwood; and the calcareous sandstones of the overlying Pilton beds.

The quartzites in the Ladock beds, Cornwall, are used for road metal.

When we turn to the Old Red Sandstone phase of this geological period, a number of excellent building stones are encountered. This formation occupies a large area in the counties bordering South Wales—in Hereford, Monmouth, Brecknock, Shropshire, and Worcestershire, and forming a belt along the north side of the South Wales Coalfield. In this tract the rocks are mainly red and purple marls in the lower subdivisions, and red gritty sandstone, with some marls and thin limestone, in the upper subdivisions. The sandstones and grits have been extensively employed for local building; they vary in colour from dark brownish-red to pale reddish-brown, and are usually felspathic, and frequently slightly calcareous; as a class they cannot be regarded as of such good quality as the Carboniferous stones from adjoining districts. Nevertheless, there are plenty of buildings to show that sound stone can be obtained. Much Old Red Sandstone has been employed in the city of Hereford, and the name 'Hereford Stone' has been used as a generic term for the sandstones from some distance around. Quarries are worked at Lugwardine, east of the town, and at Three Elms Stone; also at Cradley and Leysters Pole, north-east of Leominster, and near Monmouth and Crickhowel. The old Barbadoes Quarry near Chepstow furnished stone for Tintern Abbey. At Farlow, in Shropshire, a yellow sandstone is quarried from the passage beds between the Old Red Sandstone and Carboniferous. Certain beds of this system in this district are known as 'firestones,' and formerly

SANDSTONES AND GRITS

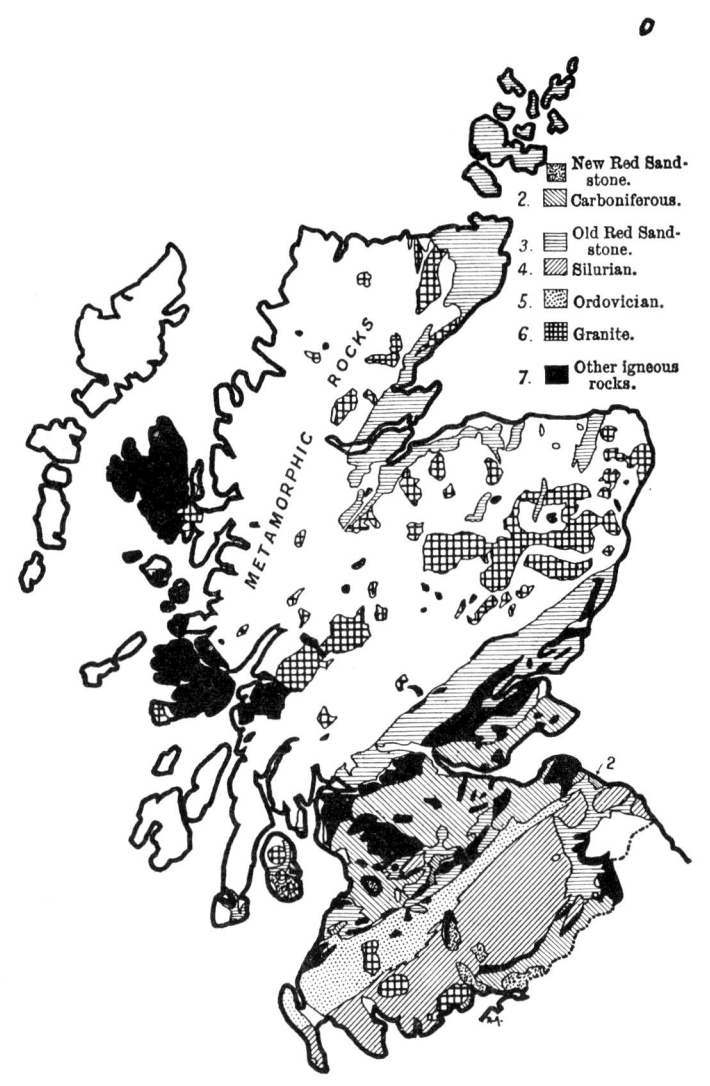

MAP IV.—GEOLOGICAL MAP OF SCOTLAND.

1. New Red Sandstone.
2. Carboniferous.
3. Old Red Sandstone.
4. Silurian.
5. Ordovician.
6. Granite.
7. Other igneous rocks.

they were often employed for hearthstones and fireplaces. They are generally hard and compact, and approach quartzite in texture.

The small patch of Old Red Sandstone at Peel, in the Isle of Man, yields stone for building, monuments, and flags

In Scotland the Old Red Sandstone of the Cheviot district does not produce many building stones of general interest; the sandstones, which are white as well as red, are confined mainly to the upper subdivision. The rocks of this age, which border the Carboniferous area of the Central Lowlands, appear as a narrow and very irregular belt between Girvan and Haddingtonshire, but on the northern side they form a broad tract from Loch Lomond on the south-west, passing through the counties of Dumbarton, Stirling, Perth, Forfar, and Fife, to Stonehaven, in Kincardineshire. Both upper and lower Old Red Sandstone are represented in this district, and the total thickness is over 5,000 feet. The sandstone and conglomerates are commonly felspathic, and the colouring is variable, including white, grey, red, yellow, purple, and dark brown varieties.

These rocks are very extensively quarried; the greatest number of large quarries occur in Forfarshire, as is indicated in Appendix B.

Farther north the Old Red Sandstone is well developed on the border of the Moray Firth and up the east coast, in the counties of Aberdeen, Elgin, Nairn, Inverness, Cromarty, Sutherland, in Caithness--the source of the well-known flags—and in the Orkney Isles. A brief statement of the subdivisions recognized in this region will give some idea of the vast thickness of this formation in north Scotland.

SANDSTONES AND GRITS

	Feet.
John o' Groat's Sandstone and Flagstone	2,000
Huna Flagstones	1,000
Gill's Bay Sandstone	400
Thurso or Northern Flagstone Group	5,000
Wick or Lower Flagstone Group	5,000
Langwell and Morven Sandstone and Conglomerate	2,000
Badbea Breccia and Conglomerate	300
Braemore and Berriedale Sandstone	450
Coarse Basement Conglomerate	50

Of course, in this great thickness—16,200 feet—of rock, although sandstones and flagstones predominate, there is a good deal of interbedded shale and some thin limestone. In this district there are numerous quarries.

Red and yellow sandstones and flagstones of this system make up the greater part of the Orkney Isles.

In Ireland these rocks are extensively exposed in the south-west and south. They include the hard, purple, and greenish Glengariff Grits of the Dingle Promontory of the lower subdivision; and the fine-grained green, yellow, and red sandstone of the Kiltorcan beds, with underlying softer red and purple sandstone of the upper subdivision. These rocks have been worked in many small quarries. In County Cork the Glengariff Grits are called 'brown stone'; the other Old Red beds 'red stone.' The Round Tower of Cloyne, of a light brown stone, is still in good preservation. In Kerry the Old Red Sandstone and some Lower Carboniferous sandstone were much used in Early Norman architecture, but now they are little employed except for rough work. The upper Dingle beds yield some good sound rough stone.

Carboniferous.—This system is by far the most important source of sandstones and grits in this country. In each subdivisional horizon of the rocks sandstones occur; it is, however, in the Coal Measures and Millstone Grit Series that most of the best-known stones are found.

Before proceeding to the examination of the rocks in

the several principal groups of the system, it will be well to note some of the more obvious changes that affect the strata as they are followed from one part of the country to another; for it unfortunately happens that the nomenclature attached by custom to the various horizons connotes very different lithological features in different districts. Moreover, the lines of demarcation between adjacent groups, which seemed natural and convenient enough in the district to which they were first applied, became forced and unduly artificial when transferred to the same series of strata some distance away.

The cause of these lateral variations in the Carboniferous rock is simply this: At the same time that organic limestones were being formed in the clear waters of one area muds and shales were being laid down in some neighbouring tracts; while sandstones and grits were being deposited in moving waters of shore and estuaries, accompanied by terrestrial and lagoonal formation of coal.

Thus we find the thick-bedded Carboniferous, or Mountain, Limestone of South Wales, North Wales, the Mendips, and Derbyshire, characterized by few and small intercalations of shale, and practically no sandstones. Northward, however, the few massive limestones gradually give place to numerous thinner beds, separated by beds of sandstone, shale, and grit, increasing in numbers and in thickness, and even including seams of coal. This is the condition of things in the Lower Carboniferous rocks of the Wensleydale (Yoredale) district. In Northumberland the clastic sediments have still further increased at the expense of the limestones, and coals are found low down in the series; in southern Scotland the so-called Carboniferous Limestone Series, with the underlying Calciferous Sandstone, is rich in coals, shales, and sandstones, while limestones play a subordinate part.

SANDSTONES AND GRITS

Sandstones and Grits of the Lower Carboniferous.—The micaceous sandstone in the Lower Limestone Shale group of the Bristol district, Forest of Dean, Mendip Hills, and north-west of England, may be passed over at once as of no economic importance. To these may be added the Coomhola Grits of southern Ireland, which also occur low down in the system. They are nearly always dark and argillaceous.

It is not until we reach the Yoredale district of Yorkshire that sandstones and grits of any thickness come into the Lower Carboniferous. Certain sandstones and grits, formerly classed as 'Yoredale' in Derbyshire, Staffordshire, and South Lancashire, are better placed in the Millstone Grit Series.

Farther north, in Northumberland, the whole of the Carboniferous Limestone Series contains sandstones associated with shales, limestones, and carbonaceous beds; none stands out with special importance except a great mass of grit, with shales, in the lower part of the series known as the Fell Sandstone Group, or Tuedian Grits. These grits form prominent features on the moorland fells about Tweedmouth, the Peel and Bewcastle Fells, on Simonside Fell (Simonside Grit), and Harbottle Fell (Harbottle Grit); also the Inghoe and Rothbury Grits and Armstead rocks: brown false-bedded sandstones.

The Carboniferous Limestone Series of Scotland contains three groups of strata: the Upper Limestone—grits, sandstones and shales, with a few limestones; a middle group, similar, but with more coal—the Edge Coal Group; and the Lower Limestone Group, again with similar lithological characters. At the top of this series the *lower* division of the Roslin Sandstone is included; this lies in the position of the Millstone Grit of the South. Grits and sandstones are quarried here and there in

142   GEOLOGY OF BUILDING STONES

MAP V.—CRETACEOUS AND CARBONIFEROUS SYSTEMS OF ENGLAND AND WALES.

SANDSTONES AND GRITS

the Edge Coal Group and in the other divisions of the series.

Below the Carboniferous Limestone Series of Scotland is a thick shaly set of beds, the Calciferous Sandstone Series, which is not so arenaceous on the whole as its title suggests, for a large part of the strata consists of clays, shales, and impure limestones; grey sandstones occur in the upper part in the Edinburgh area; and red sandstones are more generally prevalent in the lower portion, or Red Sandstone Group.

Sandstones and Grits of the Upper Carboniferous.—The rocks which constitute the younger part of the Carboniferous System — namely, (*a*) the Millstone Grit, and (*b*) the overlying Coal Measures—possess characters of great uniformity over the whole of the United Kingdom. Combined with the general uniformity, there is great diversity in the local details, as, indeed, is to be expected in formations laid down in the ever-varying circumstances of current and depth found in the shifting channels of a great estuary, and in a region subjected to frequent alternations of fresh water and marine conditions. Most of the sandstones from this series that are employed as building material have been deposited in fresh or estuarine waters.

(*a*) *The Millstone Grit.*—This series of strata—a most important source of good building material—illustrates with the utmost clearness those characteristics indicated above. Wherever this formation occurs, it consists regularly of beds of coarse grit, with occasional finer-grained sandstones, and intervening sandy or muddy shales. It is an easy matter in any circumscribed area to distinguish a sequence of shale beds alternating with grits, and this has frequently been done, the grits being numbered from above downwards, or having assigned to them distinctive names. Thus, in Lancashire, where the Millstone Grit

Series is most extensively developed, measuring in places over 5,000 feet in thickness, the following succession of beds is recognized:

1. First Grit, or Rough Rock.
   Shales.
2. Second Grit, Haslingden Flags.
   Shales with thin Coal.
3. Third Grit.
   Shales and irregular Sandstones and thin Coals.
4. Fourth Grit ⎫ Kinder Scout Grits.
5. Fifth Grit ⎭
   Shales.
6. Pendle Grit (Upper Yoredale Grit).
   Bowland Shales.
7. 'Yoredale' Sandstones.

But while some of the grit beds are very persistent, and can be followed with ease into Yorkshire, Derbyshire, and Staffordshire, such as the Rough Rock and the Kinder Scout Grit, others, the second and third, are much more variable, and the Kinder Scout Grit itself, which in places is a single strong grit, splits up elsewhere into two or even more comparatively feeble representatives. Briefly, the names and relative positions of the grit beds convey little indication of the nature of the stone as a building material when applied in different localities; the Third Grit, for example, may be a coarse massive rock, as in the Roaches, near Leek, in Staffordshire, resembling the Rough Rock of many places; while elsewhere, as in Yorkshire, it is represented by a considerable number of thinner and variable sandstones, separated by shales.

Both the base line and the upper limit of this series are arbitrarily chosen; in reality there is a gradual passage downwards into the Upper Limestone Shales, Pendleside beds, or Carboniferous Limestone Series, as the case may be, and upwards into the Coal Measures.

Distribution of the Millstone Grit. — The Millstone Grit Series in the West of England attains a thickness of between 400 and 500 feet. It is found encircling the Forest of Dean Coalfield in a narrow outcrop; small tracts of it lie at the northern end of the Bristol Coalfield, between Chipping Sodbury and Tytherington; a very narrow crop appears on the northern flank of the Mendips, between Ashwick and Mells. In the last-named locality these rocks yield a brown compact quartzite—sometimes called a firestone—used more for road metal than for building, and it is generally true of all the southern examples of the Mendip Grit that they are harder and more compact (not necessarily better stones) than those of Derbyshire and the North. Associated with the quartzitic characters it is common to find the stones inclined to be too 'venty,' and subject to irregular cracks to admit of the extraction of large sound blocks.

Round the great coalfield of South Wales the grits crop regularly; the outcrop is broader on the northern side of the field. Here the grits vary from point to point; on the Pembrokeshire coast they contain hard white sandstones and grits.

In Shropshire small patches of grits occur about Broseley, Coalbrookdale, and Lilleshall, and on the north of the Clee Hills; and a broad belt of these rocks runs northwards from south of Oswestry, by Chirk Castle, Cefn, and Minera, west of Mold, and east of the Halkin Mountains, towards Talacre. A remarkable feature of this tract is that as they are traced northwards these grits pass from the normal type into a series of cherty beds, with shale and some limestones.

In Anglesey a narrow strip extends by Trefdraeth and Llangristiolns.

On the western side of the Pennine Range the Millstone Grit fringes the north-east side of the North Staffordshire Coalfield, culminating in the fine hill, Cloud End, near Congleton; another belt extends northwards from Dilhorne, and spreads out north of Leek into a broad moorland tract, embracing the picturesque Roach Rocks (third grit) and Axle Edge; it forms the Peak in Derbyshire (Kinder Scout Grit), and, curling round the limestones and shales of the Derbyshire uplands, it builds a succession of well-marked scarps down the eastern side of the range by Hathersage, Eyam, Chatsworth (Chatsworth Grit), Stancliffe in Darly Dale—with the outliers of Stanton and Birchover—Matlock Moor, Cromford, Whatstandwell, and Belper, to Little Eaton, north of Derby.

Northward a broad tract of Millstone Grit separates the Lancashire and Yorkshire Coalfields, spreading round the former by Colne to Hoghton and Longridge, and beyond the latter by Ilkley Moor, Pateley Bridge, to Richmond.

A great tract lies between Lancaster and Settle, and many isolated outliers cap the hills in the Yorkshire Moorland, as at Whernside and Penyghent.

The same rocks form an irregular fringe to the Northumberland and Durham Coalfields.

(b) *Coal Measures.*—These rocks, in all parts of the country, consist of alternations of argillaceous shales, sandstones, and coals, with here and there very subordinate limestone.

The distribution of the principal coalfields is indicated in Map V.

Although sandstones are present in all the coalfields, they are most extensively worked in those of Lancashire and Yorkshire, where, as the list of Appendix B shows, they are of great importance. In the other coalfields mentioned below, the sandstones are quarried on a smaller scale.

The South Yorkshire, Derby, and Nottinghamshire Coalfield. —Most of the sandstone and grit beds have secured distinctive local names, which vary occasionally in different parts

SANDSTONES AND GRITS

of the field. The more important of these beds will be briefly noticed, in ascending order.

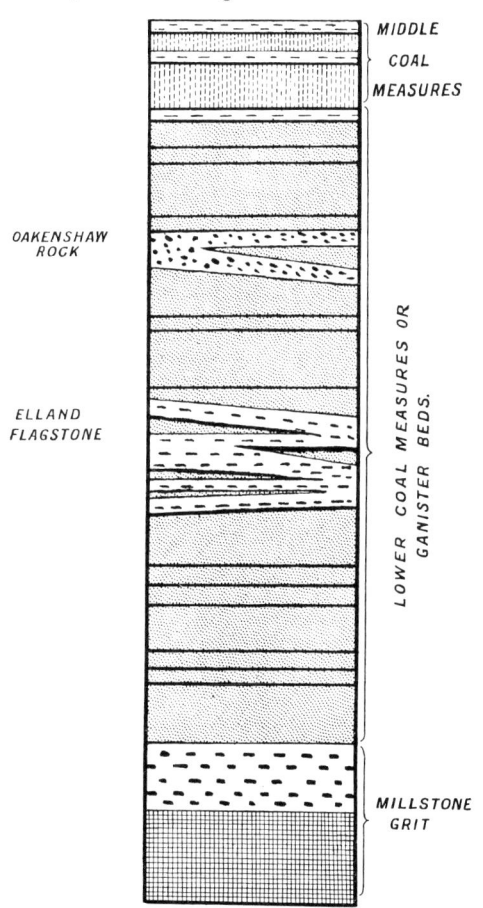

FIG. 7.—SECTION IN THE YORKSHIRE COAL MEASURES.

Crawshaw Rock = Soft Bed Flags: Generally present throughout the field; a coarse, thick-bedded grit on Onesmoor; fine-grained flaggy sandstone at Onesacre. The

Soft Bed Flags of the Huddersfield district are worked for flagstones.

Middle Rock, of the Ganister Measures: Often a very hard stone, as about Huddersfield and Halifax; a coarse grit or flaggy sandstone in other parts; numerous quarries north of Sheffield.

Loxley Edge Rock: A thick-bedded, moderately fine-grained grit, quarried about Sheffield.

Elland Flagstone = Brincliffe Edge Rock = Greenmoor Rock (Fig. 7): One of the most important of the Yorkshire stone beds. Although it varies in thickness and in characters, it can be traced in every part of the coalfield, and is equivalent to the Rochdale or Upholland flags of Lancashire. It is a thick series of beds, which are quarried where they are massive for building stone, where they are flaggy for paving stone. The best quality stone is of a pale blue colour, fine-grained and compact; some varieties are yellowish-brown. It is largely worked about Green Moor, Rastrick, Elland Edge, Halifax, Swales Moor, North and South Owram. In the neighbourhood of Leeds the stone is not so good as about Halifax. Between Thornton and Bradford the lower part is irregularly bedded and useless = rag; the middle part is a valuable building and flag stone, and can be split into slabs from 1 to 15 inches thick and several feet in length; the upper part yields stone that can be split into tile stones, $\frac{1}{8}$ inch thick, for roofing.

Greenoside Rock: Developed between Sheffield and Huddersfield, and particularly around Greenoside, where it is a good deal quarried; it is usually a coarse, thick-bedded grit.

Penistone Flags: This is a group of three beds of flaggy sandstone, separated by shales, best developed about Penistone. The lowest is fine-grained and micaceous, 40 to 60 feet thick; the middle bed is thin and variable;

the top bed is the most constant, and is often a hard and fine-grained stone.

Oakenshaw or Clifton Rock, the topmost important bed of the Lower Coal Measures, occupies a large area in the northern part of the coalfield. The stone is usually a fine-grained sandstone. In several places it yields good flags and local building stone.

Silkstone Rock = Sheffield Rock = Falhouse Rock = Slack Bank Rock, east of Leeds: A somewhat variable stone; about Falhouse it is fine-grained, compact and hard, with well-marked bedding; quarried about Christchurch and Littletown. East of Leeds the Slack Bank Rock is the principal stone; it varies from fine and stong to coarse, laminated, and crumbly; current bedding is often very pronounced.

Park Gate Rock: A thick and important bed about Sheffield; largely quarried at Bradgate; it dies out at Helsley Park. Farther north it is represented by the Birstal Rock = Cropper Gate Rock; at Carlinglow and Howley Park it is a very superior, massive, fine-grained white sandstone. At Birstal both white and brown stone are quarried for high-grade building stone; it is also sawn into paving slabs.

The Manor and Emley Rocks, respectively south-east of Sheffield and near Emley, are not of much importance.

Thornhill Rock and Birdwell Rock, southern part of coalfield = Dewsbury Bank, Morley, Middleton, Robinhood and Oulton Rocks of the northern part: Forms the most striking features of the Middle Coal Measure landscape, and is very extensively worked for building stone. The stone is usually fine-grained and compact; bluish-white when fresh, but weathers brown. When carefully selected it is a very good stone. Some varieties are coarser. At Morley and Oulton good flags are sawn from the large blocks.

The Woodhouse Rock, about Woodhouse and Handsworth; the Kexborough Rock at that place, and at Barough, and the Horbury Rock, between Bretton and Ossett, yield very fair stone locally.

The Handsworth, High Hazles, Barnsley, Kents Thick and Kents Thin, Woolley Edge, and Abdy Rocks, are sandstone and grit beds of variable character and local importance. The Woolley Edge Rock is largely quarried at Hemingfield, in Worsborough Dale, and near Barnsley, where it is thick-bedded and coarse. The 'Strong Blue' stone of Whitwood is from this horizon; it is a beautiful pale blue stone, fine-grained and compact, and a first-class building stone.

The Treeton Rock, south of Sheffield, Oaks Rock, north and south of Barnsley, and Acton Rock, of Acton, and Glasshoughton, are all about the same horizon. The Oak Rock is largely quarried between Tinsley and Barnsley.

The Lower, Upper, and Middle Chevet Rocks, Houghton Common Rock, Brierly Rock, and Ackworth Rock, are all worked in places, but they are frequently rather soft. Of the Ackworth Rock Professor Green says: 'It stands better than its appearance would lead one to expect.'

The Pontefract Rock, which was formerly classed as Permian, is thick-bedded and tolerably strong, but 'the mouldering condition of the castle ruins shows that most of the stone is unfit to resist the action of the weather for long periods.'

The Red Rock of Rotherham and Harthill is a thick mass of red, purple, and salmon-coloured sandstone; it is largely quarried for building stone and grindstones.

South Lancashire Coalfield.—In this district the principal sandstone beds are worked in the Lower Coal Measures. The more important beds include the following: The Riddle Scout Rock, Upholland Flags, Rochdale Flags, Old Lawrence Rock of Hayton, Burnley Flags, Woodhead Hill Rock. These rocks are largely worked for flagstones,

SANDSTONES AND GRITS

which are of good quality, somewhat micaceous and brown or grey in colour; as building stones they are extensively employed in the neighbourhood.

In the Middle Coal Measures of this district are the Blenfire Rock of Oldham Edge, a massive red sandstone, and the Chambers Rock of Rocher, a flaggy sandstone.

The adjoining coalfield of Cheshire yields a good strong stone from the Lower Coal Measures, Kerridge Rock, near Macclesfield.

In the *North Wales* Coalfield the sandstone of Gwespyr and Talacre are the most important; it is fine-grained and has a yellowish-green tint.

The building stone of the *South Wales* Coalfield and that of *Bristol* and *Somerset* is limited to the massive series of sandstones known as the Pennant Series. A very similar stone—sometimes called 'Forest Stone'—is worked in the coalfield of the *Forest of Dean*. The better-class stone from these districts is hard and durable; the colour is usually darker and more sombre in effect than that of most of the North Country Coal Measure stones; it occurs in various shades of dark grey, blue-grey, and pinkish-grey; this last variety is usually the hardest. The stone contains a good deal of felspar and fragments of slaty rock.

The coalfields of *Scotland* are worked at several places, as indicated in Appendix B; some of these rocks occur in the Upper Red Measures, as at Dunmore and Bredisholm; others are in the lower beds.

The *Carboniferous Rocks of Ireland*, as in England, furnish more sandstones than any other formation. They are subdivided as follows:

> Coal Measures: Grits and sandstones in shales, with coal, ironstone, and fireclay.
> Millstone Grit: Flagstones, sandstones and grits, shale and some coal.
> Pendleside Rocks: Similar beds, with more black shale and thin limestone.

Upper Carboniferous Limestone (with occasional sandstones and conglomerates).
Middle Carboniferous Limestone or Calp (with sandstones in Sligo, Tyrone, Ulster, etc.).
Lower Carboniferous Limestones.

Variable Basement Beds
{ Lower Carboniferous Sandstone.
Lower Limestone Shale and Conglomerate.
Carboniferous Slates.
Coomhola Grit. }

Few really large quarries are now working sandstone in Ireland, though there is abundance of good stone, and many small quarries are worked from time to time. The stone presents every variation of colour, texture, and cement; but, as in England and Scotland, the red stones are scarce in these formations. A more detailed account will be found in Appendix B.

The New Red Sandstone.—Under this head fall certain bright red, yellow, and white sandstones, which, with a more precise nomenclature, would be classed either as Permian or Trias. Since the stones from these systems have many points of resemblance, and the allocation of certain younger red sandstones is still to some extent *sub judice*, the more comprehensive classification may be conveniently retained here.

The distribution of the New Red rocks is indicated with sufficient clearness in Maps II. and VI. It will be observed that they extend from the south coast of Devonshire northward, and spread out broadly in the Midlands, whence the outcrops turn in two branches, one on each side of the Pennine axis.

The Lower Sandstones and Breccias of the south coast of Devon, about Dawlish, Teignmouth, Kingskerswell, Heavitree, Crediton, and Exeter, are regarded as Permian by Mr. Ussher. Their strong red colour, so well exhibited in the cliffs of Dawlish and the neighbourhood, cannot fail to arrest attention. Most of the sandstones are rather soft

SANDSTONES AND GRITS

and friable; but the conglomerates and breccias, often quite coarse, with pebbles and fragments of Devonian limestone as large as a hen's egg, form fairly strong stones, and they have been quarried on a small scale in numerous places for building purposes—*e.g.*, at Paignton, Chelston, Exminster, Heavitree, and Kingskerswell. Some good examples of the uses of this very coarse building stone may be seen in the villas outside Newton Abbot, on Milber Down; it has been used in the sea-walls at Torbay and other places.

The sandstones in Devon and Somerset are only occasionally hard enough for building stone, and they vary rapidly from place to place. They have been worked at Poltimore and North Clyst, and near Exmouth; also near Williton and Bishop's Lydeard. A pale Keuper sandstone, in thick beds, is quarried about North Carry, Sutton Hams, and Knapp. The last-named stone was used in the Wesleyan College, Taunton.

In the Mendip district a 'dolomitic conglomerate,' or breccia, locally called 'Millstone,' or 'Millgrit rock,' is a red sandy stone, full of large lumps of Carboniferous limestone. It has been used in the locality for building, and has been polished as a marble; near Clevedon, in Glamorganshire, it has been used for architectural carving. 'Draycot Stone,' quarried near Axbridge, is a variety of the rock. Some of the sandstones worked for building near Yatton (Claverham or Clareham Stone), Chew Magna, Sutton Mallet, and Pyle are so calcareous that they should properly be regarded as limestones with much sand.

Radyr Stone is the name given to the Triassic Conglomerate in the neighbourhood of Cardiff, where it is used for building and road metal. It has been much employed in railway bridges and for copings. The colour of the stone from Radyr Quarry is variable; red varieties are quarried near St. Nicholas and Vianshill.

The Bunter formation, or Lower Trias, is usually divisible

into an Upper Red and Mottled Sandstone; a middle group, the Pebble Beds or Conglomerate; and a Lower Red and Mottled Sandstone. These beds rarely yielded anything strong enough for building stone, except in Lancashire, where a reddish-brown, rather soft sandstone from the Pebble Beds has been quarried about Liverpool, Manchester, Stockport, West Derby, Kirkdale, and Walton. Stone of this kind was used in Chester Castle. The Upper Sandstones, which are here red below and yellow above, have been used about Birkenhead and Ormskirk.

The dark red sandstone of St. Bees and Corby, in Cumberland, is probably of Bunter age; it is not a very strong stone. The best Triassic building stones come from the Keuper formation, especially from the lower parts. The stone is white, yellowish, pink, or red; usually fine-grained, but occasionally coarser. In Warwickshire the stone has been very widely used. The Waterstones division, or Lower Warwick Sandstone, has been quarried at Guy's Cliff, Coton End, and Cullington; it has been employed in Warwick Castle. The Pendock stone, from near Malvern, is a Keuper Sandstone. Near Tewkesbury a white sandstone, 20 to 30 feet thick, occurs in the red marl of the Upper Keuper. The reddish-brown stone of Ombersley and Hadley, near Droitwich, has been used in the restoration of Worcester Cathedral.

In Staffordshire a soft, cream-coloured sandstone has been worked at Colwich, and used in many of the local churches and mansions; and the well-known Alton and Hollington stones are obtained from the Keuper rocks in this county. These stones, worked in several quarries, are fine-grained to moderately coarse. The colours range from cream, or 'white,' to pink, red, and reddish-brown. Large blocks can be obtained. They are known as Alton and Hollington white firsts and seconds, and 'mottled' and 'red.' They have been used in Trentham and Maer

SANDSTONES AND GRITS

Halls; the Town Hall, Derby; Croxden Abbey; St. Asaph Cathedral; and in many churches and public buildings.

In Shropshire the Keuper Beds are quarried in the Hawkstone Hills; at Weston, Belton and Muxton, and Grinshill, near Shrewsbury. The Grinshill stone from the Bridge, or Cureton Quarries, is a fairly coarse-grained stone, which occurs in large masses in the lower part and in flaggy beds in the upper part of the quarries. The best stone is white or cream-coloured; other beds are pink or red. It has been used in the old churches in Shrewsbury and in other large buildings.

In Cheshire light red stone has been quarried at Manley, near Dunham, and used in the restoration of Chester Cathedral; also at Runcorn, Weston, Peckforton (Peckforton Castle), Helsby, Storeton, and Bidston Hill, near Birkenhead. Much of Old Chester is built of local stone of this age.

The bright red sandstone of Penrith and Lazonby, in Cumberland, is of Permian age. The succession of beds here is as follows:

	Feet.
Upper Sandstone / Upper Brockram	150
Middle Sandstone (Penrith and Lazonby)	300 to 1,000
Lower Brockram	100

The Penrith stone is fine to moderately coarse in grain, and very free from intergranular matter; the quartz grains have recrystallized to a very marked degree. It is not a very strong stone, and should be selected with care; much of Penrith is built of it.

The Brockram is a coarse breccia, consisting mainly of fragments of Carboniferous limestone in a matrix of red sandstone; it has been quarried about Kirkby Stephen and Appleby for local buildings; the Romans used it in Hadrian's Wall.

In some places it is dolomitic, and it is often so calcareous that it should pass as a limestone; it is sometimes burnt for lime.

On the other side of the Pennine Range the Permian Magnesian limestone is frequently sandy, and the Mansfield red and white stones are often regarded as sandstones, with a crystalline dolomitic cement (see p. 197).

In Scotland the New Red Sandstones occur in a few small isolated areas; the more important will be briefly mentioned.

The Annan area in Dumfries has been worked at Corsehill, Cove, Annanheath, and Kirtlebridge. The stone is fine in grain, even and compact, and can be freely carved and turned. It often has a fine lamination, which makes it necessary to exercise care that it is set on its right bed for outside work. In colour it ranges from strong bright red to pale pink, and a small amount of almost white stone is also obtained. The Corsehill stone is the best known, and is deservedly popular where a red stone is required. It has been freely used in London, Edinburgh, Glasgow, and elsewhere—Jew's Synagogue, Great Portland Street; Great Eastern Railway Hotel, Liverpool Street; Town Hall, Reading; Mount Stuart Mansion, Rothesay; University Buildings, Edinburgh; State Capitol (interior), Albany, New York; American Fire Insurance Company's Building, Baltimore, Md., and in several buildings in Stockholm.

To the north-west of Annan another tract lies about Lochmaben and Corncockle Muir. The well-known Corncockle red stone is obtained here; it is bright in colour, and resembles Corsehill in texture, though some of it is rather softer.

Another tract is situated about Dumfries; the principal quarries are those of Locharbriggs. The bedding varies a good deal, and the stone is best where it is steepest.

SANDSTONES AND GRITS

The stone is rather less strongly coloured than the average Corsehill, and it is rather coarser grained; it is moderately hard but easy to work. It has been very extensively used in Glasgow.

Farther up Nithsdale is the Thornhill tract; the principal quarries are those of Gatelawbridge and Closeburn. The stone is finer grained than that of Locharbriggs and usually paler; it is strongly current-bedded, and exhibits dark spots in places, which, however, usually disappear on exposure. The Closeburn stone is dark red. "Old Mortality" worked in Gatelawbridge quarries; but much of the moulding and other dressing is now carried out on the spot by machinery.

In Ayrshire a patch of New Red rock occurs about Mauchline. The principal quarries are at Mauchline, Barskimming, and Ballochmyle. The stone varies in strength from hard and compact to friable; the former kinds are used for grindstones. It usually exhibits well-marked current-bedding; the colour is a bright brick red. The excellent weathering qualities of the stone are exemplified in the ancient castle of Auchinleck.

In the south of Arran the New Red stone is quarried at Corrie and Brodrick Wood. The stone is coarser and softer than most of those previously mentioned. Its colour is a dull dark red.

Reference may here be made to certain sandstones of Rhætic age, which have been worked for many years in Glamorganshire. The principal quarries are those of Quarella; the beds have been worked also at Llandough, Cowbridge, St. Hilary, and Pyle. The Quarella stone is a moderately fine-grained sandstone, slightly calcareous. In colour it ranges from white to grey and greenish-grey. The white and green are suitable for dressings and the grey for landings. It has been used in Llandaff Cathedral, the Colonial Institute, etc.

The New Red rocks are only very limited in their occurrence in Ireland, but where they do appear they have often been quarried, although the stone is usually poor and more friable than the best English stones. It passes under the name of 'red-free' wherever it is worked. It is quarried near Larne, in Co. Antrim, in Co. Londonderry and Meath. In Armagh a red calcareous breccia from the base of the Trias has been quarried for flagging near Red Barn, and a Permian sandstone from Grange was used for rough work in the restoration of Armagh Cathedral, 1835. In Co. Down a deep red rather soft stone has been quarried at Kilverslin, near Moira, and used in the Great Northern Railway bridges. Scrabo, near Newtonwards, is the best-known quarry in these rocks. There are several quarries, and the stone varies in colour—grey, yellow, red—but is commonly light brown, with an argillaceous or siliceous cement, and variable texture; well-selected stone behaves well. It has been used a good deal in Belfast, the Academy, Albert Memorial, St. Enoch's, and other churches, and in Stourmount Castle and other buildings in Newtonwards. The Globe quarry produces a light-coloured stone of even grain, which stands fairly well, though it is soft (Robinson and Cleaver's warehouse, Belfast). The Dundonald quarry, near Comber, is similar to Corsehill.

Jurassic.—The Jurassic strata of Great Britain may be considered to be limited to the broad tract shown on Map VI., stretching across England from the coast of Dorset to the north coast of Yorkshire. The areas of these rocks in Sutherlandshire and the Western Highlands of Scotland and those in Antrim may, for our present purpose, be disregarded, as they are of insignificant practical importance.

The Jurassic rocks are valuable repositories of limestone, clay, and ironstone; but sandstones are comparatively scarce.

Beginning with the lowest formation of the Jurassic, we find that the Lias, although it locally contains sandy beds, has no sandstones worthy of the name. A thin-bedded sandstone occurs in the Lias of Londonderry, and has been used for second-rate flags. The Inferior Oolite in the south and west of England contains sandy beds, but no valuable sandstone. The so-called 'grits,' 'pea-grit,' etc., near Cheltenham, are limestones. Farther north and east, however, hardened sandy beds do appear. The sandy development of the lower parts of the formation becomes more prominent in parts of Oxfordshire, where, however, there are no good beds of sandstone. They are quarried about Great and Little Tew. A characteristic section near Heythorpe Common is as follows:

		Ft.	in.
Inferior Oolite	Rubbly oolitic stone, with shale	1	0
	Marly shale bed, with nodules of calcareous sandstone at base	1	0
	Flaggy calcareous sandstone, with shale	1	6
	Sandy and flaggy current-bedded calcareous sandstone	6	0

At Steeple Ashton and Chipping Norton about 7 feet of soft sandstone, or 'oven stone,' occurs.

In Northamptonshire, Rutlandshire, and Lanarkshire the Inferior Oolite is more sandy. A sand rock has been quarried for building stone near the racecourse, Northampton.

Long ago the stone of Helmedon was well known, and was used in the mansions of Stowe and Woburn; at Etdon a pale sandstone (freestone), together with a red common walling stone, were worked; and a calcareous, rather micaceous, sandstone has been quarried at Thorpe Mandeville.

The Northampton iron ore passes, within a short distance, into a building stone, near Duston. The following is a list of the beds in the New Duston Stone Pit:

	Feet.
White sands } Sandy clay }	3 to 4
Dark grey carbonaceous clay	?
Yellow clay sand	1
Iron-stained yellow sandstone	10 to 12
Fissile and somewhat oolitic slaty beds, the *White Pendle*, or *Duston Slate*, more or less current-bedded	4 to 6
Yellow and Best Brown Hard, even-bedded brown calcareous sandstone	12*
Rough Rag and Hard Blue Rag, a calcareous sandstone (the blue portion is more distinctly oolitic than the brown)	6 to 7
Ironstone ... about	3

A similar sequence is found at Harleston; here the main sandstones are known as *Harleston Stone,* used for setts and kerbs.

At Kingsthorpe a yellowish sand-rock was used as a building stone in the village, and in some of the old buildings in Northampton—*e.g.,* barrack wall, union infirmary, etc. It is seldom used now, as it is not very durable, although it hardens on exposure in dry weather; formerly it was used for furnaces, as it withstood the heat well.

A hard red rock, with greenish centres, 20 feet thick or more, has been quarried near Cottingham Church for building stone. The stone has also been worked at Detborough, Rockingham, and Uppingham.

The Jurassic rocks of Yorkshire are more sandy and less calcareous than those of the south-western counties. A series of estuarine beds takes the place of the Inferior and Bath Oolites. The *Grey, or Scarborough, Limestone* has not been very appropriately named, as it really consists of several alternations of strata of different lithological character, which are not always grey, and in which there is

* *Duston freestone*, used for ashlar, tombstones, coping, paving.

little that is really limestone, the majority of the beds being calcareous shales and sandstones, and in some places even coarse grits. The general composition of this rock is very variable. To the south of Scarborough it consists principally of calcareous shales, with thin nodular ironstones, and a little calcareous sandstone. These latter towards the north become more prominent, and over the interior moorlands develop into massive beds of coarse sandstone; while to the south, in the Howardian Hills, they pass into fine-grained flaggy sandstone, and finally die out altogether. The grey limestone, where it first rises from the sea, is a very insignificant bed, of not more than from 3 to 7 feet thick; but it rapidly thickens out towards the north and west, increasing to as much as 20 feet or more near Scarborough, to 70 feet at Cloughton, and to about 90 feet at Peak.

It yields a strong, rough calcareous sandstone, incapable of being dressed, but suitable for rough work; it has been used in Scarborough Pier, and on this account is known locally as the 'Pier Limestone' (Mem. Geol. Survey, 'Jurassic Rocks,' vol. i.).

The *Middle Estuarine Beds* contain, as a rule, only thin beds of sandstone, but occasionally these thicken out, as they do at Haigburn Wyke, whence a very sound stone has been obtained, used in the clock-tower of the station at Scarborough. They are worked in a small way near Kirkby Knowle.

The *Upper Estuarine Series* of Yorkshire (=part of the Inferior and Great Oolites) produces strong rough sandstones at several places. At the base is the Moor Grit, which is well exposed in the northern and moorland districts; it varies from a fine-grained to a moderately coarse rock. There are many small quarries; those of Cloughton and Old Fold, at the head of Riccal Dale, may be mentioned. From the latter stone was taken for the railway bridges on the Helmsley branch of the North-Eastern Railway. Near Whitby and in other places good flags of large size are obtained. They have been used in paving the Whitby Quay.

The *Middle Calcareous Grit*—a series of alternating calcareous and sandy beds—is strongly developed in the western end of the Vale of Pickering, about Helmsley. In the Pickering quarries sandstones appear in the upper and limestone in the lower part.

In many of the Yorkshire rocks, as well as in those of Lincolnshire, large siliceous concretions, locally called 'doggers,' are sometimes broken up and used for rough building; they have frequently been employed for the prehistoric monuments in this district.

The Jurassic rocks of Scotland are strongly arenaceous; they are used only locally.

Cretaceous.—There are really no important sandstone building stones in the Cretaceous System of this country. In the lower part soft sands and sand-rock are of frequent occurrence, and locally they are sufficiently hard and compact for building; they are, however, very little used for this purpose to-day.

In the district known as the Weald, in the counties of Kent, Sussex, and Surrey, the following Cretaceous formations have been recognized:

Upper Greensand.
Gault.
Lower Greensand { Folkestone Beds. Sandgate Beds. Hythe Beds (Kentish Rag). Atherfield Clay.
Weald Clay.
Hastings Beds { Upper Tunbridge Wells Sand with Cuckfield Clay. Grinstead Clay. Lower Tunbridge Wells Sand. Wadhurst Clay. Ashdown Sand { Ashdown Sand. Fairlight Clays.

SANDSTONES AND GRITS

The Weald Clays and Hastings Beds are confined to the Weald district, but the two Greensands and Gault extend farther afield, as shown in Map V.

The strata of the Hastings Beds are variable, and some difficulty is experienced in correlating them in different parts of the district; hence the old names for some of the stones connote rather the kind of stone than the geological horizon.

In the upper part of the Ashdown Sands, near Hastings, a calcareous sandstone appears under the name *Tilgate Stone*. A similar stone, bearing the same name, occurs in the Wadhurst Clay, and was formerly quarried at Calverley Quarry, and a variety of this stone from the same formation was called 'Hastings Granite'; at Beech Green, near Penshurst, it is called *Beech Green Stone*. The original Tilgate Stone from Tilgate Forest is from the Tunbridge Wells Sand in the Cuckfield district.

From the Weald Clay comes the *Horsham Stone*, a calcareous sandstone, which has been used for building, and in places, when it splits readily, for paving slabs and roofing (p. 324).

Most of the Wealden Sandstones are too soft and variable to have ever permitted their employment beyond the district in which they are found;

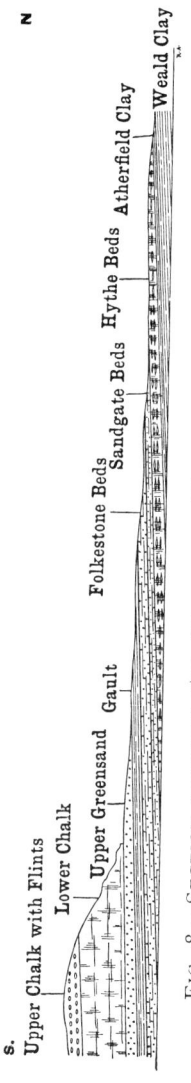

FIG. 8.—SECTION SHOWING THE CHALK AND WEALDEN BEDS, WEST OF WORTHING.

in the eastern part they are much more irregular than in the western area, where they have been extensively quarried in the past. Fair-sized blocks were obtained, and they have worn tolerably well in the old castles and ecclesiastical buildings of the district, and their warm colour is in pleasing contrast with the other paler stones.

The Hythe Beds of the Lower Greensand yield a calcareous sandstone in the upper part, known as *Bargate Stone*. It occurs over a large tract south-west of Dorking; it is still quarried about Godalming, though not so extensively as formerly. The stone-bearing beds here are about 40 feet thick, but the workable stone beds are only 6 inches to 3 feet thick, lying in sand; they are current-bedded, and very variable in composition, some being quite soft, some shelly, others argillaceous, and they occasionally pass into sandy limestones, resembling Kentish Rag. It has been used a good deal locally for buildings.

In the lower part of the Hythe Beds soft sandstones have been much quarried in the western part of the Weald; the only hard beds are found about Pulborough (*Pulborough Stone*) and Petworth. The stone was used by the Romans, and examples of later work may be seen in the keep of Arundel Castle, in the Town Hall and piers at Littlehampton, and in Pulborough Church, built in the sixteenth century.

The 'hassock' of the Hythe Beds and the Kentish Rag also are frequently very sandy and siliceous, but as they are more usually strongly calcareous they will be noticed under limestones (p. 264).

At Nutfield a very fair sandstone for building has been obtained for a long time from the Fuller's Earth pits of the Sandgate Beds. The best stone occurs above the Fuller's Earth; the total thickness of sandstone is from 12 to 15 feet. The bottom bed was formerly known as *quoin stone*, and was employed for general building. Many cottages in the village are built of this stone, and look very well; the stone has a warm yellow colour.

The Folkestone Beds yield a calcareous sandstone, usually of poor quality in the outcrop, to the west of the town.

Hard nodular beds, or concretions, are found in the Folkestone Beds, and are called *Carstone*. It is said to be the hardest and most durable stone in the district. It has been used a good deal for general building and paving. Gilbert White says of it: ' In Wolmer Forest I see but one sort of stone, called by the workmen *sand*, or *forest*, *stone*. This is generally of the colour of rusty iron . . . is very hard and heavy, and of a firm, compact texture, and composed of a small roundish crystalline grit, cemented together by a brown terrene, ferruginous matter; will not cut without difficulty, nor easily strike fire with steel. Being often found in broad flat pieces, it makes a good pavement for paths about houses, never becoming slippery in frost or rain. In many parts of that waste it lies scattered on the surface of the ground, but is dug on Weaver's Down, a vast hill on the eastern verge of that forest, where the pits are shallow and the structure thin. The stone is imperishable.

' From a notion of rendering their work more elegant, and giving it a finish, masons chip this stone into small fragments about the size of the head of a large nail; others stick the pieces into the wet mortar along the joints of their freestone walls; this embellishment carries an odd appearance, and has occasioned strangers sometimes to ask us pleasantly " whether we fastened our walls together with tenpenny nails?"' (Gilbert White, ' Natural History of Selborne,' Letter IV., 4to., 1789).

In Wilts, the Isle of Wight, and on the northern outcrop in the midland counties, the Lower Greensand is rarely hard enough to be of much value for building; but near Faringdon, in Berkshire, a few beds have been sufficiently cemented for this purpose, and for use in former times as querns and millstones. The uncemented 'Shotover Sands' of Oxfordshire are locally hardened, sometimes cherty; and siliceous concretions around Aylesbury, in Buckinghamshire, known in the district as *bowel stones,* have been used in rockeries and for copings. In the eastern counties, particularly in Norfolk, the locally hardened sands are called *Carstone, Gingerbread Stone,* or *Quernstone;* they are quarried for building at Snettisham and other places. Carstones and coarse sandstones from the Tealby Beds have been used locally for building.

The Gault and Upper Greensand (Selbornian) may well be considered together, since the former changes its character markedly as it is traced westwards, becoming in that direction more sandy.

In the Upper Greensand of the Weald district no hard beds suitable for building are found east of Westerham. About Merstham, Reigate, Gatton, and Godstone a pale, fine-grained sandstone occurs in the formation; it is usually calcareous, and contains a considerable amount of fragmentary colloidal silica (40 per cent.), and a little glauconite and mica. It is soft when first quarried, but hardens if kept for a time in a dry place. It is known as *firestone, Gatton stone, Merstham stone, Reigate stone.* Webster says of this stone: 'The quarries of Reigate Stone were formerly considered of such consequence that they were kept in the possession of the Crown, and a patent of Edward III. exists, authorizing them to be worked for Windsor Castle. Henry VII.'s Chapel at Westminster was also built of the stone procured from them, as is also the church at Reigate. These ancient quarries were situated between the town of Reigate and the Castle Hills in the north, and traces of them may still be seen in several places, as at Gatton Park, Colly Farm, and Brickland Green, which latter place is the most westerly spot where the stone has been found' (*Trans. Geol. Soc.*, v. 353).

Gilbert White describes the uses of the firestone (or Malm) in Hampshire thus: 'This stone is in great request for hearthstones and the beds of ovens; and in lining of limekilns it turns to good account, for the workmen use sandy loam instead of mortar, the sand of which fluxes, and runs by the intense heat, and so cases over the whole face of the kiln with a strong vitrified coat, like glass, that it is well preserved from injuries of weather, and endures thirty or forty years. When chiselled smooth, it makes elegant fronts for houses, equal in colour and grain to Bath Stone, and superior in one respect, that, when seasoned, it does not scale. Decent chimneypieces are worked from it, of much closer and finer grain than Portland, and rooms are floored with it; but it proves too soft for this purpose. It is a freestone, cutting in all directions; yet it has something of a grain parallel with the horizon, and therefore it should not be surbedded, but laid in the same position that it grows in the quarry. On the ground abroad this freestone will not succeed for pavements, because probably some degree of saltness prevailing within it the rain tears the slab to pieces' (*loc. cit.*, Letter IV.).

White's remarks refer to a stone very similar to the freestone of Reigate, etc., which, under the name *Malm*, or *Malm Rock*, is found in the Upper Greensand, and sometimes in the Gault, of West Surrey, Hampshire, Sussex, and Devonshire; it has often a more chalky aspect than the Gatton or Reigate Stone. The Malm Rock has been used in the masonry of the Roman Villa at Bignor, and it seems even to have been used in a polished condition in the same building. A similar grey calcareous sandstone, 20 feet thick, has been used in

Sussex, in Pevensey Castle. The softer parts of the stones are still dug for hearthstones, and the stone from Farnham has been employed in the manufacture of artificial stone.

The Malm Stone has been used a good deal on the western outcrop of the Weald anticline, particularly about Froyle, in Bentley Church, and in Lower Froyle and Isington. Farther west, it has been used about Shaftesbury, and at Salcombe and Duncombe, and other villages in the Devonshire part of the outcrop.

The Reigate Stone was used in the Asylum, Redhill, 1853-1855, along with Bath Stone, and is said to have been used in Hampton Court. It does not appear to wear well in towns; scarcely any remain of the many buildings in which it was employed in London. Even in the time of Wren it had decayed very badly in Westminster Abbey, for he records that it had mouldered away 4 inches from its original surface. For interiors the stone is eminently suitable, on account of its pleasing colour and the ease with which it may be carved. It might be used for exteriors outside large towns, where it lasts well. A good deal of the early mined stone was probably rather inferior stuff; better quality might possibly be obtained by modern methods.

A poor walling stone from the hardened Greensand of Antrim is known locally as 'mulatto stone,' a name which has also been applied to stone from the same formation in the Weald district.

The **Tertiary** formations in this country are usually in too soft a condition for the arenaceous beds to yield good building sandstone; they are mostly in the state of unconsolidated sand or loose sand-rock. From the Bagshot Beds and from some parts of the Reading Beds very hard clean quartzose sandstone or quartzites occur in irregular

SANDSTONES AND GRITS

lumps and beds, as concretionary masses in the midst of softer sand. Blocks of this stone are dug at Denner Hill and Walter's Ash, in Buckinghamshire, out of the 'Clay-with-flints,' for building stone setts, kerbs, steps, sills, and ashlar. The stone has a very clean white or slightly yellow appearance; it is very hard, and has a strong siliceous cement. It has been used at Windsor Castle. Formerly very many large blocks of this stone lay about on the surface of the ground in the London and Hampshire Tertiary districts, having been left behind by the weathering processes, which removed the less resistant material in which they originally lay. The blocks are known as 'Sarsen Stones'; they were frequently used by the old builders from prehistoric times in all kinds of structures, from primitive monoliths to more recent church and farm buildings. The *Hertfordshire Pudding Stone* is a rock of this kind, which differs from the ordinary sarsen stones in being composed of round flint pebbles, $\frac{1}{2}$ inch to 2 inches in diameter, set in a siliceous ground mass made of sand grains and angular flint sand, in siliceous cement. This stone appears occasionally in old buildings, and was often used for millstones and querns.

Occasionally the sand of the Reading Beds is locally hardened, and is then sometimes used in the immediate neighbourhood for odd building jobs. In this condition the sandstones never form thick beds, but broad slabs, 1 foot to 15 inches thick, are obtained. A stone of this kind appears in the brickyard at Upper Nately.

FOREIGN SANDSTONES.

France.—Sandstones are worked in the following formations:—*Silurian:* Vitré stone. *Carboniferous:* Reddish and white stone of St. Germain (Belfort); the grey and brown stone of Grande-Combe (Gard); the grey stone of Pégauds (Allier); the grey fine-ground stone of Mouillon (Loire). *Permian:* Tudeils and Gramont, in La Corrèze. *Trias:* Baccarat, Rhune, St. Germains (Haute Saone), Crétaux, Monestiés, Le Mentière, Lodève. *Lias:* Baissgnac, Sceauteaux, Agladières, Garde. *Upper Cretaceous:* Alet (Aude). *Miocene:* Geaune, Mugron, Porchères (Pierre de Mâne), Martigues (Grès de la Couronne), Grignan, and the Molasse of St. Juste.

Belgium.—The principal sandstones are of Devonian age, from Hainault, Yvoir, Gembloux, Wépion. Carboniferous sandstone is used a little about Rieudolte, in the Ardennes; and the Trias is quarried in Luxembourg, at Udelfangen and elsewhere.

Austria-Hungary.—Devonian sandstone is worked at Trembowla in Galicia, and Sternberg in Mähren. Culm sandstone is used in Troppau. Permian stone is used at Hohenelbe in Bohemia, and Brünn in Mähren. Cretaceous sandstones are quarried at Königgrätz, Böhm-Brod, and Chrudim in Bohemia; and about Salzburg. Miocene calcareous sandstones are very largely employed in the Vienna basin, and the Molasse sandstones in Vorarlberg.

The United States.—One of the best known of the American sandstones is the Triassic 'brownstone,' a moderately fine-grained, reddish-brown stone, sometimes called 'Portland Stone,' from the Portland and other quarries in Connecticut and New York. A similar stone, but brighter red in colour, is used in Massachusetts. A more purple or bluish-brown stone from the same formation is largely used in Pennsylvania. The 'Bluestone' of New York is a durable blue-grey stone of Devonian age; the 'Wyoming Valley Stone' of Pennsylvania is similar. The Berea grits and 'Euclid Bluestone' of Ohio are of Sub-Carboniferous age. The Cambrian Potsdam sandstone is used in Missouri, Michigan, New York, and Wisconsin (where it is sometimes known as 'Lake Superior Brownstone'). Cretaceous sandstones are worked in Montana, Nebraska, and Washington.

The Lower Carboniferous Sandstone of New Brunswick and Nova Scotia, 'Nova Scotia Stone,' of a greenish tinge, is used in New York.

Silurian sandstone is quarried in Quebec, and the Potsdam sandstone in Ontario. A Cretaceous sandstone is used in British Columbia.

In **Germany** sandstones are very extensively used for all kinds of work, including the most elaborate structures in the large cities. The most important sandstone-bearing formations are the Cretaceous and Trias. The Upper Cretaceous, 'Quader Sandstein,' is particularly well developed in the valley of the Elbe and its tributaries, where it is extensively quarried in the districts of Pirna, Zittau, and Friedberg. The stone is fine to coarse grained, and white, cream, yellow, or pink in colour. Quarries are situated at Postelwitz (annex of Brandenburg Gate, Reichstag Buildings at Berlin), Oberkerchleithen, Cotta, Weissenberg, Herrnskretschen (monolith columns in Royal Academy, Dresden), Wünschelburg, Friedersdorf, Alt-Warthau, Niederposta, etc. Lower Cretaceous ('Hilssandstein') is quarried in the Teutoburger Wald. The Wealden 'Deistersandstein' from Obernkirchen, Bückeberg in Cassel—fine-grained, white and yellow—has been used to a great extent in the Cologne Cathedral and other important structures. Other Lower Cretaceous Sandstones come from Kehlheim, Kapfelberg, Albach, etc.; many of these stones are greenish from the presence of glauconite.

Triassic stones are obtained from the Keuper Beds of Wendelstein, Schwäbisch-Hall, Zeil, Tretzendorf, Heilbronn, Mögeldorf (used with such good effect in Nuremberg), and other places in the neighbourhood of Frankfort, Stuttgart, and Würtemberg. From the Bunter formation come the sandstones of Würzburg, the Black Forest, the Pfalz district, Aschaffenburg, Nahetal, Arzweiler, Nebra, etc.

Permian stone is worked at Eggenstedt, Saxony. Carboniferous stone comes from Cudowa in Breslau, Flöha, and Gablenz, near Chemnitz. Liassic sandstones are worked in Franconia.

In **Italy** sandstones are not much used except in Tuscany, where the dark, greenish-grey 'mascigno' or coarser 'cicerchia' of the Cretaceous Flysch formation are employed.

In **Sweden** sandstones are quarried from the pre-Cambrian rocks of the Gäfle, Grenna, and Lemunda districts, from the Cambrian in Scania, from the Silurian in Övedskloster, in Scania, and Burgsvik in Gothland.

India.—Sandstones have been worked in India from the Vindhyan series, the Gondwana series, and from the Cretaceous and Tertiary formations. Of these the first is the most important source of stone; the series is widely spread in the North-West Provinces and in the valley of the Ganges. It has been quarried at Mirzepur, Chunar, Partabpur, Dekri, Gwalior, Bhartpur, etc. The best varieties are fine-grained grey, yellowish, reddish-yellow, and dark red. The stone has been used in some of the finest buildings in India, and is widely employed in Delhi, Benares, Agra, and other cities. Many ancient monoliths of great size have been made out of it, but most wonderful is the pierced work and delicate carving that has been executed in the material. The Gondwana sandstones in the Damuda Valley have been used in a similar way.

CHAPTER VI

LIMESTONES

Introductory.—The limestones that have been used for structural purposes include nearly every one of the many varieties in which the stone occurs. The essential characteristic of a limestone is the presence of a large proportion of carbonate of lime ($CaCo_3$) in its composition. In the purer varieties the amount of this compound ranges from 80 to 98 per cent. of the mass; impure forms may contain as little as 30 per cent. or less, but in these cases we are on the border-line where the rock is passing into the condition of a calcareous sandstone or mudstone, as the sandy or muddy impurity begins to be the predominant constituent. There is every gradation between the pure limestones, such as some kinds of chalk, white saccharoid marbles, and certain Carboniferous limestones, and the sandy, siliceous limestones or calcareous sandstones, like some forms of the Kentish Rag or Mansfield Stone on the one hand; also between the pure forms and the argillaceous Chalk-marls, blue Lias limestones, and Carboniferous, Devonian and Silurian mudstones, on the other hand.

Limestones vary in texture as much as they do in composition. In the Devonian and Carboniferous rocks of Britain we find massive, heavy, crystalline and sub-crystalline stone, non-porous and capable of taking a high polish readily; in the Jurassic System we have more porous, lighter stones, which can only occasionally be polished; in the Cretaceous System the most noticeable limestone is the soft, fine-grained friable chalk, and so on. In Britain, and, indeed, as a general rule, the older limestones are the more crystalline, dense, and massive; the younger ones are softer, lighter, and more friable.

The reason for this is fairly obvious; but a few illustrations will make the matter clearer. At certain spots on the British shores—but better examples are found elsewhere—great quantities of marine shells are thrown up by the sea, forming masses of loose shell sand; here and there in these banks the superficial waters have dissolved portions of the carbonate of lime from adjacent parts of the mass, and have redeposited it in patches among the shell fragments, forming thereby cemented blocks of coherent rock. The same process may be observed going on amid the oyster banks and other shelly areas on the floor of the surrounding seas. In these cases the calcreted or lime-cemented mass of shells forms a highly porous rock; the process of consolidation is yet in its infancy, and is going on day by day. On the shores of some coral islands the consolidation of loose material takes places with astonishing rapidity.

Again, upon the floors of the seas and oceans, where many calcareous deposits have their origin, there are countless organisms acting the part of earthworms on land, devouring the fragmentary shelly particles, the mud and the slime; continually reducing the size of the particles, and thus enabling the contiguous moisture, with its dissolved gases, to attack the material with greater readiness. This it does, continually taking up material in solution in one place and depositing it in another.

Borings made in coral islands have shown that in the lower portion of such calcareous deposits the mineral débris, coarse and fine, begins at an early stage to crystallize sporadically; but in many limestone formations the production of general crystallization throughout the mass is never completed until, in the course of geological changes, the beds have been covered by great thicknesses of younger strata, and have been subjected to the hardening influence of more or less powerful squeezing and folding.

The statement made above that the older limestones are the more dense and crystalline, although generally true, is subject to exceptions. The coral, shell, and foraminiferal limestones of the British Carboniferous System have undergone great pressure and strain, produced by folding, consequently we find the stone well crystallized and compact. But there are limestones in Russia of the same age as these which have not suffered the same stresses; there some of the limestone is as fresh and young in appearance as that in some modern coral islands. On the other hand, some of the Tertiary limestones, which in this country are soft and generally useless for building purposes, are represented in Alpine regions, where the rocks have been squeezed and folded in the process of mountain forming, by marbles as hard and crystalline as any of the oldest limestones known.

The Origin of Limestones.—It has already been indicated that certain limestones are formed from molluscan shell débris; and one may safely say that the majority of limestones owe their existence to the accumulations of shells or calcareous skeletons of marine or freshwater organisms. Of course, in no case is the solid rock formed directly by these agencies alone; they do but supply the raw material and framework, with which the activities of repeated solution, deposition and recrystallization have fashioned the coherent stone. In accordance with the relative magnitude of the shares taken by organic and chemico-physical agencies in forming limestones it is customary to consider these rocks under the two heads—Organically-formed and Chemically-formed limestones; but it will be apparent that no coherent rocks exist in the formation of which the chemical agencies have been entirely in abeyance. Chemically-formed calcareous rocks are known in which the direct intervention of organic life is not obvious; ultimately, however, the material from which these are constituted can be shown in many cases to have come through the mediation of some form of organism. This is not the place to discuss such problems; it will be sufficient for our purpose to enumerate some of the salient types.

Organically-formed Limestones—1. *Shell Limestones.*—The majority of these are formed from the more or less broken and decayed fragments of marine mollusca. They range from the feebly-coherent shell-sand rock (Newquay, coral islands, etc.), through the more substantial oyster and clam banks, and their equivalents in the older rocks, the heavy brachiopod limestones of the Carboniferous and Devonian. All the great divisions of the molluscan kingdom are found at one place or another acting as rock-builders; we have brachiopod, pelecypod, gasteropod, pteropod, and cephalopod limestones.

Some limestones, for instance, the Paludina limestone of the Wealden and Purbeck Beds, are of freshwater origin.

Examples of shell limestone are: Ham Hill Stone; Totternhoe Stone (in part); certain layers of Portland Stone; some Lias beds, as at Keinton; many Carboniferous limestones, lumachelles, etc.

2. *Crinoidal Limestones.*—The best-known stones of this kind are found in the Carboniferous limestone of this country and Belgium—the Birdseye Marble, Petit-Granit—Euville Stone, etc. They are formed by the disjointed fragments of the stems and 'arms' of crinoids, or sea-lilies. As limestone builders these organisms are not

confined to the Carboniferous period, for they form more or less important beds in the Ordovician, Silurian, Devonian, Trias, and Jurassic, in many parts of the world.

3. *Foraminiferal Limestones.*—The microscopic organisms known as foraminifera, many of which secrete a tiny shell of carbonate of lime, make up for their diminutive size by their countless numbers. They are still forming deposits of calcareous ooze at the bottom of some of our oceans, and they have played an important part as limestone-builders from the earliest geological periods. The most typical British example of the above is perhaps the Chalk; they enter also very largely into many beds of Carboniferous Limestone, while abroad, in southern Europe, north Africa, and central Asia, vast deposits were formed in Tertiary times by a group of large foraminifera called Nummulites—lens-shaped, disc-like bodies, ranging in size from that of a threepenny-bit, or less, to that of a halfpenny and larger. Of such a stone the Pyramid of Cheops is built.

4. *Coral Limestones.*—Everyone is now familiar with the fact that limestone is formed by corals in the reefs and islands of the warmer seas; it is not so widely recognized that many of the limestones of this class are due, not to the activity of corals alone, but to the solid remains of many other organisms, notably calcareous algæ, with molluscan foraminiferal and crustacean remains.

Freshly consolidated coral rock is often very hard and porous, and is not infrequently used in the vicinity for the construction of buildings of all kinds.

The principal coral-formed building and decorative stones of Great Britain are those in the Devonian and Carboniferous limestones (Devonshire Marble, Frosterley Marble, etc.).

5. *Bryozoa Limestones.*—On several geological horizons bryozoa have formed considerable beds of limestone; some

of the Magnesian limestones of Yorkshire have been made almost entirely by these organisms.

Chemically-formed Limestones—1. *Tufa and Travertine.*—Under this head the most important building stone is Calcareous tufa, or Travertine. This type of rock is deposited from springs or streams issuing from older calcareous rocks, or as warm springs in districts of declining volcanic activity. In some of these waters a considerable amount of carbon dioxide is held in solution, and this enables the water to carry a heavier load of carbonate of lime than would otherwise be possible. On coming to the surface some of this CO_2 is liberated—a process often very materially assisted by the action of Algæ and other water-loving plants—the deposition of a portion of the dissolved $CaCO_3$ follows, and a more or less spongy stone is formed (p. 265). A similar formation is the stalagmitic or stalactitic carbonate of lime, rarely employed as a building material, but frequently of great value as a decorative stone. Most of the tufaceous limestones used for building are of modern age, but others employed occasionally occur in the Purbeck beds of Dorsetshire, and in the Rhætic and Lias; the Burdie House limestone of Carboniferous age in Scotland is of this nature.

2. *Oolitic Limestones.*—This type of rock is in some degree akin to the tufaceous limestones, as it is formed, at least in part, by deposition from solution. An Oolite—sometimes called roe-stone, from its resemblance in texture to the roe of a female fish—is composed mainly of small subspherical bodies, consisting of concentric layers of carbonate of lime, deposited, as a rule, round a small nucleus of fragmentary shell, coral, mud, or sand-grain. The concentric layers are sometimes obliterated by a radial crystalline structure; sometimes both structures are visible in section at the same time; other oolitic grains exhibit only a granular calcite paste in the cross-fracture.

LIMESTONES

Along with the oolitic grains we may find the remains of corals, shells, echinoderms, and other organisms, giving rise to what is called a 'shelly oolite.' The oolitic grains may be clearly defined in a matrix of crystalline or pasty calcite, or they may have lost their outline, owing to a certain amount of decay and recrystallization. The whole rock may be composed of oolitic grains, or, as in a sub-oolitic rock, they may be sparsely scattered through the mass. Larger grains, the size of a pea, are occasionally found (pisolite), as well as irregularly-formed pellets and compound grains.

The typical oolite building stones in this country are those obtained from the Jurassic rocks (Portland, Bath, Ketton, etc.); but oolites have been formed in shallow seas of all ages, from the Ordovician upwards, and good examples may be seen in the Carboniferous limestones of Bristol, South Wales, Derbyshire, and Ireland; and similar deposits are being formed to-day in many places.

Dolomite and Magnesian Limestone.—The double carbonate of magnesia and lime ($CaMgCO_3$), the mineral dolomite, enters locally into the composition and alters the character of limestones of all ages. There are two principal modes of occurrence: (1) as an alteration product within a normal limestone; (2) as an original chemically deposited rock. In the former case the alteration of a wholly calcareous rock may take place while the fresh limestone or ooze is in the sea in which it has been deposited; this has been observed in the materials for borings in coral islands, and evidences of this 'contemporaneous dolomitization' may be seen in older formations, such as the Carboniferous Limestone. Or the alteration of the original limestone may have taken place long after the original material had been consolidated and uplifted; examples of this 'subsequent dolomitization' may be seen in many of the older limestones, along planes of the joints, faults, bedding, etc., and over tracts of limestone which have been subjected to solutions soaking through a magnesia-bearing cover of strata.

Magnesian limestones and dolomites of this character are not uncommon in the Carboniferous Limestone of Ireland, Derbyshire, South Wales, and elsewhere, and in limestones of all ages.

Precipitated magnesian limestones, or dolomites, seem to have been found in inland seas subject to heavy evaporation. The best-known magnesian limestones and dolomites in England—namely, those of Mansfield, etc., in the Permian System—perhaps belong partly to this class. Tufaceous and concretionary structures are very common in these rocks.

The terms 'magnesian limestone' and 'dolomite' are used somewhat loosely for any calcareous rock containing a high proportion of magnesia. There is a difference, however, for the dolomite should have a composition approximating closely to that of the mineral of that name; that is to say, it should be a dolomite rock—one formed of the mineral dolomite ($CaCO_3$, 54·35 per cent.; $MgCO_3$, 45·65 per cent.). Any limestone is magnesian if it contains an appreciable amount of the mineral dolomite. It is often possible to trace the passage of a limestone free from dolomite through one with a few sporadic crystals per cubic inch, up to a pure dolomitic rock.

Earthy, Argillaceous, and so-called Compact Limestones, Siliceous Limestones.—Many limestones contain argillaceous matter mixed with the carbonate of lime. Such rocks have been formed very often as marine muds, in which calcareous shell-bearing molluscs have lived, and it is from the decay of the shells that the carbonate of lime has mainly been derived. In this class of rock we may have a clay or shale with little lime or a limestone with little clay matter. Examples of this type of stone are to be met with in formations of all ages; the Liassic limestones are typical of the more calcareous forms, and many beds in the Jurassic System show the same characters in lesser degrees. Such stones frequently present a hard, smooth, compact appearance when freshly quarried, but they rarely weather well, and cannot be relied upon. Siliceous limestones contain more or less silica, either in the form of sand-grains or quartz fragments, as in the case of the Kentish Rag and the Mansfield Dolomite. By the accession of increasing amounts of quartz-grains these rocks pass into calcareous sandstones. The silica is sometimes in the form of minute granules of chert and flint, but stones of this kind are generally too hard and irregular in their behaviour under the tool to pay for dressing. Still another form assumed by silica in limestones is that of colloidal granules and sponge fragments, observed in some varieties of Greensand and in Beer Stone from the Chalk.

The siliceous character of certain limestones is due to included crystals of quartz, often perfect in form.

Colour of Limestones.—Limestones present a wide range of colours: we find them passing from black, through every shade of grey, to sparkling or dull white, or from a pale green tint, through yellow, brown, and orange, to a bright red. Some have a distinct pale blue colour, and more rarely a green tinge may be observed.

The cream, buff, brown, orange, and red tints are all caused by varying amounts of iron oxides. The greys and blues are caused partly by finely-divided carbonaceous matter, which is most abundant in the black limestones, and partly by iron sulphide in a very fine state of division. Many of the dark limestones emit, when struck, a more or less pronounced odour of sulphuretted hydrogen. The greenish tint seen in some limestones may be caused by grains of iron silicates. The carbonaceous limestones may carry the carbon in the form of minute plant débris; others owe their dark colour to bituminous, asphaltic, and oily matters. Such stones do not fall within the best class of limestones for building.

Marbles as Building Stones.—Marbles are limestones or dolomites that by some process of alteration or metamorphism have been completely crystallized or recrystallized. In such rocks all original structures are completely obliterated. Many calcareous and magnesian carbonate rocks capable of taking a polish are called marble in the trade, although they are not so completely altered as to be true marbles in the geological sense.

Marbles, being mainly employed as decorative stones, cannot be considered in detail here. Near their place of origin they are often used as ordinary building stones, and as such their behaviour is like that of high-class limestones. As examples of the recent use of marble in London, the buildings at the top of St. James's Street may be cited. The white stone is the well-known Pentelic marble of Greece; the pink stone is a dolomite marble, with inclusions of red jasper, from North Wales.

Chemical Composition of Limestones.—The chemical composition of limestones for building purposes is not an important character. The carbonate of lime—or in magnesian limestones and dolomites the double carbonate of lime and magnesia—should be high; much alumina, indicative of clay, is undesirable; so, too, is iron sulphide. Silica is unimportant, except in so far as it may increase the cost of working.

Table X. shows illustrative examples.

The specific gravity of limestones is about 2·7 to 2·9. The crushing strength of limestones ranges between

LIMESTONES

Table X.

	Hopton Wood.	Doulting.	Ketton.	Ancaster.	Weldon.	Portland.	Bath; Box.	Grey Chalk, Folkestone (Davis)	Chilmark.	Anston.
Carbonate of lime	98·90	95·89	92·17	93·59	94·35	95·16	94·52	94·09	79·0	54·87
Carbonate of magnesia	0·45	0·11	4·10	2·90	3·55	1·20	2·50	0·31	3·7	43·07
Alumina	—	0·79	} 0·90	0·80	{ 0·28	} 0·50	1·20	{ (NaCl 1·29)	} 2·0	0·73
Iron oxide	0·25	0·85			0·61					
Silica	0·75	2·04	—	—	0·08	1·20	—	3·61	10·4	0·56
Water and loss	0·10	0·32	2·83	2·71	1·13	1·74	1·78	0·70	4·2	0·75
	100·45	100·00	100·00	100·00	100·00	100·00	100·00	100·00	99·3	99·98

400 and 1,400 tons per square foot; the more compact varieties average about 730, and the more open ones about 640.

DISTRIBUTION OF LIMESTONES.

In every one of the great geological systems of strata there are calcareous beds of some sort, but the principal limestone-bearing formations in Great Britain and Ireland are the Devonian of South Devon, the Lower Carboniferous, the Jurassic, and the Chalk. Throughout the following the reader is also referred to Appendix C for a list of the large quarries.

With the exception of certain beds of dolomite in North Wales and in Scotland, the Pre-Cambrian rocks contain no calcareous beds of importance, and it is not until the upper part of the **Ordovician** is reached that noticeable limestones occur in that system. Here in the district around Bala appears the Bala Limestone, an impure stone, in beds 20 to 30 feet thick; of about the same geological age are the grey Rhiwlas Limestone, north-west of Bala, 30 to 40 feet thick; the Sholesbrook Limestone, near Haverfordwest, greenish and impure; and the Robeston Walthen Limestone, black in colour. A black, impure fossiliferous and pisolitic stone—the Hirnant Limestone—occurs locally at the junction of the Bala and Llandovery rocks. None of the above would rank as a building stone except for the roughest jobs.

The Coniston Limestone of the Lake District occurs on this horizon; it is a series of limestones, with shale beds some 200 feet thick. Most of the stone is dark blue in colour, compact and hard; but, as it weathers badly, it has not been much used for building. A white crystalline variety of this stone, in parts with a pinkish stain, is quarried at Keisley.

In the **Silurian** strata we have the Aymestry Limestone

in the Ludlow Beds, the Wenlock Limestone, and the Woolhope Limestone, all of which are worked in the neighbourhood of these several places and at other points along their outcrop, mainly for lime, but in part for building stone, and for this purpose they have been freely used in the neighbourhood of the quarries. The stone is usually grey in colour; some of it is hard and crystalline, and looks not unlike Carboniferous Limestone in buildings. Large blocks are rarely obtained.

The Aymestry Limestone occurs in beds from 1 to 5 feet thick, and occasionally exhibits concretionary structure. The quarries near Aymestry show a thickness of 30 to 40 feet. The stone is found in Shropshire, Staffordshire, and Herefordshire.

The Wenlock Limestone is inclined to be nodular in structure, but it varies a good deal in character, and it is broken up into beds of different thickness by shale. Its usual colour is pale grey. Near Much Wenlock the stone is more crystalline and darker in colour, and was sometimes employed as a fossil marble in small slabs of a dull grey, bluish, or pink tint. The so-called 'Ledbury Marble' is an oolitic variety of the stone found near the place of that name. At Dudley it is known as the 'Dudley Limestone,' and has been largely quarried at Wren's Nest and Castle Hill. Locally the stone is called 'White Lime.' At the former quarry 90 feet of shale separates an upper set of flaggy limestone beds, 20 to 30 feet thick, from a lower set, 35 to 40 feet thick. The stone has also been quarried—mainly for lime and as a flux—near Sedgley, Lincoln Hill and Iron Bridge, May Hill and Blaisdon Edge.

In the **Old Red Sandstone** areas there are no limestones fit for building, but calcareous beds occur in the midst of the red marls as lenticular bodies, varying greatly in size, from mere nodules up to layers several feet thick, which may thin out in a few yards in all directions, or may extend for several miles. The stone has been used for lime, and the thicker beds yield a stone very closely resembling Carboniferous Limestone in appearance. These limestones are called 'Cornstones'; they have been worked at many places between Monmouth and Abergavenny; at Creden Hill, near Hereford, and other places in the Old Red Sandstone area of the West of England and adjoining part of Wales.

Some poor kinds of limestone have been worked in the Lower **Devonian** rocks in North Devon and Somerset, about Combe Martin, Ilfracombe, Challacombe, and on the Quantock Hills; but the principal repository of limestone is the middle division of the Devonian System, in South Devon.

These limestones are often very massive, with rather ill-defined thick bedding; the masses are usually lenticular, but many attain a considerable thickness — 450 feet at Dartington. The stone is compact as a rule, but is often penetrated by small veins of calcite; in places it has been considerably dolomitized, and frequently it has been strongly stained along the joints and bedding planes with bright red hæmatite, which has been washed into the rock from the New Red strata that formerly covered all this part of the country. The prevailing colour of the stone is a cool blue-grey; occasionally it is darker grey or pinkish-grey, and near Ideford it is almost white. Sometimes oolitic structure is exhibited, but the bulk of the stone is a crystalline aggregate of calcite, with or without definite traces of fossils; in some places with many remains of corals often beautifully preserved.

The Devonian Limestone has been very largely quarried at Plymouth, Torquay, St. Mary Church, Babbacombe, Brixham, Petit Tor, Totnes, Dartington, Berry Pomeroy, Ipplepen, Ogwell, Ashburton, Newton Abbot, Chudleigh, and elsewhere. In the neighbourhood of the quarries the stone has been much used, not only in the village churches and farm-houses, but in the larger towns for all classes of building.

For ordinary work the stone is most commonly used in the form of coursed rock-faced ashlar in blocks of moderate size; for more finished work it is punched, with or without drafted margins. In other cases dressed random ashlar or rubble is employed.

Good examples of the use of the stone may be seen in Plymouth, Devonport, Torquay, Newton Abbot, etc., where it is the predominant building stone. In good-class work the effect produced by this stone is very clean and pleasant, if somewhat cold. This last result, which would be regarded by some as a defect, is sometimes overcome by introducing dressings of one of the yellow limestones, or even of buff terra-cotta. Particularly satisfactory are some of the bluish-grey varieties, which often carry a slight bloom of pink.

As a building stone it wears very well, but when employed in marine engineering work—breakwaters, piers, and the like—it is subject to rapid destruction, as is the case with most limestones, through the activity of boring organisms.

At several of the places mentioned above the Devonian Limestone has been worked and polished as a marble.

Carboniferous Limestone.—This formation is developed upon a large scale in this country. The general distribution of the Upper Carboniferous rocks has been indicated in Chapter V. In the southern districts the limestones are confined almost entirely to the lower members of the Carboniferous System, while in the north they tend to appear more and more often as isolated beds in the higher members.

Beginning with the south, we find the limestones

forming the Mendip Hills, and appearing at several small areas in Somersetshire, bordering the Bristol and Somerset Coalfields, and forming a narrow outcrop more or less completely encircling the coalfields of the Forest of Dean and South Wales. In North Wales it appears again in Flintshire and Denbighshire, in Carnarvonshire and Anglesey, and small areas occur in Shropshire and Leicestershire. North of Derby a broad tract runs northward through the Peak district, and extends over the border into Staffordshire; north of the Lancashire and Yorkshire Coalfields limestone-bearing strata come on in force, and may be followed to the Tweed, with extensions westward to Morecambe, Furness, and the Cumberland Coalfield.

In Scotland the limestones are less conspicuous members of the system, which here occupies the lowland tract between the Firths of Forth and Clyde. Small patches lie in the south of the Isle of Man, and in Ireland they cover nearly the whole of the centre of the island.

In the south of England, in Wales, and in the Peak District the Carboniferous Limestone is a tolerably compact series of beds, with only thin partings of shale. The series is finely exposed in the Avon Gorge. In North Wales Morton has grouped the beds as follows:

NORTH FLINT.	Feet.	LLANGOLLEN.	Feet.
Upper Black Limestone	200	Upper Grey Limestone	300
,, Grey ,,	150	,, White ,,	300
Middle White ,,	600	Lower White ,,	120
Lower Brown ,,	200	,, Brown ,,	480

North of the great Derbyshire masses the numerous limestone beds are separated by varying thicknesses of shale and sandstone. In different parts of Yorkshire and the adjoining counties these limestones, some of which are persistent over large areas, have received distinctive

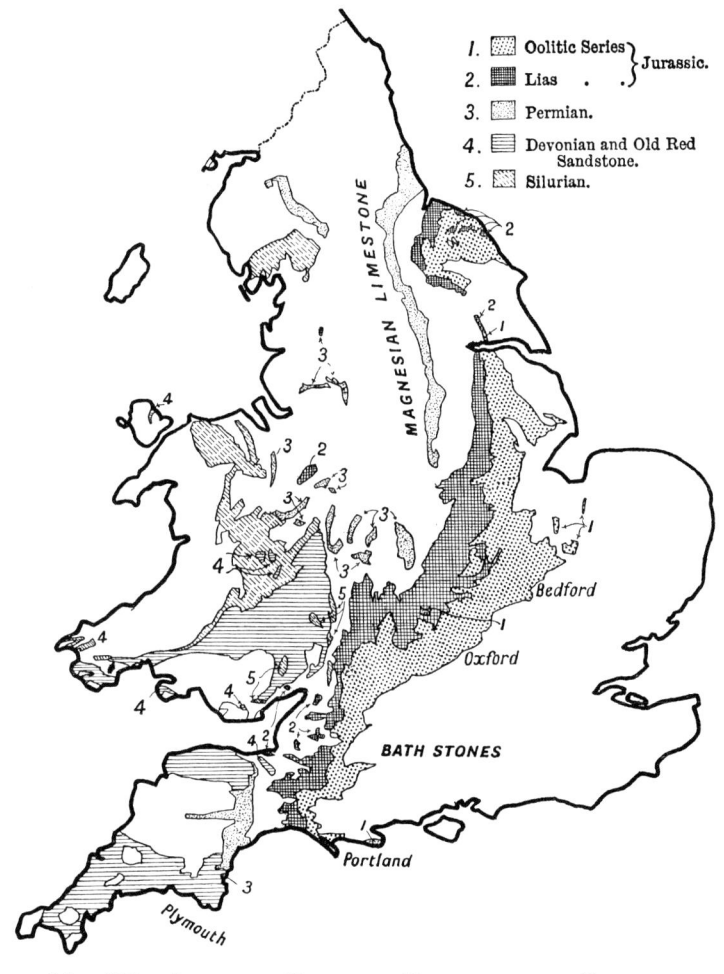

MAP VI.—JURASSIC, PERMIAN, DEVONIAN AND SILURIAN SYSTEMS OF ENGLAND AND WALES.

names; some of these are indicated below, the intervening beds being omitted:

WENSLEYDALE OR YOREDALE.	ALSTON DISTRICT.
Crow Limestone.	Fell Top Limestone.
—	Crag Limestone.
Red Beds Limestone	Little Limestone.
Main Limestone.	Great or Twelve-Fathom Limestone.
Underset Limestone.	Four-Fathom Limestone.
Third Set Limestone.	Three-Yard Limestone.
Fourth Set Limestone.	Five-Yard Limestone.
Fifth Set or Middle Limestone.	Scar Limestone.
Simonstone (Simonside) or Sixth Set Limestone.	Cockle Shell Limestone.
—	Post Limestone.
Hadra (Hadraw) Scar or Seventh Set Limestone.	Tyne-Bottom Limestone.
—	Jew Limestone.
Gayle Limestone.	—
Hawes Limestone.	—
Great Scar Limestone Series. { Melmerby Scar Limestones. Ash Fell Limestones. Ravenstonedale Limestones. }	—

In all these districts the stone has been used extensively in village dwellings, churches, and in other kinds of work, as, for example, the Midland Railway viaducts of Monsal Dale, Millers Dale, Chee Dale, etc. At the present day the bulk of the quarrying in this stone is done for the preparation of lime and for fluxing.

Although in general the Carboniferous Limestone presents great uniformity of appearance, there are many varieties of structure. The best stones for building usually come from the thicker-bedded stone of a greyish-blue colour; the thinner and more regular-bedded varieties are generally darker in colour, and contain more carbonaceous and bituminous matter. The best stones are composed of the small fragmentary remains of crinoid stems, shells, and foraminifera, and the whole is thoroughly crystalline.

LIMESTONES

The darker varieties are inclined to weather brown and dull, and they neither last so long nor look so well as the paler kinds.

Dolomitization of the limestone is of frequent occurrence; it may affect local patches or beds on a small scale, or it may change the character of the stone over a considerable area. Such is the case at Harbro Rocks, near Brassington, Derbyshire; this highly dolomitized stone has been used only locally, and its appearance is dull and sombre, but its weathering powers are very good.

The upper beds of limestone in the Midlands are frequently cherty, and are suitable in this condition neither for dressed building stone nor for fluxing.

The black limestones are sometimes called 'Lias' by quarrymen, from their resemblance to the dark beds of that formation. The Carboniferous Limestone quarried at Oystermouth, in Pembrokeshire, is so called, and the dark limestones quarried at Afon Goch, near Holywell, Flintshire, are known as 'Aberdo Limestone,' through confusion with the Lias limestone of Aberthaw, in South Wales.

In the list of large quarries in the Lower Carboniferous rocks there are few that are worked primarily as building stones. Some that yield a

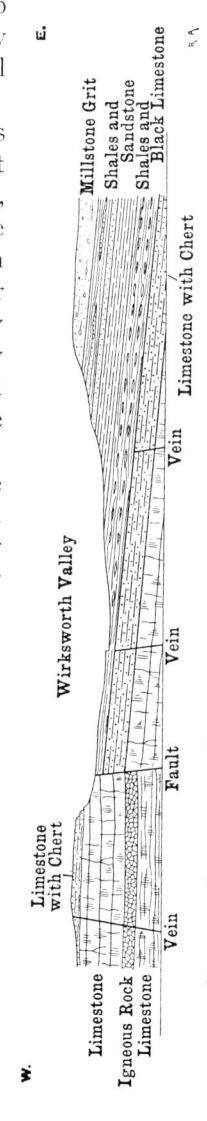

FIG. 9.—SECTION IN CARBONIFEROUS LIMESTONE AND MILLSTONE GRIT BEDS.

good dressing stone are too small to be included in the list.

Perhaps the best known of all the Carboniferous stones is that produced by the Hopton Wood Stone Company (Plate IV.), from quarries at Middleton, near Wirksworth, in Derbyshire. Several varieties of stone are worked (see Fig. 10), but the most familiar is a compact cream-coloured stone, capable of being sawn, carved, and polished. It is employed for monumental work, landings, steps, and paving-slabs. It is a very sound stone, and hardens on exposure. Out of doors, in towns, it weathers evenly, like good Portland stone; the small fossil fragments stand out in the same way in each case. It may be seen in the New Bailey Building, Victoria and Albert Museum, and Imperial Institute, London; Chatsworth House, Trentham Hall, Town Hall, Manchester, etc.

The Derbyshire and Mendip marbles are polishable varieties of the Carboniferous Limestone; so also are those of Kendal; Cleator, near Egremont; Bertham Fell, Westmorland; those from Dent and Frosterley come from the Yoredale Series.

Building stone is worked in Yorkshire at Prudham, near Hexham; Barton, near Darlington; Gilling and Garsdale; at Deganwy and Llysfaen, in North Wales (for churches, monumental and engineering work); near Wenvoe, in South Wales, for the Barry Docks; and in Ireland, at Skerries, grey (much used in Dublin); Lexlip, black; Tullamore, blue; Gillogue, Limerick, blue-black; Brachernagh, pale blue; Foynes, grey and blue, for engineering; and at many other quarries.

Permian. — The Permian limestones in this country are limited in extent, and almost without exception are magnesian in composition; hence the name Magnesian Limestone, which is applied to the more calcareous

PLATE IV

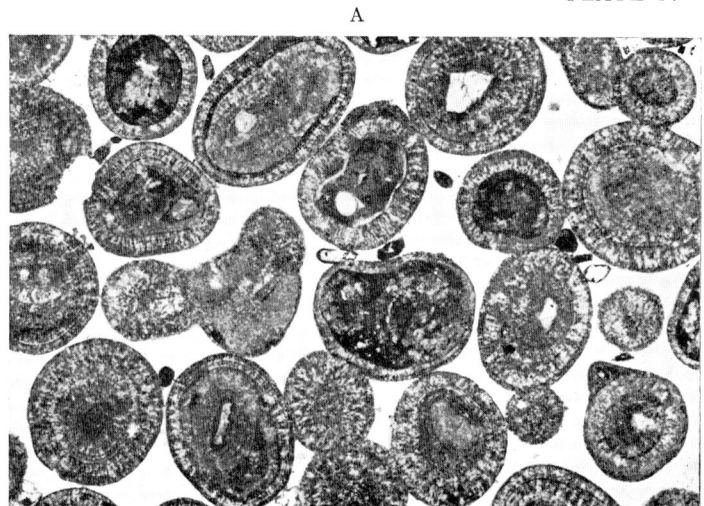

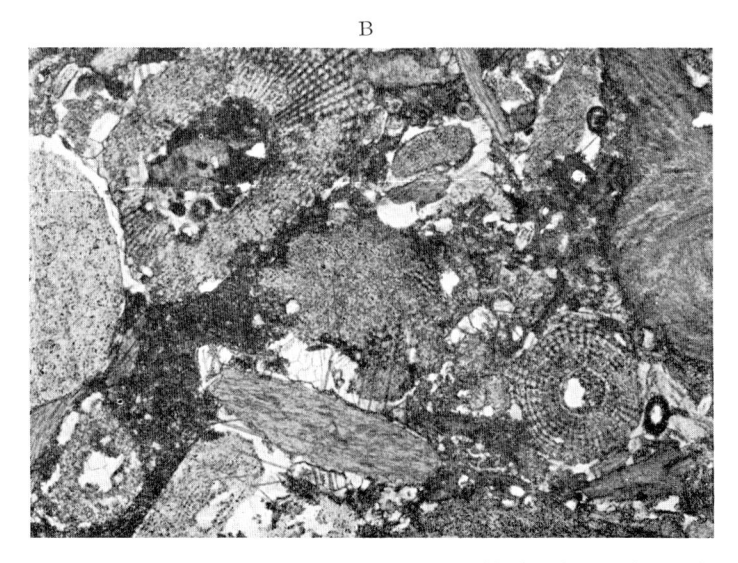

PHOTOMICROGRAPHS OF LIMESTONE. (Magnification, about 20 diameters.)
A. Oolite, Ketton Stone   B. Crinoidal and shelly, from Hopton Wood.

LIMESTONES

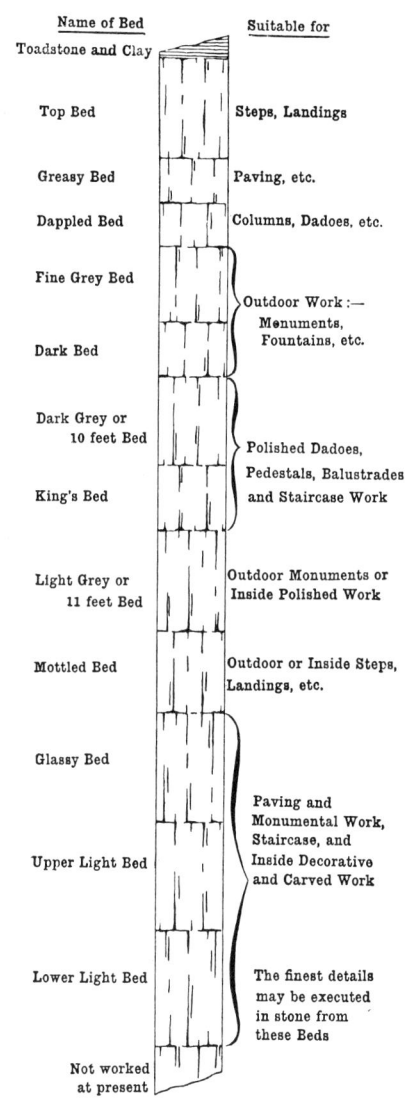

Name of Bed	Suitable for
Toadstone and Clay	
Top Bed	Steps, Landings
Greasy Bed	Paving, etc.
Dappled Bed	Columns, Dadoes, etc.
Fine Grey Bed	Outdoor Work:—Monuments, Fountains, etc.
Dark Bed	
Dark Grey or 10 feet Bed	Polished Dadoes, Pedestals, Balustrades and Staircase Work
King's Bed	
Light Grey or 11 feet Bed	Outdoor Monuments or Inside Polished Work
Mottled Bed	Outdoor or Inside Steps, Landings, etc.
Glassy Bed	Paving and Monumental Work, Staircase, and Inside Decorative and Carved Work
Upper Light Bed	
Lower Light Bed	The finest details may be executed in stone from these Beds
Not worked at present	

FIG. 10.—SECTION OF THE BEDS AT THE HOPTON WOOD STONE FIRM'S QUARRY, MIDDLETON, WIRKSWORTH.

phase of these rocks on the eastern side of the Pennine Range.

Two types of calcareous building stone are obtained from this formation: (1) Magnesian limestone and dolomite; and (2) sandstones or breccias, with more or less dolomite binding material.

The Magnesian limestone runs northward in a narrow outcrop from Nottingham, through Mansfield, Worksop, Doncaster, Brodsworth, Pontefract, Tadcaster, Knaresborough, and Darlington, where it begins to broaden out, and extends to the coast at Hartlepool, Sunderland, and South Shields.

The succession of strata changes as the outcrop is followed northward :—

NOTTINGHAM AND DERBYSHIRE.	SOUTH YORKSHIRE.	DURHAM AND NORTHUMBERLAND.
Upper Marls.	Upper Marls and Sandstones.	Upper Magnesian Limestones.
Upper Magnesian Limestone.	—	Middle Magnesian Limestones.
Middle Marls and Sandstones.	Local.	
Lower Magnesian Limestone.	Middle and Lower Limestones.	Lower Magnesian Limestones.
—	Marl Slate.	Marl Slate.
Quicksand.	Quicksand.	Sands and Sandstone.

The beds dip very gently eastward.

The upper limestones are thin-bedded and flaggy, and usually contain less magnesia than the lower and middle limestones, except in the northern part of the district; in the centre and southern part of the outcrop they are very little used.

The best building stones come from the lower limestones, which range in thickness from 40 feet in the south to 200 feet in Durham; the beds vary, however, very rapidly from place to place. In the southern part of the

LIMESTONES

outcrop, about Mansfield, etc., the best building stone is found rising up into small hillocks from out of the surrounding Permian beds.

The Permian magnesian limestones range in colour from white and cream to dull yellow or yellowish-brown. When freshly broken the paler kinds have a beautiful sparkling appearance, due to the crystalline faces of the carbonate rhombs; but on exposure they soon become grey, and lose their sugary aspect. Very common in some of the stones is the presence of minute black specks of iron or manganese oxide.

All these magnesian limestones feel rather rough and harsh to the touch, somewhat like fine-grained sandstones.

Mansfield Woodhouse (Nottinghamshire).—This stone has a warm yellow-brown colour, and exhibits a crystalline appearance on the fresh fracture; frequently minute specks of black oxide are abundant. The beds are massive but irregular, and some parts are more coarsely crystalline and porous than others. The microscopic section shows an aggregate of rather coarse and well-formed rhombs of dolomite; here and there are small irregular pores, which in places are filled with calcite. This stone has been used in Southwell Cathedral and in the lower part of the Houses of Parliament, also in the Martyrs' Memorial at Oxford.

Bolsover Moor (Derbyshire).—This is almost identical with the above in appearance and texture; it contains a few sparsely scattered angular grains of quartz. Stone from these quarries was selected by the Commissioners for the Houses of Parliament, but it was not employed; the irregular and oblique bedding and frequency of joints make it impossible to obtain large blocks; the average size of block does not exceed 4 feet by 2 feet by 9 inches.

Steetly (near Worksop, Nottinghamshire).—The stone from these quarries varies in colour from white or cream to grey and yellow. In texture the paler stone much resembles the Woodhouse Stone. The 'Yellow Bed' has a creamy appearance when freshly worked, and black specks are visible. The 'White Bed' is paler and rather coarser. Some of the beds are finer in grain than the above. Fairly regular blocks, 2 to 3 feet thick, are obtained, and the stone takes a fair polish, though it is rather porous. St. George's Church, Doncaster, was built of Steetly Stone.

North Anston (Yorkshire).—This is a warm yellow stone, fairly compact and hard. The bedding and jointing are more regular than in most of the other quarries, and blocks up to 8 feet by 3 feet by $1\frac{1}{2}$ feet are obtainable. The stone is duller and less crystalline in appearance in the fresh fracture than that from Bolsover, and the dolomite rhombs are rather smaller and contain more dull colouring matter. Minute irregular cavities are abundant, but many of these are filled with calcite.

Roche Abbey (Yorkshire).—A compact pale grey stone, white when freshly broken, with sparkling sugary appearance. The dolomite rhombs are small and crowded closely together; minute irregular pores are present, some being filled with calcite, or with a yellowish isotropic or cryptocrystalline substance. The bedding is irregular; some large blocks may be obtained, but the supply is uncertain. It has been much used for building and carved work, but it is rather soft, especially in the lower beds.

The stone from *Huddlestone*, Yorkshire, is like that from Roche Abbey, but often exhibits whitish specks. It is composed of an aggregate of even-sized dolomite rhombs.

From *Brodsworth*, Park Nook, Robin Hood's Well, and Cadeby, in Yorkshire, pale yellow, brownish, or cream-coloured dolomitic limestones are obtained, which resemble the foregoing in external appearance, but when examined with a lens or in section in the microscope they are seen to be entirely composed of fragments of bryozoa. The fossil fragments are themselves dolomitized, and they lie in a crystalline matrix of the same mineral. The Cadeby stone is softer than the others. At Smawse, Bramham Moor, Yorkshire, a fine-grained, even-textured dolomitic building stone is quarried, which is similar in appearance to those previously described, but structurally it is entirely different. It is composed entirely of small spherical oolitic or pisolitic grains, about 0·1 mm. in diameter; many of these appear in section to be filled with a yellowish isotropic substance. The matrix is a crystalline dolomitic aggregate.

Quarries in the Magnesian Limestone have been worked at Shire Oaks, Brancliff, Kiveton Park, Sutton-in-Ashfield, Pleasley, and Stoney Houghton; the stone from the last-named is cherty and more calcareous than the others.

Near Cresswell Crags, which are formed of the Magnesian Limestone, and at Shireoaks, stones have been taken for rockeries on account of the strong development of the structure known sometimes as 'Stylolite,' which gives a curiously irregular surface to the stone.

Two well-known stones are obtained from the Lower Magnesian Limestone near Mansfield, which differ from the ordinary dolomite in that they contain a considerable amount of quartz sand; for this reason they are often classed as sandstone. They will be described here because they form a special local phase of the Lower Magnesian Limestone formation, and because the magnesian and calcareous binding material is present in much higher proportion than is usual with sandstones.

Mansfield White 'Sandstone' is quarried at the south end of Mansfield in three places; of these, Lindley's is the largest: the depth of the quarry is 30 to 40 feet. The stone is yellowish-white at first, but bleaches to clean white on exposure.

The bedding is very irregular—current-bedding—in all

the quarries; but the thicker beds are in Lindley's quarry. The beds are known by the quarrymen as 'Upper,' 'Middle,' and 'Column,' in descending order; but it is obvious from what has just been said that they are not sharply defined.

The 'Upper Beds' are regarded as the 'best beds'; they are coarser than the others; the largest blocks, up to 6 feet square by 2 feet thick, are obtained here. This stone is used for facings, but not for steps. The lower part of the 'Upper Beds,' and sometimes the upper part of the 'Middle Beds,' is called the 'Second Bed'; it is used for monumental work.

The 'Middle Bed' is the most compact and hard; it is used for steps, landings, and paving.

The 'Column Bed' resembles the best beds in appearance, but it is a little darker in colour, and like the 'Middle Bed,' it contains streaks of blue marl in places. It is used for dressings, steps, and general work.

Mansfield Red 'Sandstone' is quarried in the Rock Valley and on the Chesterfield Road, Mansfield; the former is about 50 feet and the latter about 70 feet deep. The bedding and other characters resemble those in the White quarries, but the individual beds do not receive special names; the quality of the stone is said to improve with depth.

The Red and White Mansfield Stones differ mainly in colour; in composition and texture they are practically identical. The small grains of quartz are imbedded in an obscurely crystalline dolomitic matrix. The quartz makes up about half of the rock.

Both Red and White Stone are used in the pavement of the terrace in Trafalgar Square, and in the following structures examples may be seen: The Temple Bar Memorial; St. Pancras Hotel and Station; The Hippodrome, Cranbourn Street; Claridge's Hotel, Brook

Street, London; many of the London and County Bank buildings; King's College, Cambridge; Magdalen College, Oxford; Ely Cathedral; Chichester Cathedral; Newstead Abbey; Clumber Park, etc.

Lias.—The outcrop of the Lias extends from the coast of Dorset at Lyme Regis northwards between Bath and Bristol, and by Gloucester, Tewkesbury, Rugby, through the western part of Leicestershire and eastern side of Lincolnshire, to the coast at Redcar, in Yorkshire. There is also a small area in Glamorganshire.

The Liassic rocks consist largely of clay, with thin-bedded limestones, the latter predominating in the lower part of the series. Important ironstones occur in the Middle Lias.

The limestones form thin regular beds, but nowhere do they produce a really first-class building stone, though excellent paving stones are obtained, suitable for internal use, and the genuine Blue Lias Lime is an important product of this formation.

As a building stone the Lias Limestones are used only locally, and many of the older buildings in the districts indicated above are constructed of this stone; the best examples are now to be found in the country places, for in the towns—Gloucester, Stratford-on-Avon, Northampton, Grantham, etc.—its place has been taken by brick.

The characteristic colour of the stone is a bluish-grey or slaty tint, and the general effect, except in bright sunlight, amid grass and foliage, is one of sombreness approaching dulness. Moreover, it is not durable; the frost readily attacks it and causes it to splinter, and even for rough work, embankments and the like, it is not to be relied upon. In the old houses built of this stone the flues were almost always formed of brick.

The Lower Lias Limestones have been extensively quarried in Somersetshire, Worcestershire, and Warwickshire for building and paving stones. One of the best-developed quarry districts is that at Keinton Mandefield, and the following section of one of these quarries conveys a good idea of their general character:

SECTION OF LOWER LIAS AT STIPSTONE QUARRY, KEINTON MANDEFIELD.

	Ft.	in.
Posts: Limestones and clays used for road metal	3	0
Thin Yellow: Limestone in three beds	0	9
Shale	0	9
Thick Yellow: Limestones with fossils (Ammonites)	0	6
Shale	1	4
Thin Corner or *Cornstone*: Limestone with bivalve shells	0	6
Shale	0	2
Thick Corner: Limestone	0	7
Thick White: Shelly limestone	0	7
Thin White: Limestone	0	3
Cream: A poor bed of limestone	0	6
Red Liver: Shelly limestone	0	4
Thin Black: Limestone used for outdoor paving	0	6
Thick Black: Limestone used for outdoor paving	0	5
Thin Cover: Limestone used for paving	0	3
Thick Cover: Limestone used for paving	0	5
Clog: Limestone used for building	0	5
Hearthstone: Limestone used for kerbs	0	5
Thin Firestone: Limestone used for kerbs	0	3
Thick Firestone: Grey limestone used for kerbs	1	2

Thin partings of shale separate the limestone beds, many of which are employed for building stone, and those below the Thick Yellow are burnt for lime.

Most of the quarrying is now done for paving slabs (paviours) and kerbs. Some of the slabs obtained are of large size (12 feet by 12 feet). The best of the paving stones comes from the lower part of the limestone beds, usually in the zone of the Ammonite (*Ægoceras planorbis*). The slabs are usually cut square, and are either tooled or rubbed. Occasionally the surface is polished for mantelpieces, but the polish is not good, and the appearance is dull and uninteresting, except in the so-called Ammonite Marble, which has never been found or worked on a large scale.

While most of the Lower Lias Limestones are bluish-grey, there are exceptions: the Wilmscote quarries, for example, furnish white paving slab, from the layer known as the 'White,' used in the Houses of Parliament and the Law Courts, and make a very clean-looking indoor paving, which wears well. Some of the layers in the same quarry are buff in colour.

Very different in appearance and texture is the *Sutton Stone*, from the Lower Lias at Sutton, near Bridgend, Glamorganshire. It has a creamy tint, and is a compact free-working stone, either with a fine cellular texture (resembling the tufaceous limestones) or dense and smooth, like a lithographic stone. The lower beds contain the freestone; the upper ones are employed only for rough walling. According to Mr. H. B. Woodward, it is the most durable of the Lower Lias Limestones. It has been used in Llandaff Cathedral, Neath Abbey, Corby Castle, and other old castles in South Wales.

A similar stone, but more porous, the *Brockley Down Stone*, occurs at Downside, near Wrington, and on the northern side of Shepton Mallet. It resembles Downton Stone in appearance, but it is not so strong.

In the Middle Lias the irregularly developed limestone band known as the Marlstone or Rock Bed has been quarried for rough building in several parts of Somersetshire; at Stinchcombe and Dursley, in Gloucestershire; and at several places in Northamptonshire and Rutlandshire. This stone when freshly quarried from under cover has a bluish-grey or greenish tint, but the exposed portions of the outcrop, those not covered by the clays of the Upper Lias, have been weathered brown, and in this condition the stone is preferred for building. Whether it is really any better is doubtful, but it certainly will not change colour as the unweathered green stone will do after a short exposure. The Marlstone of Byfield was formerly squared for interior paving, and that from Bagpath, near Dursley, has been polished. The Marlham Stone, used for rough work near Ilminster, comes from this horizon.

Hornton Stone, from the Middle Lias of Edge Hill, has been extensively quarried, and has a high reputation locally. It is a very soft stone, easily cut with a handsaw, and it is liable to crack under pressure; for this reason the quoins of the better buildings in which it is used are usually of some other stone. Two colours are obtainable: a warm brown and a sage green, with splashes of brown. It has been used for ashlar, steps, paving slabs, sinks, and tombstones. In many of the old churches and other buildings in this district this stone has been utilized; good examples exist in Banbury, in the Catherine Wheel Inn, and in the doorway of the Mechanics' Institute.

Although it does not strictly belong to the Lias, the so-called 'White Lias,' pale and cream-coloured limestones of the Rhætic formation, may be mentioned here. It has been quarried at Sparkford, near Castle Cary, Somersetshire, where it has been employed in the local railway bridges; and at Radstock and Load Bridge, in the same county; also at Chesterton Quarry, near Warwick. The last-named stone was used in the Chesterton Wind and Water Mill, designed by Inigo Jones, and in the hall at Compton Verney. It is usually a very smooth-textured stone, compact, and not very hard; for internal carved work it is very suitable, as it can be cut with great sharpness and precision.

Inferior Oolite.—Rocks of this geological age crop out in the coast of Dorset, and continue northwards and eastwards diagonally across the Midlands into Lincolnshire and Yorkshire. From this formation the well-known limestones of Doulting, Dundry, Painswick, Casterton, Ancaster, Haydor, Ketton, and others, are obtained. Like all shallow-water deposits, the members of the Inferior Oolite are subject to frequent and rapid variation in the thickness of their beds and the nature of the sediment.

From Dorsetshire to the Cotteswold Hills the regular sequence is Midford Sand at the base, forming passage-beds with the underlying Lias, then Inferior Oolite Limestones, capped by the Fuller's Earth (Great Oolite), where it has not been denuded. East of the Cotteswolds, the lowest of the Inferior Oolites is represented by the ferruginous sandstones, white sands, ironstones, and clays of the Northampton Sand and Lower Estuarine Series; the Fuller's Earth above has thinned away. Still further east these beds are overlaid locally by the sandy Collyweston Slate (p. 318), and by another strong calcareous series of beds—shelly and oolitic limestones—the Lincolnshire Limestone. In Yorkshire the Inferior Oolite is represented in descending order by the following formations: Grey Limestone, Middle Estuarine Series, Millepore Bed, Lower Estuarine Series, with Ellerbeck Bed, Dogger, and Blea Wyke Beds. 'Thus,' as Mr. H. B. Woodward says, 'the Inferior Oolite in its course across England exhibits almost every variety of stratified rock. In some places, as on the Cotteswolds, we have fine false-bedded oolites furnishing excellent freestone; also beds of soft oolitic marl, and layers of coarse oolite and pisolite. In other places the oolite becomes very ferruginous, and almost an oolitic iron ore. Again, near Lincoln we find beds of compact limestone, with scattered oolitic grains, and beds of shell limestone. Conglomeratic beds are met with on the eastern borders of the Mendip Hills, and tiny quartz pebbles occur in some beds in the Cotteswold Hills.'

The following section is shown in a quarry north of Frome, at Oldford:

	Ft.	in.
Rubbly Oolite	3	0
Pale shelly Oolite	3	6
Pale chalky Oolite	3	0
Pale Oolite (four layers)	13	0
Brown sandy Oolite and sparry Limestone, resembling Doulting Stone	2	0

In the neighbourhood of Frome the limestone in places becomes distinctly siliceous, hard, and cherty.

Around Bath the Great Oolite quite overshadows the Inferior Oolite as a building stone, and the latter was known as the 'Bastard Freestone.' Here, indeed, it is inferior in quality to the Great Oolite, and much of it is soft, and often contains holes like the roach of Portland. (It may be well at this point to remind the reader that the term 'Inferior Oolite' connotes the relative position of the formation, and not the quality of the stone.)

The building stone of Dundry Hill composes the upper part of the Inferior Oolite; it forms a series of hard, flaggy brown limestones, resting on more massive beds of sandy limestone, the whole being more or less oolitic, and in places sparry, like the Doulting Stone. This series of freestones is 15 or 16 feet thick; it rests upon about 40 feet of 'Rag Beds,' which in turn cover 5 or 6 feet of ironshot limestones.

Passing northwards to the Cotteswold Hills, near Dursley, at Selsley Hill, Rodborough, Stroud, Haresfield, Randwick, Painswick, Birdlip, Crickley, Leckhampton, Lineover Hills, and Cleve Cloud, many local sub-

LIMESTONES

divisions have been established, the names of which are as follows:

			Feet.
Ragstones	Upper	White Freestone	5
		Clypeus Grit	6 to 15
	Lower	Upper Trigonia Grit *	2 to 12
		Gryphite Grit † } Lower Trigonia Grit	2 to 12
Freestones		Upper Freestone	6 to 20
		Oolitic Marl	5 to 10
		Lower Freestone	45 to 130
Pea Grit Series		Pea Grit	3 to 20
		Lower Limestone	10 to 25
Midford Sand (Cotteswold Sands).			

At the base of the Lower Limestone there are 5 to 9 feet of ferruginous oolite and sandy limestone, forming a kind of ragstone, which is occasionally used for rough masonry. The Lower Limestone is worked for building stone at the Ruscombe quarries, Randwick Hill; here the beds consist of 10 feet of freestone with few oolitic grains, over an oolite with bands of shelly detritus. This is a good weather-stone, but difficult to work. The Pea Grit itself is not a 'grit' at all, but a very coarsely oolitic or pisolitic limestone. At Crickley and Leckhampton the Pea Grit Series is about 38 feet thick, and is indivisible into Lower Limestone, etc.; the beds of different types alternate one with another. The harder beds are worked for building stone.

The Freestones are pale oolitic, and sometimes shelly limestone. The lower, or 'Building Freestone,' is the more important, but good stone is obtained locally from the upper division, and in some cases from hard bands in the oolitic marl. At Leckhampton there is a bed of flaggy oolite at the base of the Lower Freestone, which has been much used for rough work.

The *Ham Hill* (Hamdon Hill) stone occurs as a local

---

* From the fossil *Trigonia*. † From the fossil *Gryphæa*.

calcareous development in the upper part of the Midford Sands. Indications of similar intercalations may be observed near Yeovil Junction, North Perrott, and elsewhere in the neighbourhood. The following may be taken as a representative section of the Ham Hill beds:

		Feet.
'Ochre Beds' { Sand and thin soft stone ... Sand with thicker beds of stone		40
Ham Hill Stone { Main mass of freestone, indistinctly jointed and false-bedded. Good stone obtained 7 or 8 feet down, and thence to the bottom in the following sequence:	Feet.	about 50
Yellow beds (chief part)	35	
Coarse bed		
Grey beds (most durable)	8	
Stone beds	6 to 7	
Yellow Sands	about	80

The quarries north of Doulting and near West Cranmore are excavated in a series of very markedly current-bedded oolites, belonging to the upper portion of the Inferior Oolite. The sequence of beds is approximately as follows:

	Feet.
Inferior Oolite { Oolitic, shelly, and slightly sparry limestone, and pale oolite on very sparry oolite ...	7
Brown sandy limestone, shelly, and slightly oolitic	10
Oolitic freestones	
Massive beds of oolitic, sparry freestones ...	20
Thick-bedded sandy oolitic limestones and sparry limestones	10
Sandy and sparry limestones and decomposed ironshot limestones	10

The organic fragments in the stone are mostly crinoidal, and much of the material may have been derived from the Carboniferous Limestone, upon which the Inferior Oolite rests in the eastern borders of the Mendip Hills.

The Ragstones of the Inferior Oolite of the Cotteswold district have been subjected to much subdivision by

PLATE V

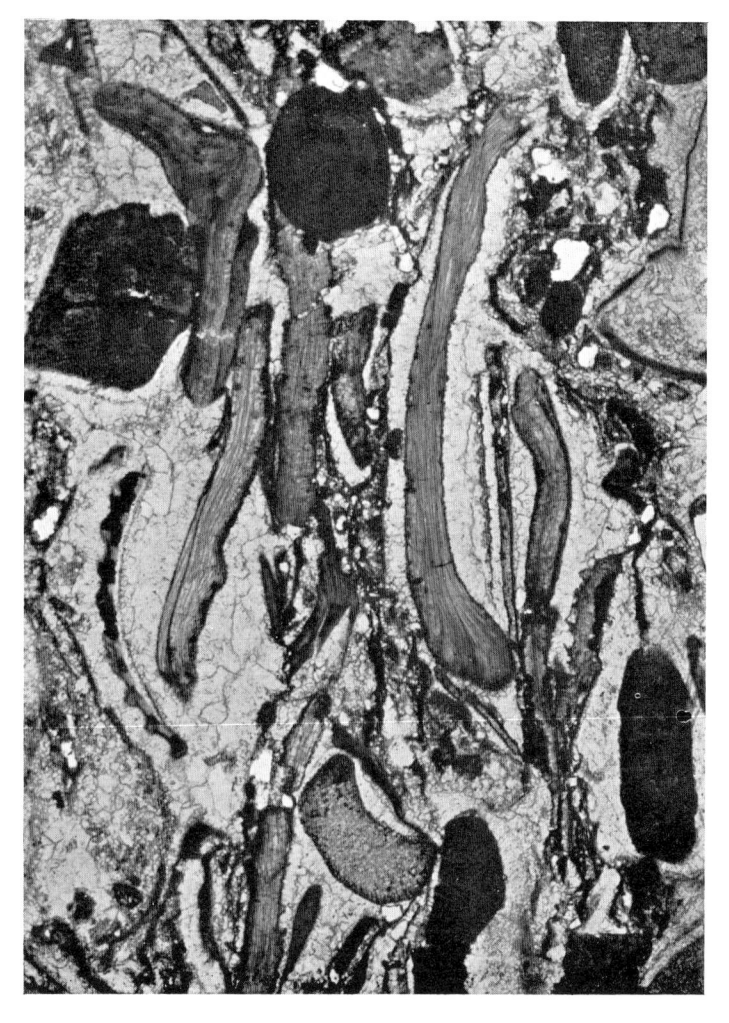

PHOTOMICROGRAPH OF LIMESTONE. (Magnification, 30 diameters.)

Ham Hill Shelly.

geologists on palæontological grounds, but these minor—and local—subdivisions are of little or no economic value. The Ragstones are for the most part earthy and ferruginous oolites and shelly limestones, only occasionally yielding blocks suitable for building purposes. Here and there calciferous sandstones make their appearance, but there are no grits in the true sense (see p. 205). The term is in this case an unfortunate misappellation which has become sanctioned by use.

The White Freestone, a fine-grained white oolite, is only locally developed; it appears on Stroud Hill and Rodborough Common, but it does not seem to have been employed as a freestone.

Exposures of some or all of the above beds may be seen in

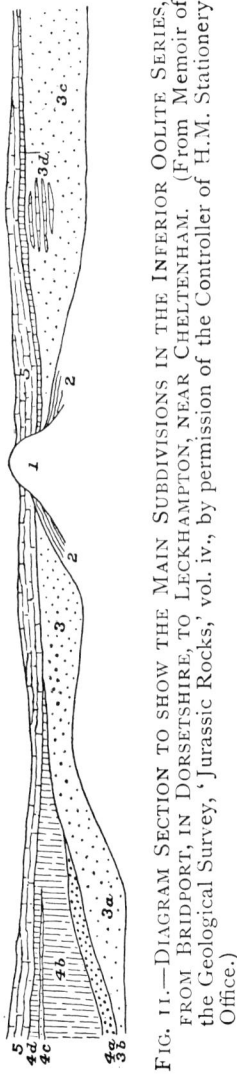

FIG. 11.—DIAGRAM SECTION TO SHOW THE MAIN SUBDIVISIONS IN THE INFERIOR OOLITE SERIES, FROM BRIDPORT, IN DORSETSHIRE, TO LECKHAMPTON, NEAR CHELTENHAM. (From Memoir of the Geological Survey, 'Jurassic Rocks,' vol. iv, by permission of the Controller of H.M. Stationery Office.)

1. Palæozoic Rocks.
2. Lias, etc.
3. Midford Sand.
   3a. Cotteswold Sand.
   3b. Gloucestershire Cephalopod Bed.
   3c. Yeovil and Bridport Sand.
   3d. Ham Hill Stone.
4. Inferior Oolite: Lower Division.
   4a. Pea Grit Series.
   4b. Lower Freestone.
   4c. Oolite Marl.
   4d. Upper Freestone.
5. Inferior Oolite: Upper Division. Ragstones, etc.

GEOLOGY OF BUILDING STONES

the quarries, which are numerous in the district. Quarries are situated on Stroud Hill; near Brinscombe; at Nailsworth, where the Ragstones yield the best weather stone; at Ball's Green, near the latter place, where the Lower Freestone has been extensively quarried in underground galleries; at White Hill (Whiteshill), the Horse-pools, Quar Hill, Painswick Hill, Kimsbury Castle.

The nature of the stone beds in this part of the country may be exemplified by the exposure in the quarries on Leckhampton Hill.

		Ft. in.	Ft. in.
Upper Trigonia Grit	Hard, irregular, earthy, shelly, and oolitic limestone	5 0	
	Hard oolite, passing down into hard shelly oolite	3 6	
Gryphite Grit	Hard, brown, rubbly, and gritty oolitic and ironshot limestone	5	0 to 7 0
Lower Trigonia Grit	Rubbly limestone	5	0 to 6 0
	Marly and shelly oolite	1 6	
	Brown marly layer	0 6	
Upper Freestone	Current-bedded oolite, with marly layers	20	0 to 25 0
Oolitic Marl	Soft marly oolite or rubbly marl	7	0 to 10 0
Lower Freestone	Pale oolite, with occasional marly layers; the upper part much false-bedded and shattered, the lower part more or less shelly	130 0	
Pea Grit Series	Rubbly, ochreous, shelly, and pisolitic limestones	12 0	
	Coarse iron-stained oolites and pisolitic limestones	15	0 to 20 0
	Brown and grey limestones	6 0	

Midford Sand.

North-east of Cheltenham, on Cleeve Hill, there are numerous quarries working different parts of the Inferior Oolite Series, including the Pea Grit pisolitic beds, the Ragstones, and the Lower Freestones (60 feet); the Upper Freestone here is represented by markedly sandy beds.

LIMESTONES

Near Stanway the following beds are shown in the Jackdaw Quarry:

	Feet.
Well-bedded Oolite, much jointed and false-bedded in places	25
White Oolite	3
Brown Oolite (best freestone)	6
Yellow Oolite	10

Quarries have been opened in the outlying masses of Inferior Oolite at Bredon Hill and Stanley Hill, and at Longborough, Broadway, and Westington Hill, about two miles south of Chipping Campden and Blockley.

In the direction of Ebrington the Lower Freestone becomes very sandy. At Bourton-on-the-Hill the freestone quarries show:

	Ft. in.	Ft. in.
White Rock: Current-bedded oolite	11 0	to 12 0
Earthy sand	2 6	
Red Bed: Hard and rough oolitic limestone, with irregular iron-stained bands	5 0	
Yellow Bed: Buff oolite, good freestone	6 0	
Ferruginous oolitic limestone, with irregular cavities	10 0	
Bottom Beds: Brown, sandy, oolitic limestones	5 0	to 6 0

(Lower Freestone)

About Notgrove another freestone makes its appearance above the Gryphite Grit; it is known as the Notgrove Freestone (10 feet thick).

The tendency for the Lower and Upper Freestones to become sandy northward and eastward of the Cotteswold Hills has already been noticed. On entering Oxfordshire this tendency is found to have affected the Ragstones also; moreover the Pea Grit, the Freestones, and the Gryphite Grit have disappeared as recognizable beds.

The most important calcareous formation in the Inferior Oolite of this part of the country is the Chipping Norton

Limestone, a series of oolitic and sandy limestone beds appearing at the top of the Ragstones. They are, however, of little use except for rough building and dry walling; their maximum thickness is about 30 feet. Sometimes the Chipping Norton Limestone is shelly, crinoidal, and oolitic, or it may pass into a fine-grained oolite and calcareous sandstone. There are several small quarries in the limestones around Chipping Norton.

Other limestone beds which may or may not be the equivalents of the Chipping Norton Limestone are worked in places, as at Heythrop, where there are 20 feet or more of coarse-grained shelly oolites; they have been employed in many of the buildings in the neighbourhood.

In Northamptonshire, Rutlandshire, and Lincolnshire the Inferior Oolite Series is divided as follows:

Upper Estuarine Series (Great Oolite).
Inferior Oolite
- Lincolnshire Limestone.
- Collyweston Slate.
- Lower Estuarine Series } Northampton Beds.
- Northampton Sand

As we are at this moment concerned only with the calcareous rocks, attention must be paid to the topmost division alone, the Lincolnshire Limestone. (Certain calcareous sandy stones, such as the Harleston Stone and the 'Yellow and best Brown Hard' stones of Duston will be found mentioned on p. 160.) This formation is a thick lenticular development of marine limestones. The beds vary greatly in character; at one point they are composed almost wholly of shell débris, at another they are built up largely of the remains of corals, with or without an accompaniment of oolitic grains. As Professor Judd has pointed out: 'The rocks of the two facies of the Lincolnshire Oolites do not maintain any constant relations with one another; at some places, as at Barnack and Weldon, beds of the shelly facies occur almost at the base of the

LIMESTONES

series, while at others, as about Geddington and Stamford, the strata with the Coralline facies occupy that position. Sometimes, as at Ketton and Wansford, we find beds in the Lincolnshire Oolite entirely made up of fine oolitic grains, and these constitute some of the most valuable freestones. Very rarely the grains of which the rock is composed are very coarse, and it becomes a pisolite.'

The old freestone quarries of Helmedon, long ago disused as a source of building stone, are in the Lincolnshire Limestone.

The good freestones and rags which occur in this formation lie upon divers horizons; thus the Barnack Stone is found near the base; the Weldon Stone somewhat higher; those of Casterton, Clipsham, Ancaster (Haydor and Wilsford Quarries), Ketton and Stamford near the top of the series (Fig. 12).

At Great Weldon, where the Lincolnshire Limestone has been quarried to a considerable extent, the following series of beds is exposed:

	Ft. in.	Ft. in.
Soil		
Rubbly and decomposed rock	11 0 to 12	0*
Fissile limestone ...		
Current-bedded oolitic limestone with holes		

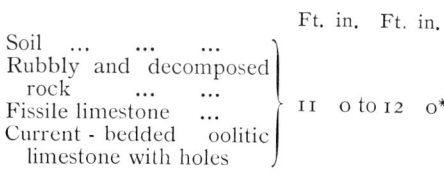

* Forming the overburden.

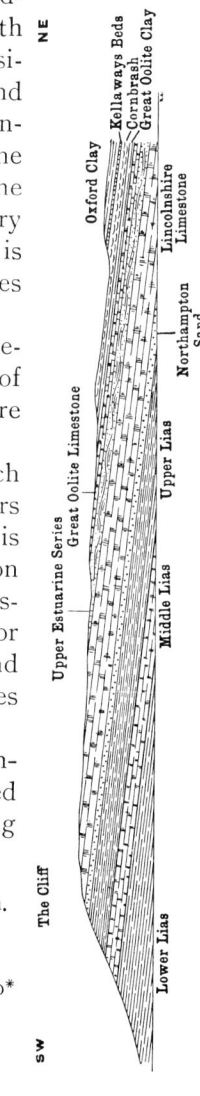

FIG. 12.—SECTION IN LINCOLNSHIRE OOLITE: LINCOLN TO SUDBROOK.

		Ft. in.	Ft. in.
Lincolnshire Limestone	Hard, shelly, blue-hearted limestone, and Weldon Rag or Weldon Marble	2 0 to	3 0
	Oolitic freestone, current-bedded in parts (the 'A' bed)	2 6	
	Oolite ('A 1') bed bottom freestone	3 6 to	4 0
	Fine oolite, local ('B' bed)	1 6	
	Rough, shelly limestone, with moulds and casts of shells	3 0	
	Fine oolitic freestone, pink	4 0	
	'Marble bed,' fossiliferous	2 0	

Northampton Beds: Ironstone and sandy beds.

The limestones at Wakesley are about 30 feet thick; a shelly bed, 6 to 8 feet near the base, resembles the Weldon Rag.

A hard blue siliceous limestone in the Lower Estuarine Series is quarried at Morcott, and similar hard blue stone is obtained from the Northampton Beds at Uppingham. The extensive quarries about Barnack, whence the Barnack Rag was formerly obtained, are all abandoned, and the supply of stone has been supposed to have been exhausted, but a similar stone is still obtained for local use near the village. The celebrated Ketton Stone is obtained at Ketton Heath, from the upper part of the Lincolnshire Limestone. A section in this neighbourhood shows the following beds:

		Ft. in.	Ft. in.
Upper Estuarine Series: Clays and sands		28 0	
Lincolnshire Limestone	'Crash bed': Red-stained freestones	2 6	
	On oolitic freestone	3 6 to	4 0
	'Rag': Irregular, tough, sparry, blue-hearted oolite, passing down into good yellow freestone—lower part of Ketton Stone	6 0	

Near Stamford about 12 feet of good stone lies low down in the Lincolnshire Limestone; the upper part of this is a hard compact blue-hearted limestone, with scattered oolite grains—the 'Stamford Marble'—which passes down into a buff limestone, with brown oolitic grains; the paler portions of this stone once passed under the name of 'Stamford Stone.'

The oolitic freestones of Casterton lie in the higher part of the Lincolnshire Limestone. At Clipsham the oolitic freestones and shelly rag beds lie, as at Ketton, beneath a considerable thickness of Upper Estuarine beds.

The compact subcrystalline limestone worked about Thistleton and Grantham is not suitable for building; and the same remark applies to the coarse oolites, shelly oolite, and earthy limestone at Waltham, though some of the coral-bearing oolite was formerly quarried for this purpose.

At Ponton and Houghton the Lincolnshire Limestone is about 100 feet thick; the best freestone, a rather marly oolite, lies near the base.

The freestone of Ancaster has long been famous; here the best beds occur near the top of the series; the general section of strata in this district is as follows:

		Feet.
Upper Estuarine Series: Blue clays, etc.		
Lincolnshire Limestone	Rag	10
	Freestone	20
	Rag	2
	Freestone	2
	Pale limestones, sub-oolitic, current-bedded in lower part	50
	Grey, earthy, sub-oolitic limestones ⎫ Shelly oolitic limestones, with ochreous galls ⎬	10
	Calcareous sandstones and irregular clay bands	2
Lower Estuarine Series: Sands and clays.		

The stone is quarried on Wilsford Heath, and formerly in the Old Quarry north of Copper Hill Farm, and at the Castle Quarries. At South Ranceby about 16 feet of current-bedded oolite is found, which improves in quality as the depth increases, and has yielded building stone.

Several varieties of stone, but none of much value, have

been quarried from the lower beds of the Inferior Oolite escarpment east of Caythorpe, Fulbeck, and Leadenham, and again between Washenborough and Dunston. Immediately around Lincoln the limestones are about 80 feet thick. The Greetwell Road Quarry shows the following beds:

		Ft. in.	Ft. in.
Lincolnshire Limestone	Marl and rubble.		
	Top and Bottom Nerlys: Freestone, with white marly sub-oolitic limestone	5 0	
	Upper Silver Bed: Fine-grained fissile oolite	2 0	
	Sink Stone: Sub-oolitic limestone	1 0	
	Bottom Silver Bed: Compact sub-oolitic limestone	1 6 to	2 0
	Marl and sub-oolitic limestone	1 6	
	Variable sub-oolitic and coarsely oolitic limestone, with *Walling Bed* in lower layers; the more compact beds are regarded as good *Weather Stone*	4 6	
	Oolitic limestone (not used)	2 0	

Northampton Sand.

Hard grey limestone (2 feet) and blue shelly and argillaceous oolite and compact limestones are worked in the Dean and Chapter Pit; the lowest beds are fine-grained, buff, more or less oolitic shelly limestones, which include a 'Silver Bed,' as at Greetwell Road; some of the beds resemble Stamford Marble. A series of unimportant quarries in cream-coloured oolite, with clay layers, lie along the escarpment north of Lincoln.

In Northamptonshire the Lincolnshire Limestone has been divided by W. A. É. Ussher into an upper group, the Hibaldstow Beds, 20 feet; and a lower group, the Kirton Beds, 30 to 40 feet. The former consist of buff and cream-coloured oolites, some fine and some coarse; the latter are grey limestones interstratified with clay bands; they yield no important building stones.

The limestone building stones in the Inferior Oolite Series of Yorkshire are only of local significance. The

LIMESTONES

Grey Limestone, or *Scarborough Limestone*, is in reality a variable set of strata. 'To the south of Scarborough it consists principally of calcareous shales, with thin nodular ironstones, and a little calcareous sandstone. These latter towards the north become more prominent, and over the interior moorlands develop into massive beds of coarse sandstone; while in the south, in the Howarden Hills, they pass into fine-grained flaggy sandstone, and finally die out altogether' (Fox-Strangways). It is sometimes known locally as the 'Pier Stone,' from the fact that it was largely employed in the construction of the harbour pier. This variety is a hard calcareous sandstone, obtainable in large rough blocks, which are incapable of being dressed.

In this region the rock most closely resembling the Inferior Oolite of the West of England is the *Whitwell* or *Cave* Oolite, of South Yorkshire and the Howarden Hills, 20 to 30 feet thick, worked for many years in the Mount Pleasant Quarries, near Whitwell, principally for lime and road metal. This stone, which is a soft white sandy oolite, is blue-hearted when freshly exposed. It has been used in the construction of the dock at Hull, and for the monasteries of Holderness; also for interior work in the neighbourhood of Broughton. When employed for this purpose it was called 'Cave Marble.' The Millepore Bed is of the same age as the Cave Oolite, but is a hard, highly calcareous, siliceous rock; it has been used in the harbour pier at Scarborough.

The main building stones of the Inferior Oolite and corresponding strata are as follows:

Ham Hill Stone is yellow-brown, current-bedded limestone, composed almost entirely of fragmentary shell débris — pectens, oysters, etc. — bound together by a strongly ferruginous, calcareous cement: the metallic iron has been said to reach as high a figure as 14 per cent.

The total thickness of the stone is about 50 feet, but individual beds rarely exceed 2 feet, though some of them attain a depth of 6 feet. Blocks up to 5 tons may be obtained; the 'Grey Beds' are regarded as the best weather stone. The existing quarries have been carried to a much greater depth than the old ones, which date back to Roman times. It has been proved a very satisfactory stone in the country air, and it has been extensively employed in the neighbourhood of the quarries for many miles around for churches and mansions and other buildings, as at Brympton and Montacute. Its employment for paving has been tried, but with unsatisfactory results. The New Travellers' Club is a good example of the behaviour of the stone in Town; at Daly's Theatre it has not behaved so well.

Similar stone, obtained near Yeovil Junction in thin beds, has been used locally by the London and South-Western Railway Company.

Dowdeswell Stone, Dowdeswell, Cheltenham, Gloucester; used in Christ Church, St. Luke's, and other churches in Cheltenham.

Doulting Stone, Doulting, near Shepton Mallet, Somersetshire, is a light brown, buff, or grey stone, uniform in tint and even in texture. At first it is fairly soft, and has an earthy appearance in the fresh fracture; but it becomes paler and harder after exposure, and weathers well. There are two principal varieties of the stone, obtained from separate quarries: the 'Brambleditch,' 'Fine Bed,' or 'Brown Bed'; the other, known as the 'Chalynch,' 'Weather Bed,' or 'Grey Bed,' is harder, and has frequently a purple tinge not seen in the 'Fine Bed.'

Doulting Stone has been used in Wells Cathedral, eleventh and twelfth centuries; Glastonbury Abbey during the same period; also in the restorations and additions to

Exeter, Bristol, Canterbury, Llandaff, and Winchester Cathedrals; Naval Barracks, Portsmouth; Brazenose and Hertford Colleges, Oxford; Holy Trinity Church, Sloane Street, London; the Cheshire Cheese, Surrey Street, Strand, London; Grand Hotel, Cromer; and Paddington Station.

Dundry Stone.—This is a buff or cream-coloured, somewhat shelly, sub-oolitic limestone. The oolitic structure is not evident in the hard specimen, and the calcareous matrix is distributed in irregular, cloudy masses, causing the stone to be sometimes more dense and sometimes more open in texture. It is quarried in the open and mined on Dundry Hill, south of Bristol, where it has been extensively used in the past. The maximum thickness of bed is 6 feet. The beautiful church of St. Mary, Redcliffe; the Mayor's School, Bristol; St. Mary's Church, Cambridge; and St. John's Church, Cardiff, have been built of the stone; and it has been employed in the restoration of the Cathedral, Bristol, and Llandaff Cathedral.

Painswick Stone, from the Painswick Quarry, Avening, near Stroud, Gloucester. It is a rather soft pale cream-coloured oolite. The grains are somewhat smaller than those of Ketton Stone, and there being a good deal of calcareous matrix between them, the surface of the rubbed stone feels smoother. It is a strong, durable stone, appropriately employed for interior work generally — for steps, hearths, chimneypieces, carvings and mouldings. It should not be in exposed or damp situations. It has been used in the staircases of the Houses of Parliament (1852).

Ball's Green Stone, from the place of this name, near Nailsworth, Gloucester, is an oolite very uniform in grain.

The stone is obtained from underground workings

which have been in operation for many years. The maximum depth of the beds is 3½ feet, and blocks up to 7 tons may be obtained. It is recommended for staircases, and is employed in internal church work: altars, screens, and fonts, and for tracery work in windows, etc. It has also been used for chimneypieces and for floors.

Clipsham Stone, from Clipsham, near Charlbury, Oxfordshire. This is an oolite of medium grain, and of a buff or deep cream tint. On the rubbed surface the grains appear opaque and solid, within a fairly well crystallized matrix.

Blocks up to 50 cubic feet can be obtained. It is suitable for copings, sinks, troughs, and for monumental work. It has been employed in Ely and Peterborough Cathedrals; in New College and the Municipal Buildings, Oxford; in Ashley, Exton, Melton Mowbray, Uppingham, and other churches in the neighbourhood; and in several buildings in Stamford.

Bourton Stone, from Bourton-on-the-Hill, Gloucestershire. This is not considered such a good stone as the Painswick or Cheltenham Stones. The 'Yellow Bed' is the best oolite freestone from this quarry; the bed is 6 feet thick, and blocks from 10 to 40 cubic feet are obtainable. The 'Red Bed,' 5 feet thick, is a very hard, coarse, iron-stained stone, suitable for foundations. The 'White Rock' is used for rough inside walling.

Besides the stones mentioned above, the Lower Freestones are quarried at Birdlip, Broadway, Brookhampton, Stroud, Quar Hill, near Horsepools, Haresfield, Silsley Hill, Uley Bury, Temple Guiting, Stanway, Seizincote, Longborough, etc.

The Upper Freestones are quarried at Nailsworth, and the Pea Grits, near Haresfield, Chickley, and Cleeve.

The Chipping Norton Limestone is a very variable stone, employed only locally for building purposes.

Chipping Campden Stone, Westington Hill Quarry. The upper bed, 5 to 6 feet thick, is a hard, brown current-bedded shelly oolite, with closely-packed grains; it is used for planking, covering drains and culverts, and similar rough work.

The 'Yellow Bed,' 5 to 7 feet, is an oolite freestone used for carving.

The 'White Post,' from immediately above the latter, is used for general building purposes; it is 6 to 7 feet thick.

Blockley Stone a fine-grained pale buff stone of close texture; oolite structure is not discernible on the worked face.

Casterton Stone, from Casterton, in Rutlandshire, near Stamford (Lincolnshire) a fine-grained buff or warm cream oolite. It differs from many oolite freestones in that the rubbed surface shows the individual grains intact—not fractured, nor with the centre of the grains dragged out, as is usually the case—this makes the stone harsh to the touch and gives it a matt surface. It has been incorrectly described in some works on this subject as 'Stamford Marble.'

It is employed for general building purposes, and was largely used in ecclesiastical structures from the fifteenth to the eighteenth centuries. It is suitable for tracery, mullions, etc., and may be placed without reference to its original bedding.

It has been used in the New Record Office, Fetter Lane, E.C.; Ely Cathedral; and churches at Barnet, Southgate, Daybrook Church, Nottingham; in various University Buildings in Oxford and Cambridge; in the Technical School and other buildings in Stamford.

Ketton Stone, Ketton, Rutlandshire. Three very distinct varieties of stone are obtained from the Ketton Quarries, but there is only one to which the appellation 'Ketton

Stone' can be properly applied; the others are the Ketton Rag and the 'Crash.'

Ketton Stone is an oolite of moderately fine grain, even texture, and warm cream or buff colour. As in the Casterton Stone, the oolite grains are clearly visible on the rubbed surface; but there is this difference—the individual grains are united by small pegs of limestone with their neighbours, and there is little of the nature of a matrix. These peg-like connections bind the whole mass of grains in a strong stone. It is a splendid stone for strength and durability. The thickness of the bed is 3 to 4 feet; blocks 2 feet to 2 feet 9 inches on the bed and 9 feet long, or from 1 to 150 cubic feet, may be obtained. From the nature of the stone it must obviously be very porous; it is easy to work and hardens on exposure. It is suitable for dressings, mullions, stairs, etc. It has been used in the upper part of St. Dunstan's Church, Fleet Street (1833); in the modern portions of Peterborough and Ely Cathedrals, and parts of the abbey at Bury St. Edmunds, and for the restorations of York Minster; also in Sandringham Hall; Lord Rothschild's enclosure in the Jews' Cemetery, Willesden; and St. Pancras Railway Station (dressings, 1872); No. 75, Jermyn Street (1854); and some of the colleges at Cambridge.

The Ketton Rag is an oolite similar to the Ketton Freestone in colour and size of grain, but very much denser and heavier, owing to the complete development of a crystalline calcite cement, which fills all the interstices between the grains. This calcite cement or matrix has formed large crystals, each embracing many hundreds of the grains of oolite; and it is to these large crystals that the characteristic appearance of shimmering splashes of light (lustre mottling) are present on the smoothed surface of the stone. Thin veins of calcite frequently traverse this stone, and it is to be observed that the oolitic grains are cut through on the rolled surface and do not stand out as in the freestone. The bed is about 3 feet thick. Ketton Rag is said to have been largely employed for monumental work, but the quarry-owner says it is unsuitable for building on account of its behaviour in frost. It may be seen in the Tower of London and York Minster.

The 'Crash Bed,' a coarse ferruginous oolite, full of shell débris, is very soft when first quarried, but hardens extremely on exposure. The colour is purple-brown or reddish-brown, in many shades. It is employed locally for rough fences, etc.

Weldon Stone, Weldon, Northamptonshire, cannot be described in general terms, because the different beds yield stones which vary considerably (p. 211). The Weldon Rag, or Weldon Marble, is a blue shelly limestone, yellow or brown when it has been long exposed; it is very hard and capable of taking a polish. It is used principally for rough work, and for steps and landings. The 'A' and 'A1' beds are both warm cream-tinted, medium-grained oolites, with a tendency to become more or less shelly in places; the former is coarser than the latter, and the current-bedding planes in it are more pronounced. The shell fragments are usually small. The 'B' bed is local in its occurrence, but it is a good fine-grained oolite, soft and easy to work. One of the lower oolitic freestone beds has usually a pink tinge. The lowest bed at the base of the freestone series is shelly, and is known as the 'Marble Bed'; it resembles the Rag. All the freestones are soft, and are sawn out of the bed without the aid of water; they harden on exposure. Blocks may be obtained up to 10 tons.

The stone has been used in the restoration of Rochester Cathedral and of the Chapter House, Lincoln; also in the New University Library, Cambridge, Geddington Cross, and in numerous churches and mansions in the neighbourhood.

Barnack Stone, Barnack Rag, or *Barnack Mill Stone*, formerly obtained in considerable amount from Barnack Mill Quarry, near Stamford, Northamptonshire, is a strong shelly oolite of light brown colour. It occurs in beds 3 or 4 feet thick, and blocks of 3 tons have been obtained. The original quarries have not been open for very many years, but it was employed in very early times in ecclesiastical buildings: King Wolfere is reputed to have used this stone, A.D. 664, for Peterborough Abbey. It was employed also in the abbeys of Bury St. Edmunds, Romsey, and Crowland, Peterborough Cathedral, Burleigh House, and Peterloo College. Many churches in Lincolnshire, Cambridgeshire, and North Suffolk were built of this excellent stone, including Boston, Collyweston, Holbeech, Kettening, Ketton, Moulton, Spalding, etc.

Ancaster Stne (including *Wilsford Stone*, from the old Castle Quarries), Ancaster, near Sleaford, in Lincolnshire. This stone is an oolite that varies somewhat in tint from pale to warm cream or buff, and is occasionally pinkish. Similarly, although it is usually a very fine-grained oolite, some of the beds are coarser and contain comminuted shells. The oolitic grains are not very readily distinguished on the freshly-rubbed face. The total thickness of workable freestone is about 13 feet; the beds are thin, seldom reaching more than $1\frac{1}{2}$ feet, and they tend to run one into the other. Blocks weighing 5 tons can be obtained. The stone works easily and hardens with exposure.

Ancaster Rag, or *Weather Bed*, is a coarse shelly oolite, containing a mixed assemblage of organic fragments and small pellets. It is warmer in colour than the freestone, and tends to have a shelly, mottled appearance. It contains a fair amount of crystalline calcite in patches. It dresses with a good face. The Ancaster Stones are suitable for all the ordinary building purposes; the whiter kind is considered the best for carving and moulded work. In all cases the best stone is that which is obtained from beneath a good cover of clay. It has been employed in Lincoln Cathedral, Belvoir Castle, Wollaton Hall, Belton House, Boston Church, restoration of Balmoral Castle, the newer parts of Lincoln's Inn, St. Pancras Station (dressings), St. Albans Abbey, Whitehall Cellars. The Weather bed has been used at the Surveyors' Institute and in the gate-posts, Lincoln's Inn.

Haydor Stone, Haydor Quarry, Grantham, Lincolnshire, is a buff fine-grained oolite, akin to the Ancaster Stone in general character and in the uses to which it may be put. The matrix in places is formed by large crystals of calcite, which embrace many oolite grains; in this it resembles Ketton Rag. It has been used in Belvoir Castle, Culverthorp House, Lincoln Cathedral, and in numerous churches, including those of Boston, Grantham, and Newark.

The Great Oolite Series.

The Great Oolite Series in the south-west of England includes, in descending order, the following: Cornbrash, Forest Marble and Bradford Clay, Great Oolite and Fuller's Earth. The lowest of these formations is a greenish-grey clay or marl, with one or more inconstant bands of soft limestone near its upper limit. The Fuller's Earth Rock, a blue marly limestone, is of little economic significance. The Forest Marble is an irregular calcareous and argillaceous deposit, extremely variable from point to point and of little commercial value. The more noteworthy quarries will be mentioned later.

The middle member, the Great or Bath Oolite, as the

source of all the varieties of the well-known *Bath Stone*, merits considerable attention. It consists of beds of hard shelly limestone, oolitic freestone, limestones with more or less calcareous mud, marl, and thin sandy flags.

Before examining the characters of the Great Oolite in detail, it may be well to return for a moment to the Great Oolite Series, and note the conditions under which it was formed. It is clear from the nature of the fossil contents of these rocks, as well as from the stratigraphical evidence, that in the south-west in Dorsetshire, Wiltshire, Gloucestershire, Oxfordshire marine conditions prevailed; and the oolite and shelly limestones and their accompanying marls and sandy layers were laid down on the floor of a shallow sea, in which currents played an important rôle. When, however, we trace the strata north-eastward, across Northamptonshire and Lincolnshire into Yorkshire, we find them passing more and more completely into estuarine formations. Thus the beds which are well developed in one locality thin out and disappear as they are followed across the country, or they pass gradually into others possessing very different characters. The Fuller's Earth is traceable from Dorsetshire to Gloucestershire. The Great Oolite is found only in Wiltshire, Somersetshire, Gloucestershire, and Oxfordshire; eastward it passes into the Great Oolite Limestone. Similarly, the Forest Marble and Bradford Clay begin to change in parts of Oxford and Northants into what is called the Great Oolite Clay. The Cornbrash alone appears constantly at the top of the series, from Dorsetshire to Yorkshire, varying in thickness from 10 feet or less up to 25 feet. It yields a rubbly iron-stained limestone, which is rarely of the least use as a building stone, although numerous small quarries have been dug out of it; at Upton Noble, Somerset, hard limestone is obtained suitable for rough building.

LIMESTONES

There is everywhere an intimate relationship between the Fuller's Earth and the Great Oolite, and a gradual passage from the one to the other. The sporadic appearance of the Fuller's Earth rock was an early indication of the great calcareous development that was soon to follow. The Great Oolite generally seems to have commenced with somewhat sandy calcareous deposits, which now very often exhibit fissile, slabby, or shaly characters. These fissile beds in Oxfordshire, about Stonesfield, and in neighbouring parts of Gloucestershire, form an important series, which has been dignified by the special name Stonesfield Slates.

The overlying Forest Marble is less closely related to the Great Oolite. In some places it overlaps the latter, and in others it has clearly been formed, in part at least, at the expense of the Great Oolite, which has been worn away by the waters which deposited the Bradford Clay at the base of the Forest Marble.

Returning now to the Great Oolite, we find it best developed in the neighbourhood of Bath, whence it derives the appropriate name of Bath Oolite. Its presence here as a massive formation has been the determining factor in the shaping of all the hills around the city, and strong features are made by it northward and westward in the direction of Minchinhampton and Cirencester; but within seven miles south of Bath, a little beyond Bradford-on-Avon, it has entirely disappeared. The southerly disappearance may be explained partly by the thinning out and modification of the beds, as they are traced laterally, and partly by a local erosion of the beds in this direction during and prior to the deposition of the overlying Forest Marble and Bradford Clay.

In the Bath and Bradford districts the following broad divisions may be recognized:

                                                                Feet.
Forest Marble and underlying Bradford Clay.
             ⎧ Upper ⎧ Upper Ragstones: Oolitic and
             ⎪   ⎨   shelly limestones ... ... 12 to 50
  Great   ⎨   ⎩ Oolitic Freestones ... ...   8 to 30
  Oolite   ⎪ Lower ⎧ Lower Ragstones: Shelly and
             ⎩   ⎨   marly limestone, and beds of
                     ⎩   more fissile oolite ... ... 10 to 40
Fuller's Earth.

No general section, however, can convey an adequate idea of the variability that is to be met with in the numerous quarries and mines; a few representative sections will, therefore, be given from different parts of the district.

The most southerly-situated workings in the Great Oolite are those immediately south and east of Bradford-on-Avon.

At Upper Westwood (Avoncliff or Ancliff) the section is as follows:

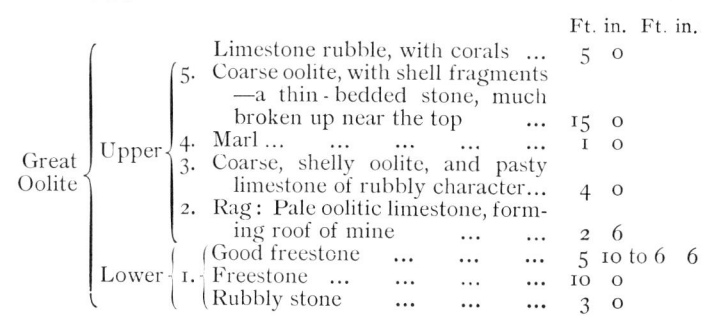

                                                    Ft. in. Ft. in.
           ⎧   ⎧   Limestone rubble, with corals ... 5 0
           ⎪   ⎪ 5. Coarse oolite, with shell fragments
           ⎪   ⎪   —a thin-bedded stone, much
           ⎪   ⎪   broken up near the top   ... 15 0
           ⎪ Upper ⎨ 4. Marl 1 0
  Great   ⎨   ⎪ 3. Coarse, shelly oolite, and pasty
  Oolite   ⎪   ⎪   limestone of rubbly character... 4 0
           ⎪   ⎩ 2. Rag: Pale oolitic limestone, form-
           ⎪   ing roof of mine 2 6
           ⎪   ⎧ ⎛ Good freestone ... ... 5 10 to 6 6
           ⎩ Lower ⎨ 1.⎨ Freestone 10 0
                     ⎩ ⎝ Rubbly stone 3 0

Here the freestone contains ochreous galls in places in the lower part, but it is on the whole very uniform and free from open joints.

It is worthy of remark that quite near the above locality, in the limekiln quarry on the east side of the Farm Road, the freestone is only 6 to 7 feet thick, and

LIMESTONES

is so markedly current-bedded as to be of comparatively little service as a building stone.

At the old quarry known as Murrel (or Murhill), near Winsley, the upper division consists of about 24 feet of mixed limestone and clay bands resting on 7 feet of Rag; the lower division has 10 feet of freestone (soft in the lower part), and below this 43 feet of Rag beds, with layers of freestone and occasional partings of clay.

On Farleigh (Farley) Down, north-west of Monkton Farleigh, we find beneath a capping of Bradford Clay:

		Ft. in.	Ft. in.
Upper Division	Hard, shelly, oolitic limestone	15 0	to 20 0
	Flaggy, current-bedded, white oolite, passing down into the shelly oolitic Rag bed	3 0	to 4 0
	Hard, compact, and oolitic limestone, with ochreous and sandy galls	1 0	
	Roof Bed: Coarse oolite	1 0	to 2 0
Lower Division	Pale oolitic freestone ... about	24 0	

The freestone here used to be divided by the quarrymen in descending order as follows:

	Ft. in.	Ft. in.
Capping (fine-grained oolite, used for carving)	1 6	to 1 8
Grey bed	1 8	to 3 0
White bed (used for carving)	10 0	
Hard weather bed	2 6	to 3 0
Red weather bed	5 0	to 9 0

Some of these beds show curved bedding, others do not.

On Combe Down we find:

		Ft. in.	Ft. in.
Upper	Ridding or rubble of limestone	10 0	to 12 0
	Rag beds		
Lower	Picking beds (used for ashlar)	22 0	
	Cockly bed (with fossils)		
	Freestone	10 0	to 12 0

The Rag, Picking, and Cockly beds are here smooth-jointed limestone, more or less oolitic. The freestone is current-bedded on a small scale, and in places is divided into three weather beds and a bottom bed. At Odd Down 2 to 3 feet of ridding rest directly on the lower division, which consists of 10 to 15 feet of Best Stone (freestone), in five or six beds, on 15 to 20 feet of Bastard Stone (Lower Rags).

In the vicinity of Box, south of the railway, the general arrangement is as follows:

		Feet.
Upper	Rubble	
	Fine, pale, and fissile oolite, much current-bedded and occasionally marly, with harder beds known as *Scallett* at or near the base	20 to 35
	Coarse, shelly oolite, including a bed known as the *Scallett Rag*, or White Rag; in some places this group is divided into Black, White, Malmy, and Red Rags	5 to 10
	Corn Grit: Current-bedded oolite	15 to 20
	Roof Bed: A hard, coarse, shelly oolite ...	3 to 5
Lower	Ground Bed: The principal oolitic freestone	12 to 14
	Stone Beds: Various (not worked here) ...	30 to 40

Fuller's Earth.

At Corsham, where the freestones are worked on an extensive scale, they are approached by inclined shafts or tunnels through the overlying Forest Marble. In different workings the freestones—in several beds—vary in thickness from 10 to 25 feet; they are struck at various depths from the surface down to 100 feet; and the 'Corsham Stone' is believed to be beneath the horizon of the Box 'Ground.' The foregoing details will suffice to show the uniformity in general character and the variability in local phases of the Great Oolite of the Bath district.

LIMESTONES

Following the Great Oolite no farther northward than Tetbury and Minchinhampton, considerable changes will be observed to have set in, as the following generalized statement will show:

			Feet.
Upper	Kemble Beds	Thin, even beds of oolite, with alternating bands of mud. False-bedded oolites, becoming freestones in places, with marly and sandy layers	30
	White Limestone	White oolite and sub-oolite limestone, occasionally current-bedded; some beds of Dagham Stone	10 to 24
Lower	Bath Freestone and Stonesfield Slate	Freestone, current-bedded oolite, and ragstone 'black rock' ... ... Grey limestones, with white oolite grains ... ... Sandy beds, passing in places into 'Stonesfield Slate'	40

In this district the top of the Great Oolite and the Forest Marble show so many features in common that difficulty is experienced in separating the two formations. The freestone of the lower division is a fine-grained buff rock, 12 feet or more in thickness, in six or seven beds, near Tetbury Station; while in Vaze's Quarry, northward of the village, the Kemble Beds yield 10 to 12 feet of shelly oolite, employed for rough walling; and beneath them, still in the same group, hard and soft beds of buff and white oolite—current-bedded oolitic freestone (10 to 12 feet).

On Minchinhampton Common, or Amberley Heath, between Minchinhampton and Stroud, the Great Oolite has been quarried for many years. The following section must be regarded as only approximately correct in detail, since the beds are rather variable:

			Feet.
Upper = White Limestone	Rubble		5 to 6
	Dry-Wall Stone: Thin-bedded, pasty limestone and sandy beds		
	Hard, pale, smooth limestone, with scattered oolite grains		1
	Planking: Current-bedded oolite, pale buff or white; rather shelly		8
	Thin, shaly paving		
	Soft White Stone: Current-bedded, white, shelly oolite		4 to 6
Lower	Weather Stone: Current-bedded, shelly limestone, made of comminuted shells, with beds full of whole shells (Lima)		12 to 15

It is noteworthy that here the presence of shells in the stone is a favourable feature.

In the neighbourhood of Cirencester the sequence of beds is usually as follows:

		Feet.
Upper	Kemble Beds: Current-bedded oolites and marly beds	10 to 30
	White Limestone: White, pasty, and oolitic limestone	8 to 10
Lower	Current-bedded and fissile oolite, with calcareous sand at the base.	

Numerous quarries, none of importance, exist in this district. Between Northleach, Burford, and Shipton-under-Wychwood the Great Oolite presents the following general sequence:

			Feet.
Great Oolite	White Limestone	White, shelly limestones, more or less oolitic, with occasional beds of marl and brown limestone	12
	Marly Beds	Marl, with bands of oolite and clay	18 to 25
	Freestone	Pale, flaggy, and shelly oolite, and current-bedded, white or blue limestone and marl ...	12 to 20
		Freestone: Obliquely-bedded oolite, coarse and slabby in places.	
	Stonesfield Beds.	Stonesfield Slate: Fissile, sandy limestone and marl ... about	4

LIMESTONES

Some years ago the stone-mines and quarries of Windrush, two and a quarter miles west of Taynton, produced a freestone of excellent quality, and doubtless much good stone is still available in the neighbourhood, although it is little worked at the present time. Here the upper division consists of about 15 feet of rubbly limestone, sandy marl, clay, and marlstone; and the lower division is a series of beds of firm white oolitic freestone, alternating with shelly ragstones. The thickness of the beds is 15 feet, but the workable freestone is not more than 11 feet; this stone can be obtained in large blocks. Good blocks may also be obtained from a quarry north of Barrington Park, from a current-bedded oolite freestone about 7 feet thick. North-west of Burford Signett a hard brown, shelly oolite, from the upper division, has been employed for building purposes; and north-west of Swinbrook, east of Burford, 9 feet of white false-bedded oolitic freestone occurs in thick, much-jointed beds.

The old shallow quarries on Taynton (Tainton) Down are now disused; formerly much valuable stone was obtained therefrom. The freestone here belongs to the lower division of the Great Oolite, and at Milton-under-Wychwood it lies beneath a thick cover (36 feet) of the upper division—White Limestone and Marl beds—some of which are employed for lime and roadstone. The quarries are situated about midway between Burford and Milton (p. 238).

Near Chadlington, Sarsden, and Chipping Norton there are several small quarries in which the Great Oolite Beds may be seen resting on those of the Inferior Oolite.

It has been previously pointed out that differences exist between the strata representing the Great Oolite Series in the south-west and in the central and eastern outcrops. The series in Northamptonshire, Buckinghamshire, Bedfordshire, and Lincolnshire consists, in descending order, of the Great Oolite Clay (Blisworth Clay), Great Oolite Limestone, and Upper Estuarian Series. The total thickness of the series in this area ranges from 30 to 100 feet.

With the upper and lower of these divisions we are not concerned, since they contain no limestone of importance. The Great Oolite Limestone presents a very variable assemblage of calcareous beds; many are soft, white, and marly; some are shelly and laggy, resembling the Forest Marble. *Oolitic structure is rare*, although it is occasionally met with. The limestones are thickest and best near the base, where they are very hard and blue in colour when unweathered. In parts of Lincolnshire some of the beds are so iron-stained as to resemble the Cornbrash. Near Bedford the Great Oolite is 25 to 30 feet thick; near Northampton it is 25 feet; and in Lincolnshire it varies from 12 to 20 feet. There is no sharp limit to either the top or bottom of the Great Oolite.

No first-class building stones are obtained from the Great Oolite here; but for local purposes in rough walling sufficiently good stone is got from a number of small quarries—*e.g.*, an oolite and earthy limestone at Deanshanger; thin beds of earthy and oolite limestone at Great Linford; thin blue limestone at Cox's Quarry, Bedford; oolite and non-oolitic beds in Warrington Quarry, north of Olney; oolite and shelly beds at Stowe-Nine-Churches. From Blisworth is obtained a flaggy limestone, 'Pendle,' 5 feet; an oolitic limestone; 'blocks' used for window-sills, etc., 2 feet 9 inches; and several thin beds of hard blue stone.

The section exposed in the quarries at Moulton Park, near Kingsthorpe, may be noted here, although the stone is mainly used for lime; there are several beds of rough grey limestone, some 3 feet thick. This series is found in Franklin's Pit, Bedford.

LIMESTONES

		Ft.	in.
	Soil and boulder clay	3	0
Bastard Beds	Rubbly, fossiliferous marl	1	2
	Pale, earthy, slightly oolitic limestones	0	5
	Marly clay	1	2
	Earthy and shelly limestone	0	2
Great Oolite	*Pendle:* Earthy and marly limestone	1	6
	Clay	0	4
	Paving Stone: Shelly, earthy, oolitic limestone	4	0
	Jubs: Marly and oolitic limestone, with veins of clay and marl.		

Upper Estuarine Series: Blue and greenish marly clay.

The Paving-stone Bed used to be sawn up for hearthstones, flues, flooring, window-sills, chimneypieces, etc. Near Oundle several beds have been employed for building purposes; most of them are hard, blue, shelly stone, of the nature of Rag.

Nine feet, or thereabouts, of good stone is quarried at Benefield, where it resembles Forest Marble. An interesting feature of the Great Oolite at Alwalton and Castor is the presence of oyster-beds. These beds, with a hard blue limestone, form the steep escarpment of Alwalton Lynch. The 'Alwalton Marble,' as this stone was called, was formerly dug along the scarp; it took a good polish, but was not very durable.

West of Saxby, Owmby, and Normanby the Great Oolite Limestone furnishes fairly thick beds, capable of being used for building stones; but there is a tendency for the stone to develop irregular fissibility, a character which is accentuated as the outcrop is followed northward.

When the Great Oolite Series is followed northward into Yorkshire, it is found to be represented by the Upper Estuarine Series and Cornbrash alone, and in this there are no limestone freestones to be noticed. (See Sandstones.)

The building stones of the Great Oolite Series are as follows:

The Bath Freestones.—

Box Ground or *St. Aldhelm Stone.*—This stone, known for long under the former title alone, has been quarried for centuries on and about Box Hill. The second name has been introduced by the owners more recently on the strength of a legend associated with St. Aldhelm, mentioned by John Aubrey in his 'Description of Wiltshire,' who says: 'Haselburg Quarrie is not to be forgott; it is the eminentest freestone quarrey in the West of England, Malmesbury and all round the country of it. The old men's story that St. Aldhelm riding over there, threwe down his glove and bade them digge, and they should find great treasure, meaning the quarry.'

Most of the stone at Box is now mined.

Box Ground is light brown in colour when fresh; it turns paler on exposure. It is somewhat finer-grained than many of the other Bath stones, and the oolitic grains themselves are fairly close together (Plate VI.). The main body of the Box Ground stone is 12 to 14 feet thick, and good beds range from 9 inches to 4 feet in thickness. Blocks weighing 6 tons may be obtained in good lengths. The weight per cubic foot has been given as 123 to 127·9 pounds, the average being probably about 126 or 127. It is, perhaps, the best of the Bath weather stones, and is recommended for exposed positions, sills, strong courses, plinths, cornices, and damp places near the ground.

In London it has been used at Apsley House, Hyde Park; Passmore Edwards' Hall, Clare Market, Strand; Albert Mansions, 90 to 120 Victoria Street; and at University College and Lloyd's Bank, Oxford; Longleat House, near Warminster (sixteenth century); Witley Court, Bow; Corsham Court, Waddesdon Manor, Aylesbury; Lacock Abbey (thirteenth century); Malmesbury Abbey (eighth century); Christ Church, Richmond Road, Surrey.

PLATE VI

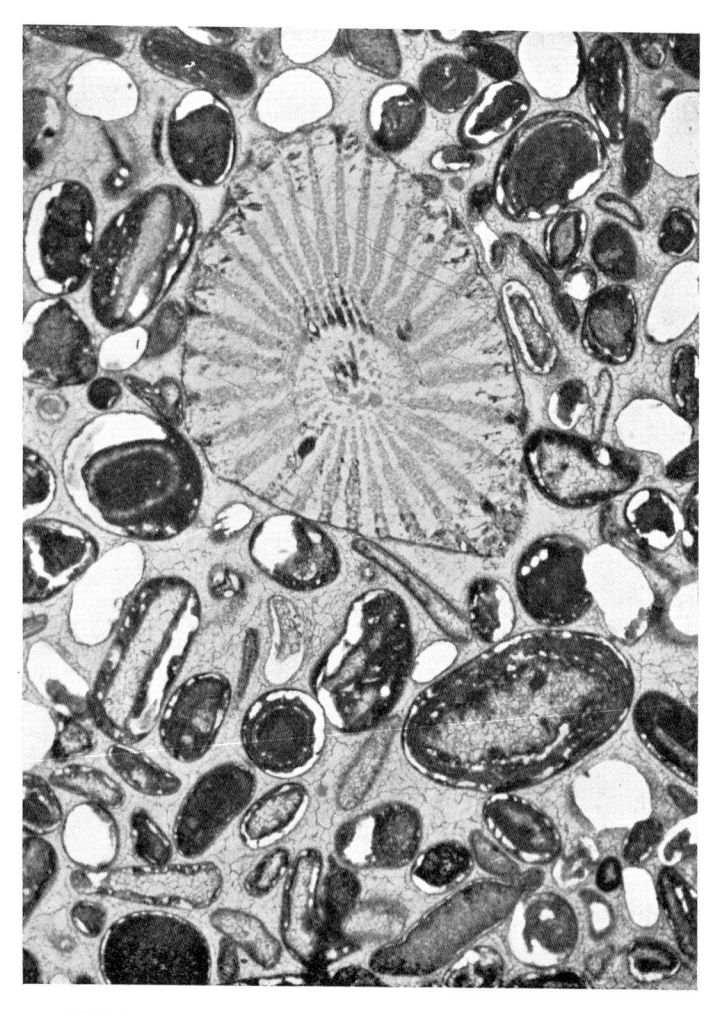

PHOTOMICROGRAPH OF LIMESTONE. (Magnification, 30 diameters.)

Bath Oolite, Box Ground.

The Scallett Bed at Box is not much worked, on account of the expense, but it is said to take carving very well. Overlying the hard Rag Beds, which form the roof of the Ground freestones, and below the Scallett, is the Corn Grit.

The *Corn Grit* is a rather earthy and shelly oolite, in colour resembling Corsham Stone; its maximum thickness is 4 feet. Although it is a very sound, strong stone, it is incapable of withstanding exposure; but it is suitably employed for interior landings, steps, columns, staircases, and for engine beds. The cost of labour on this stone is greater than that for Corsham. The weight is 130 pounds per cubic foot, and the crushing resistance 110 tons per square foot.

Corsham Down Stone, from Corsham Down and Corsham Ridge, and *Monk's Park Stone* are both moderately fine-grained oolites of even texture, and cream-coloured or 'light-stone'-coloured. These freestone beds attain a total thickness of from 15 to 25 feet; individual beds reach 4 feet, and blocks up to 12 feet can be obtained from Corsham, and somewhat longer from Monk's Park. Blocks of 100 cubic feet, weighing about 7 tons, are sometimes hauled out of the mines (about 16 cubic feet to the ton). These stones are suitable for all kinds of inside work; they carry a sharp arris very well, and may be employed in outside facings, mullions, jambs, tracery, and heads.

The Corn Grit bed is present at Corsham.

Corsham Down Stone has been used in the Marlborough Hotel, Bury Street, St. James's, W.; the London and South-Western Bank, Finsbury Pavement; Brook House, 10 to 12, Wallbrook Street; St. Matthew's Church, Fulham; St. Matthew's Church, Hull; the G.P.O., Lowestoft; the Priory, Dover; Truro Cathedral; the Central Free Library, Cardiff.

Monk's Park Stone has been employed in St. James's Church (R.C.), Spanish Place (interior and exterior); the Polytechnic Institute, Borough Road, Battersea, S.E.; Salisbury House, Finsbury Pavement, E.C.; Municipal Buildings, Reigate; Town Hall, Ilford, E.; Post Offices, Bournemouth and Bristol; St. Agatha's Church, Stratford Road, Birmingham; Town Hall, Cape Town, South Africa.

Hartham Park Stone, from the mines at Upper Pickwick, near Hartham Park, is a fine-grained oolite of even texture and warm yellow colour, which has a very pleasing effect. It has been employed in the Parkside Flats, Knightsbridge; in Harley House, Marylebone Road, W.; and in the Art Gallery, Bristol.

Farleigh Down Stone, from Monkton Farleigh (Farley), is a fine-grained, even-textured oolite of a warm cream tint. It is the cheapest of the Bath Stones. As its weather-resisting properties are not of the best, it should not be employed outside, except for flush work, and thennever near the ground, but for all inside work it is quite satisfactory. The best stone can be got in blocks 4 feet 6 inches thick, but it does no run to such good lengths as Corsham or Monk's Park. Its weight is given as 122 pounds, and crushing resistance 62·5 tons. This stone has been used in Harrington Buildings, Queen Street, Cardiff; Taff Vale Railway Station; and in many small churches and chapels.

Combe Down Stone, which has been quarried for many centuries at Combe Down and Odd Down, south of Bath, is a pale brown oolite of medium grain. It is liable to contain thin veins of calcite and small iron-stains, but these, though they interfere slightly with the ease of working, do not detract from its weather-resisting properties, which are good; it is therefore suitable for plinths and projections, and similar outside work. The weight per cubic foot is 128·6 pounds, and the crushing resistance is 117 tons. There are from 10 to 12 feet of freestone here, and the beds are from 1 foot to 4 feet 6 inches thick; only moderate lengths can be obtained.

Combe Down Stone has been used in Buckingham Palace (along with Portland); in the Roman Baths, Bath (second and third centuries); Prior Park Mansion, Bath; St. Mary's Parish Church, Torquay; Church of Our Lady of the Assumption, Cambridge; St. George's Church, Kidderminster; and the restoration of Henry VII.'s Chapel, Westminster (1819).

Bradford Stone, from Bradford-on-Avon, is a light brown medium-grained oolite, obtainable in beds up to $3\frac{1}{2}$ feet. Like the other stones from this part of the stonefield, the *Winsley Ground*, *Stoke Ground*, and *Westwood Ground*, it is a fair weather-stone; and it may be noted here that the stones from this part of the district are all warmer in tint than those from farther north. Bradford Stone has been employed at Manchester, in the Guildhall Chambers, Lloyd Street, and the Chambers, 16 to 20, Kennedy Street; in the Nurses' Institute, St. Thomas's Hospital, and the Congregational Church, Caterham.

Stoke Ground Stone, from Limpley Stoke, is a light brown oolite of medium grain and somewhat open texture. Weight, 120 pounds; crushing resistance, 90 tons. The best bed is 6 feet thick, and yields blocks of moderate length up to 40 cubic feet. This stone will take carving.

It has been used in the corner block of offices at 9, Billiter Square, E.C.; the Hospital, West Heath, King's Norton; Grand Hotel, Wentworth Street, Peterborough; Gwyn Hall, Neath.

Winsley Ground.—The best bed at Winsley is a medium-grained oolite of a warm brown tint. The beds run up to 4 feet in thickness, giving blocks of medium length. The stone is not much worked at the present time.

Westwood Ground.—The best bed from the Upper Westwood and Avon Cliff (Ancliff), is a light brown oolitic weather-stone of close texture. The weight is 130 pounds; crushing resistance, 110 tons. The thickness of the bed is 6 feet, and good blocks can be obtained up to 40 cubic feet.

Other Great Oolite Stones. — Stone was formerly quarried on a considerable scale on Bathampton Down, east of Bath; it was used in parts of Windsor Castle and in Bowood House. It was known as the *Baynton* or *Hampton Stone*, but it must not be confused with the Hampton Stone from the same geological horizon at Minchinhampton. The Minchinhampton 'Weather Stone' is a hard, non-absorbent stone, oolitic and shelly, and in places sandy. It is somewhat coarse in texture, and liable to variation, but when carefully selected and dried in the open before use it is a very durable stone.

Windrush Stone, of which there are two principal varieties, known as Windrush Hard and Windrush Soft, is a cream-tinted, even-grained oolite. The hard beds are more shelly that the soft ones and are considerably heavier. This stone is practically unworked at the present day; it was formerly mined. Examples of its use may be seen in Windrush House and Barrington House, Gloucestershire.

Taynton Stone, from Taynton (Tainton, Teynton), near Bourton-on-the-Water, Gloucestershire, was formerly worked in a number of small quarries, but it is now employed only for the estate. It is a soft, coarse-grained, shelly oolite, with strongly marked current-bedding. It is obtained in two colours— No. 1, cream; and No. 2, warm orange-yellow. The same stone is now worked near by, at Milton-under-Wychwood and Swinbrook, in Oxfordshire, whence it is still sold as Taynton Stone, though it is sometimes called *Milton Stone*. The depth of this bed, both Nos. 1 and 2, is $6\frac{1}{2}$ feet, and blocks can be obtained up to 100 cubic feet (50 to 60 cubic feet is the average large-size block). It can be sawn as desired at the quarries. The weight per cubic foot is about 136 pounds, and the crushing resistance 150 tons.

Many of the oldest buildings in Oxford, of the twelfth, thirteenth, and fourteenth centuries, still remain to show the good qualities of this stone. It has been used in the ancient parts of Christ Church Cathedral, in Magdalen College and School, in Merton College and Chapel, Mansfield College, St. John's College, and Blenheim Palace. It is also to be found in London, Birmingham, Southampton, etc., interior of St. Paul's Cathedral, Byland Abbey.

Guiting Stone, quarried at Temple Guiting and Guiting Power, Gloucestershire, is a soft, porous, shelly oolite of a strong orange-yellow tint, and with prominent current-bedding. The weight per cubic foot is 130 pounds. The beds run to 6 or 7 feet in depth, and blocks of any reasonable size may be obtained. The stone hardens on exposure, and it weathers very well. Care should be taken, however, to select blocks without soft layers in the current-bedding, for portions of a building exposed to friction. It has been considerably used of late as a dressing in conjunction with red brick, and examples will be found in Sudeley Church and Queen Catherine's tomb in the interior; Sudeley Castle; the parish church, Winchcombe, Hayles Abbey, and other churches in the neighbourhood.

Puddlecote Stone, *Charlbury Stone* (Oxfordshire).—These two oolitic freestones are practically extinct; formerly they were largely employed in the neighbourhood. In many twelfth-century buildings in Oxford the Puddlecote was used. These stones are medium-grained oolites, of a warm yellow or buff tint.

Denton Stone (Denton Blue and Denton White), obtained from Denton, south-east of Northampton. The 'White' is a yellowish shelly limestone, without any evident oolitic structure, but with a mottled appearance, caused by irregular nodular masses. The 'Blue' is a similar stone, of a bluish or greenish-grey colour, with darker and lighter irregular confused markings.

The other sources of Great Oolite building stone may be briefly recapitulated here: they are Tadmarton, a close-grained oolite; Shalstone, a hard white shelly oolite; Blisworth, an oolite sawn into sills, jambs, chimneypieces, etc.; Oundle; and Gedlington Chase.

Building Stones of the Forest Marble.—Except for purely local purposes in rough walling, flooring, pitching, and the rougher farm-buildings, this stone is not employed. These limestones are shelly and oolitic, often bluish in colour, and appear in irregular thin beds. 'Stone planks' or slabs, up to 4 feet square, and from 3 to 5 inches thick, are obtainable in some places, but the stone is always difficult to square. There are many small quarries in the Forest Marble, and it is frequently exposed in quarries worked for the underlying Great Oolite. The more noticeable quarries are those at Bothenhampton, 6-feet bed; Long Burton, 12 to 15 feet, in beds from 2 to 9 inches thick; Bowden, 20 feet of very shelly and slightly oolitic limestone, the so-called *Bowden Marble;* Wincanton and Redlinch; North Cheriton; Charleton Horethorpe; Bratton; Wanstrow. About Frome there are numerous quarries working 7 to 15 feet of shelly current-bedded oolitic limestone. It is locally known as 'Frome Stone,' and many houses in the town are built of it; it weathers well. At Norton Bridge the Forest Marble formerly yielded a good deal of building stone, and the quarries were probably worked in Roman times. The section here is of interest. It is as follows: Beneath a cover of 5 feet of Cornbrash the Forest Marble consists of 2 feet of clay and marl, followed by 12 to 15 feet of grey and buff shelly and oolitic limestone and close-grained, gritty, calcareous rock, current-bedded, and very hard in places and irregular in thickness; these are recognized, in descending order, by the quarrymen as Slatt, Strong Lime, Planking Vein, Soft Bed, Freestone, good Weatherstone, Building Stone, and Paving (ripple-marked). Other quarries are at Cirencester, Kirtlington, Bicester, Lillington Lovell, Bletchington, Epwell.

Building Stones of the Cornbrash.—As already indicated, the Cornbrash is by nature unfitted for building purposes; nevertheless, it has been used at Radipole, East Coker, near Templecombe, and Upton Noble and a few other places, for local rough work. The stone so employed is usually a tough shelly limestone, incapable of being dressed.

Corallian.—The Corallian beds do not yield any limestone building stone of great importance; nevertheless, there are several calcareous freestones that deserve attention. Some are very fair stones, that might be better known; others have been employed more extensively than their qualities merit.

This series is represented by beds of oolite, marl, ragstone, clay (Lincolnshire), sandstone, and sand. The only noteworthy freestone in the southern outcrops are the Osmington Oolite of South Dorset, the beds on the same horizon at Marnhall and Todbere, in the northern part of the county, and at Calne, in Wiltshire.

The *Osmington Oolite* is best exposed on the coast between Ringstead Bay and Osmington. Here it consists of nodular limestone beds, with intercalated layers of oolite, more or less shaly; at the base is a stout bed of hard calcareous grit. The same formation is seen in the form of current-bedded oolites at Black Head and Sandsfoot, near Weymouth. It has been quarried in a small way at Linton Hill, south-east of Abbotsbury; here blocks can be obtained up to 3 feet by 2 feet by 1 foot.

The stone is a medium-grained, warm yellowish-brown oolite, the oolitic grains being cemented in a matrix of crystalline calcite and calcareous paste. It has been used a good deal in the village of Abbotsbury, where its warm colour has a very pleasing effect, and from the appearance of St. Catherine's Chapel, which stands on a very exposed position, it seems to wear very well.

Marnhull Stone.—This is a buff marly oolite, in parts shelly, occurring in regular beds 1 foot to 1½ feet thick, with marl partings between; 15 feet or more of the beds are shown in the quarries. The stone hardens and whitens on exposure. Blocks up to 20 cubic feet have been obtained. It has been used in Gillingham Church, Hinton St. Mary, Sutton Waldron, etc.

Todbere Stone.—The character of the beds in this quarry is shown by the following section:

		Ft.	in.
	1. Rubbly oolite, shelly in places	6	0
	2. Buff, flaggy, oolitic limestone, shelly in places	5	0
Corallian	3. Hard, blue, shelly, oolitic limestone, weathering brown	2	6
	4. Rotten oolite and marl	0	6
	5. Current-bedded oolites	9	0

Beds Nos. 2 and 5 furnish good building stones; the full thickness of the lower freestones is 14 feet.

Calne Stone—This is a coarse oolite, frequently shelly and current-bedded; it comes from the upper portion of the Corallian.

In the quarry at Staple Ashton, Wiltshire, 6 feet of marly oolite and pisolite rests on 6 feet of current-bedded oolite, which is used for building, and may be obtained in large blocks.

Oolitic freestone has been worked in a small way at Goat Acre, north of Hilmarton, and at Preston and Purton. At Faringdon a blue-hearted oolite, 3 feet thick, is used as a building stone.

LIMESTONES

Headington Stone ('Shotover Limestone').—The quarries at Headington and Bullingdon, near Oxford, in the Upper Corallian, illustrate very well the difficulty of obtaining regular supplies of good stone from a series of very valuable beds. Here all the beds are very changeable; current-bedded shelly limestone passes laterally within a short distance into rubbly coral rock, or the hard blue or grey oolite limestone passes into an oolitic and shelly sand. The hard grey limestone is the best; the beds range from 5 to 12 feet in thickness, but are very inconstant. The upper earthy limestone, mainly composed of shell débris, without good cementing matrix, has been the most extensively employed in the past, with bad results, in many of the college buildings in Oxford. Wadham seems to have stood the best.

Wheatly Stone is exposed in large quarries north of Wheatly, near Oxford. The best stone is a hard shelly limestone, occasionally oolitic, and not unlike some varieties of Cornbrash. Fragments of mixed shells and other organic remains, together with pellets of calcareous mud, are held in a matrix of crystalline calcite. Like the Headington Stone it is very variable; the hard beds in places give place to others, in which the shell débris is almost entirely devoid of matrix. It was formerly used in Oxford with rather better results than in the case of the Headington Stone.

At Wicken, near Upware, Cambridgeshire, a pale cream, blue-hearted, current-bedded oolite is quarried from a face of about 12 feet.

A coarsely oolitic rock from the Upper Corallian Beds has been used for building purposes from a quarry northwest of Highworth.

The Corallian building stones in Yorkshire are mainly siliceous in character; some are highly calcareous sandstones, others are siliceous sandstones, and there are some curious intermediate forms.

The Passage Beds, or Greystones, and some of the harder limestones in the Lower Limestones are used for rough work. This is true also of some of the beds in the Upper Limestone, in which there are many quarries; but the most important limestone in this division is the Hildenley Stone.

Hildenley Stone is a peculiar local bed in the Upper Limestone division; it is an extremely fine-grained cream-coloured, or greyish limestone, not unlike a compact form of chalk. It is quarried at Hildenley, near Malton. For carved work it is by far the best stone in this district, more especially for interior work. It was employed in the interior of Kirkham Abbey, where, as Mr. C. Fox-Strangways says, ' The fine state of preservation of the mouldings in these ruins, which have been exposed to the weather for centuries, testifies to its durability.' In Malton Old Abbey and in the chapel at Castle Howard the same stone was used. The thickness of the bed is about 4 feet. Occasionally a small oyster-shell in a silicified condition interferes with the free use of the tool in carving.

The Upper Calcareous Grit contains thin-bedded argillaceous limestones in the southern part of its outcrop; these are quarried in the neighbourhood of North Grimston, Langton, Birdsall, and near Kilburn, principally for cement, but also for farm-buildings and similar rough work, for which it is convenient, as it breaks into short slabs. A rather sandy variety of this stone has been used in Wharram Church; but it is not to be recommended for better-class buildings.

Portlandian.—Portland Stone is obtained from the formation known by geologists as the Portland Beds or Portlandian Series, from their great development and clear exposure in the Isle of Portland.

The rocks of this age are not confined to Portland, for they are splendidly exposed in the fine cliffs of the 'Isle' of

LIMESTONES

Purbeck, in South Dorsetshire, and the outcrop may be followed into Wiltshire, Oxfordshire, and Buckinghamshire. They occur beneath younger rocks in North Dorsetshire, Sussex, and Kent, and probably through parts of Bedfordshire and Cambridgeshire. Representatives of these rocks appear in Lincolnshire and East Yorkshire, but with features very different from those which characterize those on the south coast.

The Portlandian Series is divided into an upper and a lower group of strata; in both of these the character of the beds varies somewhat rapidly from point to point; the horizon of a good limestone in one district may be a useless stone in another, or a clay deposit may take the place of a sandy one. It is in the upper division that the famous building stones are found; the lower group consists of sands or clays, with little stone (Fig. 13).

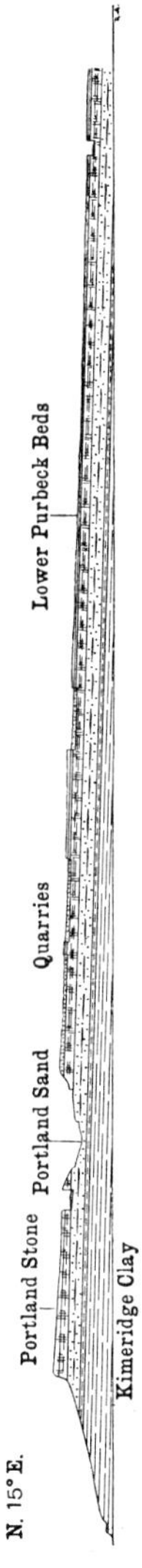

The Portland Beds are essentially shallow marine accumulations, though here and there are indications of estuarine conditions; the building stones are all marine.

Portland. — In the Isle of Portland the rocks are well exposed in the cliffs, and in the numerous quarries. On Verne Fort, at the northern end of the island, and in the neighbourhood of some of the quarries where the Purbeck Beds have been stripped off, the Portland Stone comes to the surface; elsewhere it is hidden under a cap of Purbeck Beds, and nearly all the quarry faces show the one resting in the other. The general

FIG. 13.—SECTION OF THE ISLE OF PORTLAND.

dip of the beds is towards the south-east. In many of the quarries the overburden of Purbeck Beds is very thick—from 10 to 20 feet. The Upper Portland Building Stone Series comprises the following beds, in descending order:

	Ft. in.	Ft. in.
Roach	1 6	to 4 0
Whit Bed	4 0	to 9 0
Curf and Waste	4 0	to 9 0
Best (or base) Bed	5 0	to 9 0

These broad divisions hold good generally throughout the island, but local variations occur, so that even adjacent quarries show different sections. Building stones are rather thicker on the west than on the east side. Blocks are obtained of 10 to 15 tons.

The *Roach* is a pale buff oolitic limestone readily distinguished from the other building-stone beds by the large number of holes scattered through the rock; these are the hollow moulds and casts of fossils, *Trigonia gibbosa* and *Cerithium Portlandicum*, known to the quarrymen respectively as 'horse-heads' and 'Portland screws.' Casts of other fossils also occur.

The coarsely vesicular or 'roachy' character is not strictly confined to the top bed or true roach, for the other beds assume this phase irregularly and in patches; thus south of Weston there is an upper brown and lower white roach; in a quarry at Kingsbarrow a Base Bed Roach, or Little Roach, occurs between the Curf and the Base Bed; near Rufus Castle roachy patches are found at different horizons in the freestones; while in the south, about Portland Bill, the whole series is 'roachy,' while the top bed is the most compact. In places the Roach Beds contain cherty nodules.

The Roach Beds are used locally for heavy engineering work, in the harbour, etc.

The *Whit Bed*, or Top Bed of Freestone, or Brown Bed, lies immediately beneath the Roach; the two beds are often literally one, and the separation has to be artificially made, the roachy portions of rock adhering to the freestones being chipped away.

This bed is a fine-grained oolitic limestone, of a warm cream to pale brown tint when fresh, but becomes white on exposure. In a rather irregular manner it is filled with shelly detritus, which tends to make it harder than the other beds under the tool. The thickness varies from 4 to 15 feet, but the thickest developments of the bed are seldom good throughout. Occasional flint nodules occur in this bed locally. The Whit Bed sometimes (Stewards Quarry, Kingsbarrow) has a separate lower member, the Bottom Whit Bed, 3 to 4 feet thick.

Current-bedding is sometimes present, and a mottled effect is produced in some of the stone after weathering by an irregular mixture of the Whit Bed and Base Bed types (Plate VIII.).

The *Curf*, with or without flints, is a band of poor stone, usually from 4 to 9 feet thick, which is waste, lying between the Whit Bed and the Base Bed; in the quarry south of Weston it is absent, and consequently the former rests directly upon the latter.

The *Base Bed*, 'Best Bed,' Lower Tier, White Bed, or Bottom Bed, is the lowest of the freestones worked; all the beds beneath—the Cherty Beds, 60 to 70 feet thick—being too full of siliceous concretions. The Base Bed is a fine-grained oolitic limestone, varying from 5 to 19 feet in thickness, paler in colour and softer than the Whit Bed, and not so markedly oolitic.

Portland Stone is so well known and so largely employed, particularly in the London district, that examples may be seen in every direction. The product of the quarries varies from time to time; at the present moment the Wakeham

Quarry stone is one of the most uniform in texture and tint (new War Office building).

Isle of Purbeck.—On the south-east coast of Dorsetshire, between Durlston Head and St. Alban's Head, the Portland Rocks stand out prominently in the upper portion of this fine range of cliffs; they are usually capped by a thin remnant of lower Purbeck Beds.

The clearest section of the strata is seen in the Winspit (Windspit) Quarry. It is as follows:

OVERBURDEN OF PURBECK BURR AND OTHER LIMESTONES AND CLAYS.

		Feet.
Upper Portland	*Shrimp Stone:* A fine-grained, white or cream limestone, formerly burnt for lime	8
	Blue Stone: A hard, grey, shelly, oolitic, sometimes roachy, limestone	9
	Top, Upper, or *Pond Freestone:* The best stone (oolitic)	7
	Flint Stone: Limestone, with white flints	4
	Nist Bed: A soft freestone	2 to 4
	House Cap: Coarse, shelly, oolitic limestone	5 to 6
	Under Picking Cap: A hard, cherty stone	2 to 3
	Under Freestone: Good oolitic stone	6
	Inland or *Cliff Beds:* A cherty series.	

A similar section is exposed in St. Alban's Head. The well-known caves in the cliff at Tilly Whim are old underground workings, which were in active operation up to 1811. The principal stone worked here was an 11-foot bed of oolitic freestone, occurring in the midst of more broken and cherty beds. This freestone was apparently the equivalent of the lower freestone of the other quarries. Another old quarry at Dancing Ledge shows about 30 feet of freestone, the upper part very much current-bedded and separated from the lower by a cherty layer.

At Seacombe are the Headbury and Cliff Field Quarries on the eastern side, and the Halsewell Quarry on the western side—all mainly underground works. In the last-named important quarry there are five or six beds of

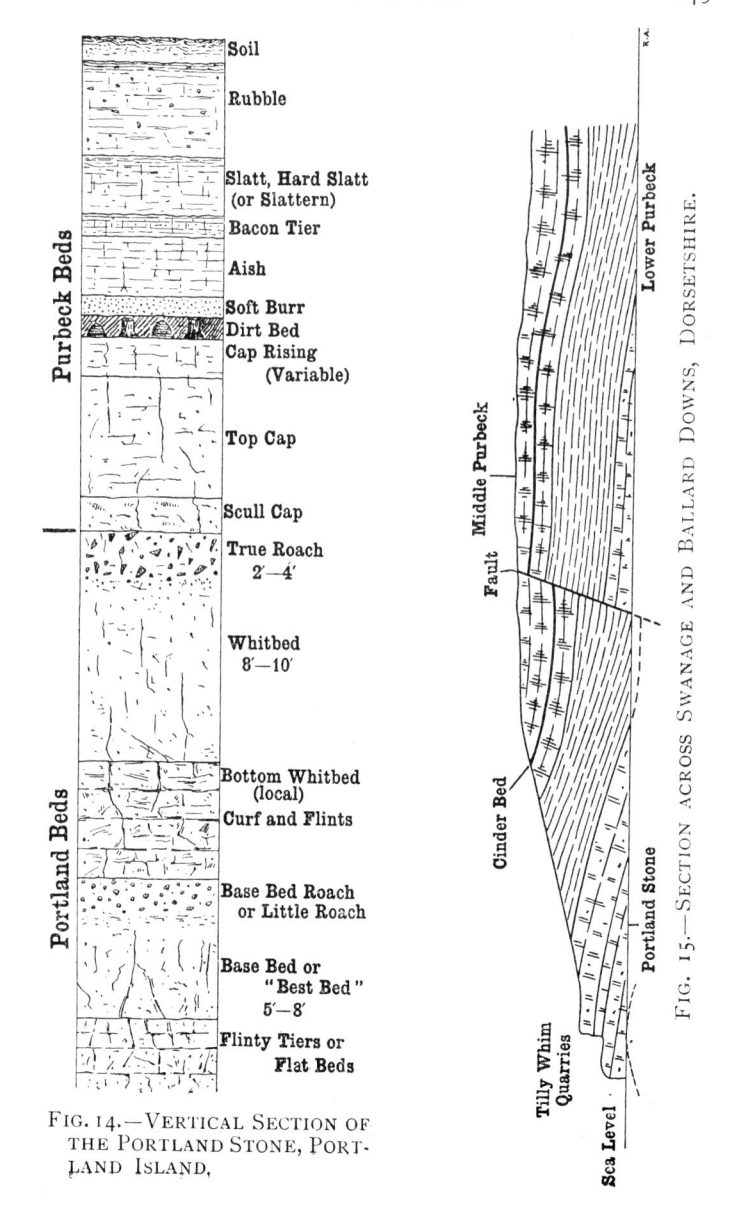

LIMESTONES

Purbeck Beds
- Soil
- Rubble
- Slatt, Hard Slatt (or Slattern)
- Bacon Tier
- Aish
- Soft Burr
- Dirt Bed
- Cap Rising (Variable)
- Top Cap
- Scull Cap
- True Roach 2′—4′

Portland Beds
- Whitbed 8′—10′
- Bottom Whitbed (local)
- Curf and Flints
- Base Bed Roach or Little Roach
- Base Bed or "Best Bed" 5′—8′
- Flinty Tiers or Flat Beds

FIG. 14.—VERTICAL SECTION OF THE PORTLAND STONE, PORTLAND ISLAND.

FIG. 15.—SECTION ACROSS SWANAGE AND BALLARD DOWNS, DORSETSHIRE.

massive oolite, amounting to about 15 feet, occasionally cherty, lying between an upper and lower cherty series of limestones.

At Renscombe the Pond Freestone is obtained by open works, but the under freestone is taken out from underground galleries. At London Doors Quarry, north of Encombe House, the upper freestone is a white limestone, and the beds are much shattered.

The freestone series in Purbeck is from 40 to 50 feet thick; the underlying cherty beds are from 60 to 75 feet thick.

The Portland Stone from this district is sometimes called the 'Purbeck-Portland' to distinguish it from that obtained from Portland itself; that from Seacombe Quarries is known as 'Seacombe Stone.' On the whole, the stone here is harder and denser than that from Portland, and this may have been the reason for abandoning the Tilly Whim works, whence the hardest and most durable of all the stones was obtained. Nearly all the quarries are situated on the sea-cliff, but owing to the nature of the coast, the loading of the stone into ships is a matter of difficulty except in very calm weather.

The under freestone at Winspit is used for kerbs, channels, and sinks. The House Cap is a coarse-grained stone, suitable for breakwaters and other rough work. The Blue Stone is a hard and durable stone, used locally for lintels, gate-posts, etc. The Shrimp Stone is so named by the quarrymen from the presence of a small fossil crustacean. The Pond Stone is the best for general building purposes. 'Cliff Stone' is a name sometimes applied in a general sense to any of the stones from the cliff quarries in this district.

Seacombe Stone varies slightly, according to its origin in one or other of the several beds; usually when fresh its colour is bluish-grey, but it assumes a pale yellow tint on drying. The best quality stone from these quarries is

LIMESTONES

thoroughly good material, both in colour and durability; and its hardness makes it suitable for employment when the softer rocks of Portland would fail—namely, in floors, landings, steps, and copings. Its weight is quoted as 149·31 and 151 pounds per cubic foot. It is employed at the Lighthouse, Margate; the Prison, Winchester; Clock House, Dover; West India Docks (fifty years); Law Life Assurance Company, 185, Fleet Street, 1854; and in many churches.

The stone from Tilly Whim was used (with others) in the construction of Corfe Castle.

Vale of Wardour.—In the Vale of Wardour, as in the Isle of Portland, the Purbeck Rocks form the overburden or cover of the Portlandian Beds.

The general section is as follows:

			Ft. in.	Ft. in.
Upper Portland Beds.	Upper Building Stones	Buff, sandy, and oolitic limestones. Compact limestones and occasional cherty seams in the lower part	10 0	to 16 0
	Chalky Series	Soft chalky limestone, with nodules and veins of black chert	4 0	to 24 0
	Ragstone	Brown, gritty, and shelly limestone, divided in places by seams of rubbly marl	4 0	to 5 6
		Pale, shelly, and oolitic limestones, with rubbly, shelly marl at base	3 3	
	Main Building Stone	Trough Bed (sometimes called 'Whit Bed'): Hard, buff, sandy, and oolitic limestone merging into glauconitic and sandy limestones, divided locally into—	2 8	

Main Building Stone divisions:

	Ft. in.	
Green Bed	5 0	
Slant	1 0	
Pinney	2 0	15 4
Cleaving or Hard Bed	1 0	
Fretting Bed	3 4	
Under Beds	3 0	

Resting on Lower Portland Beds.

The Upper building stone is disposed to be 'roachy' —that is, full of hollow casts of the shell of *Cerithium Portlandicum*, etc., as in the Isle of Portland. The chalky series has been supposed to correspond to the Base Bed of Portland; the soft stone has been used for rubbing hearthstones.

In the main building-stone series a proper understanding of the value of the individual beds is a matter of some difficulty, owing to the rather complicated local nomenclature.

The Trough, Hard, or White Bed is a sandy oolitic stone, buff or yellowish-brown in colour, said to weather the best of the Chilmark stones; the workable thickness is about 2 feet or a trifle more, and random blocks average 16 cubic feet. It is used for steps, paving, cornices, plinths, sills, and engineering work.

The Green Bed (Scott or Brown Bed?) is variable in texture and often shelly; it is both sandy and glauconitic. Blocks may be obtained of the same dimensions as in the Trough Bed. It is used for general purposes: ashlar, random rubble, carving, and mouldings. The Pinney Bed is more crystalline than the others and is harder at the bottom; its colour is yellowish-brown; it works freely and is said to weather well. The Fretting Bed is more sandy than the others, and the Slant Beds are very friable.

When dry and after exposure, most of these stones assume a grey, often a greenish-grey, tint; and it is very difficult to recognize any difference between the products of the Chilmark, Wardour, or Tisbury Quarries, or the stone from individual beds. The quarries have been worked for hundreds of years, but there is less activity now.

The stone has been used in Salisbury Cathedral (thirteenth century), where, except in the west front, it has worn wonderfully well; the Upper building stone was employed to some extent in this edifice. In the restoration of Rochester Cathedral, 1872, it was used in the exterior of

LIMESTONES

the north transept, the east end, pinnacles, and clerestory windows. The Trough Bed was used in Westminster Abbey, exterior of the Chapter House, in 1867. Other structures in which it has been employed are the Harnham Bridge, Salisbury (thirteenth century), and Tisbury Church, of the same period; Wardour Old Castle (fourteenth century); in Chichester and Truro Cathedrals; Wilton, Fonthill, and Romsey Abbeys; Christ Church Priory and Balliol College, Oxford; London and County Banks, Hastings and Banbury; Mutual Life Assurance, 101, Cheapside, 1860; Post Office, Westminster Bridge Road; and Postal Sorting Office, Hampstead.

The appearance of Portland beds here is due to slight undulations in the strata, which have brought these rocks nearer the surface, where they have been exposed by the cutting of the Chilmark Ravine by the River Nadder. The inclination of the Jurassic beds as a whole is towards the east-south-east. Both upper and lower building stones are worked on the east side of the ravine; only the lower building stones are worked on the western side. Most of the workings are underground.

In the Teffont (Chilmark) Quarry the following arrangement of beds was noted by Mr. H. B. Woodward in the lower or chief building stones:

	Ft. in.	Ft. in.
Chalky Series.		
White Bed: Gritty limestones used for hearthstones	1 6	to 4 0
Rubbly marl (rag)	0 6	
Shelly limestones	3 6	
Trough Bed: Pale shelly, oolitic, and sandy limestone	1 3	
Rubbly marl, passing into roach	0 6	
Green Bed: Hard buff or pale greenish-grey sandy oolite, merging into bed below	2 6	to 2 9
Pinney Bed: Brown glauconitic and oolitic sandy limestones in three or four 12-feet layers. This bed derives its name from the presence of a small *serpula*.		

The local names mentioned above might be supplemented by others used in different parts of the district.

The upper building stones are mainly in the condition of roach—that is, they contain hollow casts of the Portland Screw and other associated fossils, and occasionally layers in the lower series assume the same characters.

The Chalky Series, which resembles certain beds in the Portland rocks between Portisham and Upwey, in Dorset, is of interest from its very close similarity to Chalk, from which it is difficult to distinguish except by the aid of the fossils.

On both the north and south side of the River Nadder, by Chicksgrove Mill, there are quarries and underground workings in Portland beds, all somewhat sandy in nature.

At Wockley, south-east of Tisbury, the series is thinner, and the Ragstone is absent. Here, beneath about 20 feet of Purbeck beds, we find:

	Ft. in.	Ft. in.
Roach ... ... } Chalky limestone ... ... }	10 0	to 15 0
Buff and greenish glauconitic limestone ...	2 0	to 4 0
Compact and very shelly limestone, passing down into sandy limestone (quarried for freestone)	4 0	to 5 0

A similar section is exhibited in the quarry south of Tisbury Station.

Swindon and Bourton.—A small area of Portland beds is exposed around Swindon. Here, beneath Purbeck beds, the succession of beds is approximately as follows, though, as they vary rapidly, the measurements differ from time to time:

		Ft. in.	Ft. in.
Upper Portland	Chalky beds.		
	Oolitic limestone (Roach)	5 0	to 6 0
	Irregular grey clay.		
	Sands with beds of hard sandy limestone	2 0	to 3 0
	Sands with indurated, false-bedded bands (*Swindon Stone*)	20 0	to 25 0
	Fossiliferous limestone ('Cockly Bed') ...	4 0	
	Bluish-grey limestone, with black chert and pebbles	3 4	

LIMESTONES

Oxfordshire.—Around Shotover Hill, Garsington, and Cuddesden, the Portland beds form a large outlier. The rocks in this district are much more sandy and glauconitic than the typical series in the Isle of Portland; but some of the calcareous sandstones and irregular limestones have been dug for building purposes. Quarries containing calcareous beds exist at Great Milton and Great Hazeley. At the latter quarry a very fair quality of limestone has been obtained in limited quantities from ancient times, but the freestone is rather spoiled by layers of shells. Above the workable stone is a hard, splintery, gritty limestone with fossils, called by the quarrymen 'Curl,' which is suitable for rough walling, and was formerly used for chimneypieces.

Upper Portland beds yield rough building stones of variable quality from the quarries at Thame, Brill, and Aylesbury. The high wall bounding Hartwell Park, at Aylesbury, built of this stone, presents an interesting appearance from the frequent introduction of large Ammonites into the structure. Some of the Aylesbury beds are 'roachy' in character.

Outliers of this rock are quarried at Oving and Whitchurch, north of Aylesbury, where a thin bed of hard, grey, blue-hearted limestone, weathering brown, and one or more thin shelly beds have been used in the villages.

Purbeckian—*Purbeck Stone, Swanage Stone.*—The principal source of this stone is the Middle Purbeck formation on the hilly ground south of Swanage. It consists of a series of thin beds of grey shelly limestone, with intercalated shales. It is evident from the fossils, which these rocks contain abundantly, that they were formed under estuarine and marine conditions, while those of the Upper and Lower Purbeck were laid down in completely freshwater conditions.

The numerous beds of limestone are recognized and

named individually by the local workmen, and the whole series is grouped by them into what they call 'veins'; but, as is only natural, there are differences of opinion as to the limits of the veins in different parts of the area. The sequence of the stone beds, in descending order, is as follows:

		Ft.	in.
White Roach	...	4	6
Laning, Lane End, or Leaning Vein	...	5	6
Royal	...	5	0
Red Rag	...	2	2
Rag	...	1	0
Under Rag	...	1	6

Freestone Vein
- Top Shingle
- Shingle
- Under Picking
- Grub
- Roach
- Pink Bed
- Grey Bed
- Thornback
- Freestone Bed
- Blue Bed

16 0*

Downs Vein
- Lias
- Lias Rag
- Laper
- Under Picking
- Upper Tombstone Bed
- Brassy Bed
- Lower Tombstone Bed

13 0*

This formidable list by no means exhausts the catalogue of local names, for we find some quarrymen recognize a White Bed, Black Bed, a Bottom Bed, Thick Bed, etc.

The Laning Vein is used for sea-walls, ordinary building, paving, and tombstones. The Freestone Vein includes beds used for all kinds of work; the Freestone Bed is the one most generally useful. The Thornback is used for paving slabs and kerbs, the Grey Bed for steps and kerbs, and the others for setts and road metals. The Downs Vein yields good paving stone and tombstones. The

* Including shales.

Feather Bed is used for sea-walls and for paving and kerbs; it is regarded as the best of the stones for the latter purpose. The New Vein is employed for steps, paving-slabs, and tombstones.

In general the Purbeck Stone is trimmed into kerbs, paving-slabs, steps, setts, gate-posts, window-sills, sinks, troughs and landings, and building blocks. It may be provided with a rubbed face, or rock face, or may be draughted. Owing to the thinness of the workable beds of limestone and their steep northerly dip, there are practically no open quarries in the tract occupied by the Middle Purbeck rocks near Swanage. Their place is taken by shallow mines 40 to 70 feet deep. Sloping shafts, 5 feet square, are made, so that the stone can be dragged up in trucks by means of a capstan and rope, by horse or donkey power. The beds are worked in the direction of the dip; the stone is extracted by means of cross-bars and wedges. The shafts are paved with stone for the trucks, and rude steps are cut alongside for the use of the workmen. As the stone is worked away the galleries are supported sometimes by timber, but more usually by blocks of useless rock. Early in the nineteenth century the stone was actively worked in the cliffs in Durlston Bay. The process of getting the stone here was very simple: a slight undercut was made in the cliff, and props were inserted. After a time the props gave way; a fall of rock occurred, and from the débris the masons selected and trimmed the suitable stones.

The stone is dressed in sheds at the surface, and then sent by cart to Swanage, where the blocks, or 'bankers,' are piled, ready to be shipped or sent away by rail. Open quarries were formerly worked on the hills about Durlston Head. The stone mines—some hundreds of shafts, each with its surrounding heaps of 'ridding' or 'scar heaps'—cover the northern slopes of the hills south of the Vale of Swanage from that town towards Kingston beyond which the stone is apparently too much crushed by the uptilting movements to make it payable.

The quarries, or pits, are kept distinct, and the area belonging to each is frequently marked out on the surface by stone walls. The mines are sometimes owned by the workers, and there is little grouping under large firms. Many curious customs still prevail in connection with the labour, and it has been suggested that the quarrymen may be descendants of a colony of Norman stone-workers. They form a distinct class of people, owing to their curious customs and intermarriage. The men constitute the 'Company of Marblers and Stone-Cutters of the Isle of Purbeck,' and maintain the privilege of confining the stone trade to themselves and their sons, rigorously excluding strangers.

The stone is much used in the south-east of England, being sent to Brighton, Bournemouth, Portsmouth, Winchester, Weymouth, Salisbury, etc., and formerly it was extensively used in London for paving.

The Upper Purbeck Beds of Swanage yield the well-known Purbeck marble with *Paludina* shells, but no building stone of importance is now worked on a large scale. Formerly a coarse, rather soft, shelly limestone—the *Soft Burr*—occurring at the base of the Upper Purbeck Series, was a good deal used, and would seem to have been fairly durable. It was only obtainable in small blocks, which may be seen in the walls of Corfe Castle—along with other beds from the Middle Purbeck—in the old tower of Swanage Church, and in the restoration of the church at Kingston, in Lord Eldon's estate.

Beyond the Isle of Purbeck beds of the same age are found resting upon the Portlandian rock of the Isle of Portland, but they are either treated as waste or used only locally. There the Cap Beds (see Portland Section,

p. 249) are too hard and irregular in their behaviour. The Soft Burr is only a rough stone, employed in small local buildings, especially in chimneys, because it stands fire well. The Bacon Tier is used only for rough boundary walls. The Hard Slatt is put to similar purposes; it is too hard to square, as a rule, but it has been employed for paving, as in the brewery at Wyke.

Again, the Purbeck beds produce a very good stone at Upwey, near Weymouth; but here it is the lower and not the middle division that is worked.

The spire of All Saints' Church, Dorchester, was built of the '*Cypris* Freestone.'

At Teffont Evias, in the Vale of Wardour, and near Chicksgrove, smooth-grained, pale, hard limestones are worked in a small way for building purposes.

Purbeck limestones, known as 'greys' and 'blues,' were formerly worked in Sussex.

Cretaceous Chalk.—Notwithstanding the large area occupied by the Chalk in England, it has been put to little use as a building stone. The principal reason for this is its softness and inability to withstand the weather when cut into blocks and exposed in buildings. The Upper Chalk in the South of England is always soft and friable, but in Yorkshire it is much harder as a whole, and contains many more interbedded layers of marl. This horizon of the Chalk is quarried for building blocks at Boynton, about a mile and a half west of Bridlington; and stone of a very fair wearing quality could be obtained in the Yorkshire Chalk District should there be a demand for it. In the southern and eastern counties practically all the chalk used in buildings is from the Middle or Lower Chalk horizons. It has been employed in the outer walls of cottages and small houses in Cambridgeshire, Norfolk, Suffolk, Hertfordshire, Lincolnshire, and in the northern parts of Hampshire. The upper part of the Middle

Chalk has been quarried for building blocks at Medmenham, Bucks; in Norfolk the lower part of the same horizon is usually employed.

The treatment of chalk during and after quarrying varies in different districts and with variations in the stone. In some cases it is quarried in large masses, which are allowed to weather for some months, a process which reveals irregular joints and bedding planes that are invisible in the freshly-exposed stone, and then they are squared into blocks of about 1 foot long by 6 inches square. In other cases the stone is squared without a prolonged weathering process. The summer is usually regarded as the best time for squaring the blocks, which should be thoroughly dry before being used.

In building cottages with chalk the method in common use is to face the inner and outer surface of the wall with squared blocks, and fill in the spaces between them with chalk rubble. The walls of Castle Acre, Norfolk, are built of chalk, faced with flints.

Chalk freestone has been employed in St. Albans Cathedral, in parts of Windsor Castle, and in the interior of Isleham Church; good examples of carving in chalk are to be seen in Ely Cathedral, St. Pancras Priory (ruins, 800 years).

I am indebted to Mr. E. L. Lutyens for the information that he has employed chalk from the Medway Valley, from near Romsey, and from Stockbridge, Hants, for ashlaring, window mullions, and door-heads.

The Chalk Freestone of Amberley, in Sussex, is obtained from about the same horizon as the Beer Stone; it is a compact greyish-white rock, obtainable in large blocks. It has been used in vaulting and forming groined ceilings at Arundel Castle, and for similar work in St. Saviour's, Southwark; in the new Roman Church, Norwich; and in the restoration of the groined roofs in Chichester Cathedral.

LIMESTONES

Totternhoe Stone, or *Clunch Stone*, is a pale greenish-grey rock, almost white when freshly broken; when examined closely it has a distinctly granular appearance, and to the touch it feels harsher than ordinary chalk. The last-named character has probably led to the stone being described incorrectly as a 'sandy limestone.' This it certainly is not, for the greater part of the rock is composed of the minute polygonal cylinders of calcite, which result from the breaking-down of the large oyster-like shells of *Inoceramus*; these are set in a rather muddy calcareous matrix. Scattered throughout the mass of the stone are small grains of glauconite, which are responsible, in part at least, for the greenish tinge.

The principal quarries are at Totternhoe, near Dunstable, Bedfordshire, where it was formerly gotten by adits driven into the best beds at various places along their outcrop. Here the total thickness of stone is from 15 to 17 feet. The upper 9 feet are softer and of poorer quality; the best stone is obtained from two beds about 4 feet thick. Blocks up to 1 ton in weight are said to be obtainable, but the frequency of jointing and bedding planes keeps down the size of most of the output. The not infrequent appearance of small nodules of hard iron pyrites makes the practice of cutting this stone with a hand-saw usual; the blade is changed when a nodule is encountered.

The Totternhoe Stone is a local phase of the Lower Chalk; in Bedfordshire and Buckinghamshire it is from 6 to 15 feet thick, and the beds have been quarried at Ivinghoe, Barton Hill, and Kersworth Hill. In Cambridgeshire it is 12 to 15 feet thick, quarried at Burwell; and it has been traced into Norfolk, where it may be seen at Roydon and Stoke Ferry. Marham Church, in the county, is built principally of Lower Chalk, probably from near the same horizon; the tomb with recumbent figures within the church is carved from this same stone.

Though it is a soft stone when quite fresh, it hardens on exposure, but it cannot be safely used for exterior mouldings or carving. Being light (116 pounds) and readily worked, it is suitable for interior work and vaulted roofs. It has been employed in the west front of Dunstable Priory Church (where it has deteriorated badly), in the organ screen at Peterborough Cathedral, at Ashridge, and many churches in Bedfordshire, Buckinghamshire, and Hertfordshire. The parish church at Luton is faced principally with alternate cubes of Totternhoe Stone and black flints. It has been used also in Fonthill House, Wiltshire, and Woburn Abbey.

Beer Stone is the best known and most valuable of the Chalk building stones. It is a cream-coloured stone when fresh, turning grey upon exposure; it never has the greenish tinge often observed in Totternhoe Stone, and it may be readily distinguished from that rock by microscopic examination, for the Beer Stone is made up of minute irregular, corroded fragments of shells, foraminifera, echinoderm plates, and bivalve mollusc shells, which are cemented together by a calcareous paste, largely crystalline in character. The weight has been recorded as 131·7, 140, and 158 pounds per cubic foot. The crushing strength (Kirkaldy) is about equal to that of Portland Stone.

The quarries are situated at Beer, near Axminster, in Devonshire. Several old quarries were formerly worked here as far back as Norman times. Owing to the great thickness of overburden, the stone is now worked by adits driven into the hillside. The total thickness of the best stone is about 12 feet, the thickest bed being 4 to 5 feet; blocks can be got up to 8 tons. The stone, when fresh, holds a good deal of water, and it is very soft and is easily cut with a hand-saw; on exposure it hardens considerably. Beer Stone is on the geological horizon of the Chalk known as the *Rhynconella Cuvieri* zone—that is, it lies at

PLATE VII

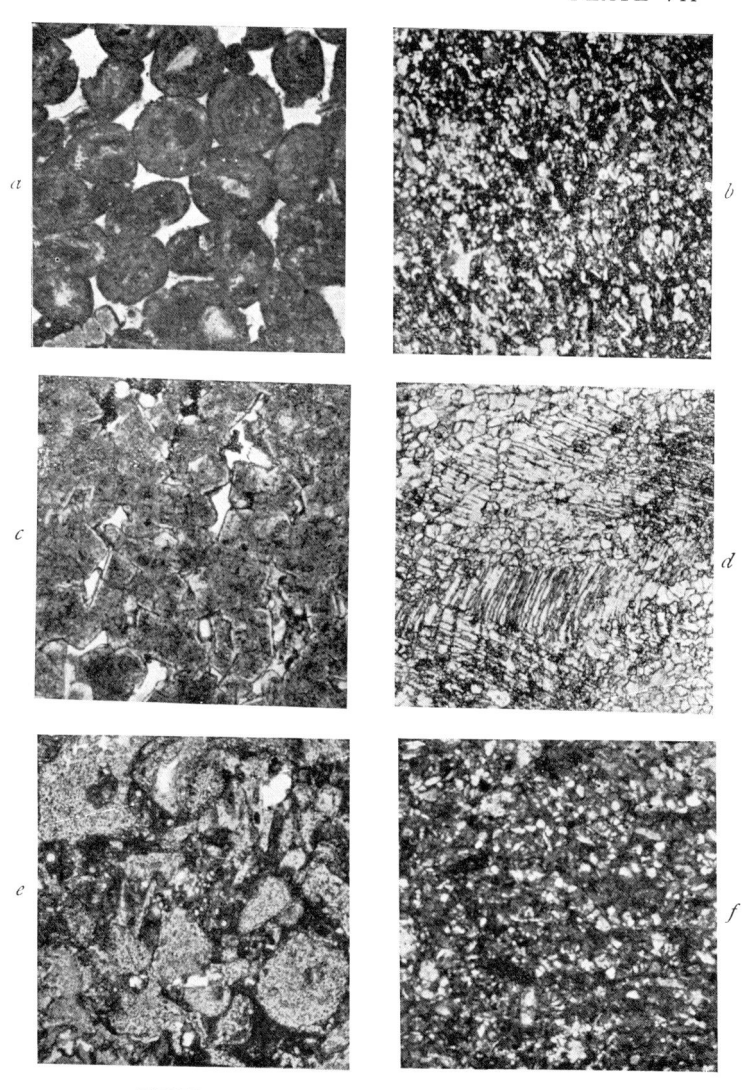

PHOTOMICROGRAPHS OF VARIOUS LIMESTONES.
(Magnification, 30 diameters.)

a, Portland Stone, Waycroft; b, Black limestone, Poolvash, Isle of Man; c, Dolomitic limestone, Bolsover; d, Sheared and partly marmorized limestone, Devonshire; e, Beer Stone; f, Totternhoe or 'Clunch' Stone.

LIMESTONES

the base of the Middle Chalk. The peculiar qualities of the stone which render it valuable for building are very local; the same beds traced inland are found to change rapidly in character. A quarry (now obsolete) at Sutton, near Widsworthy, eight miles north of Beer, at one time yielded a 'Sutton Stone' from these beds. A similar stone was formerly dug at Ware, near Lyme Regis.

Beer Stone is well adapted for internal work, plain or carved, but it cannot be recommended for exteriors in large towns. In Ludgate Square and the Carlton Club it has not stood at all well, though Mr. P. E. Masey has mentioned that in St. Stephen's Chapel, Westminster, its behaviour has been very satisfactory. Masey records its use in the following buildings: Exeter Cathedral (mostly for interior work, from the Norman period downward; the beautiful series of piscinæ in the chapels, of early Decorated date, are of this stone); St. Pancras Church, Exeter; and the churches at Whimple, Tallaton, St. Lawrence Clist, Clist-Hydon, Awliscombe, Payhembury, Buckerell, Feniton, Musbury Seaton, Branscombe, Axmouth, Axminster (Norman doorway), Lyme Regis, Ottery St. Mary (thirteenth-century work), Honiton, Colyton, Chard, Broad Clist, Uplyme, Combpyne, Charmouth, and Shute. 'In all these churches Beer Stone had been used for exterior dressing and interiors, generally at dates ranging from the eleventh to the sixteenth centuries. The condition of the stone I found to be generally good; mullions of windows in some cases have failed through not having had the stone set in its natural bed.' There is a remarkably perfect rood-screen of this stone in Oulesford Church, fifteenth century. It was possibly used in Winchester (William of Wykeham's work at the close of the fourteenth century).

Kentish Rag constitutes the main part of the Hythe Beds in the lower greensand of the Weald district. These beds are composed of many alternating layers of limestone (Rag) and soft calcareous sandstone (Hassock); the layers range in thickness from 6 inches to 3 feet. The thickness of the whole is about 60 feet. Both kinds of sandstone are prone to much local variation; the Rag may contain over 90 per cent. of $CaCO_3$, or it may carry much sand; grains of glauconite are usually abundant. The Hassock is commonly soft, sandy, and earthy, but occasionally it is sufficiently hard and calcareous to form a building-stone. The Rag is hard and impervious, grey and greenish-grey in colour, and weighs about 166 pounds per cubic foot. It has been extensively used in churches in the home counties in the form of random and regular coursed work; it cannot be dressed, and it is not suitable for interior walls. The best stones weather well, but from its variable character adjoining blocks often exhibit great differences in this respect. In weathering it tends to flake gradually, particularly in the sandy varieties. It is largely quarried about Maidstone for roads. Hassock is warmer in colour, and more porous; when sufficiently hard (= Calkstone), it is suitable for lining Rag walls.

Sussex or *Bethersden Marble*, a greenish-grey stone, full of a large species of *Paludina*, from the Weald Clay, was formerly much used in the Perpendicular churches of the district, and is often confused with Purbeck marble, which is said to be more characteristic of the Early English style. This stone is used in the altar-steps of Canterbury Cathedral.

Tertiary Limestones.—In this country limestones are very poorly represented in the Tertiary rocks; none is found in the London Basin and in the Hampshire Basin; the only one that requires notice is the Bembridge Limestone, which makes up the lower 12 to 25 feet of the Bembridge Beds in the Oligocene System of the Isle of Wight.

LIMESTONES

This limestone consists of several beds of slightly different texture and colour, but usually of a pale cream or greenish-white; compact, concretionary, or marly. It is of freshwater origin, and contains fossil land or freshwater shells. Years ago the Bembridge Limestone was extensively quarried as a building stone at Binstead, near Ryde, in the Isle of Wight.

Travertine.—In districts where limestones abound, in all parts of the world, spring waters, more or less heavily charged with carbonate of lime, ooze or flow out of the hillsides, and deposit some of this material on the ground. The rock so formed is usually porous, and is often found enclosing the casts of leaves and twigs; also bones of various animals, and land and fresh-water shells, which have been covered by the deposit very much in the same way that all sorts of miscellaneous articles are coated in the so-called petrifying wells of Matlock Bath, Naresborough, and elsewhere. The colour of this rock is usually buff, cream-tinted, or yellowish when freshly cut, but on exposure it turns to a pale grey colour; sometimes it is coloured red, brown, or black by iron oxides or enclosures of clay and soil.

The process of formation is essentially simple; it depends primarily upon the liberation of carbon dioxide, by which the calcium carbonate was held in solution as the bicarbonate. The carbon dioxide may be freely dissipated in the air, or its removal may be accelerated by the action of algæ and mosses. Newly-formed Travertine is soft, friable, and rather mealy in texture; so long as it is full of moisture it is soft, so that it may often be cut with a spade. On exposure it hardens to a surprising degree, and forms a very strong, light rock; older portions of such deposits are generally harder, and the microscopic structure is largely crystalline.

The chemical composition may be illustrated by the

analysis by L. Brognard of the Travertine of Lillebourne, France:

	Per Cent.
Carbon dioxide	42·43
Lime	54·32
Insoluble in acid	2·03
Iron and alumina	0·91
Phosphoric acid	0·31
Sulphuric acid	trace
	100·00

Samples taken from the rock and from blocks in the Roman Theatre near by (Juliobona) gave practically identical results.

Travertine is found in small deposits at several places in Great Britain: in Derbyshire, at Matlock Bath; the Via Gellia, near Cromford; Alport and Monsal Dale; in Wales, near Harwarden, Llangollen, Wrexham, Prestatyn, Caerwys, and Pwll Gwyn; at Kepwick, in Newton Dale, near Saltergate; in Troutdale and in the Forge Valley, along the base of the Tabular Hills in Yorkshire; at East Malling; Boxley Abbey, in Kent; Totland Bay, in the Isle of Wight; Osbournby, in Lincolnshire; Dursley, in Gloucestershire; and in parts of Scotland and Ireland.

Practically all the travertine now quarried in this country is used for rock gardens, aquaria, and similar purposes; but formerly it was employed for buildings in the neighbourhood of the quarries, and it was particularly favoured, on account of its lightness and strength, for vaulted cellars, arches, etc. Portions of Berkeley Castle are built of the Dursley travertine, and dwelling-houses may still be seen in parts of Wales and Derbyshire in which it forms the preponderating material.

Large deposits of travertine occur at Hammam-Meskhouten, near Constantine, in Africa, and they almost obliterate the site of Hierapolis, in Asia Minor; in America they are formed by the well-known Mammoth Hot Springs

LIMESTONES

of Yellowstone Park; in short, they are irregularly distributed in all countries where limestones occur or where volcanic activity manifests itself. The most familiar European localities are those in Clermont and Normandy in France, and Tivoli and other places in Southern Italy. In Rome from earliest times it has been the most important building stone; it has been employed in the Cloaca Maxima, the exterior of the Colosseum, the Theatre of Marcellus, and many other ancient buildings; while among more modern structures, the Church of St. Peter, the Castle of St. Angelo, the Church of the Lateran, and many other churches and private mansions.

It is desirable to point out that much confusion exists in the descriptions and nomenclature of varieties of travertine. The older Latin name for the stone was *Lapis Tiburtinus*, from which we get the modern Italian *Travertino* and *Tiburtino*; in England we speak of *travertine, calcareous tufa, calc-sinter*, or simply *tufa*; these correspond with the French *tufs calcaires* and *travertin* (*tuffeau* is a siliceous rock), and with the German *travertin* and *kalktuff*. All these terms are applied to the porous rock described above; but it happens that in certain places instead of a porous rock suitable only for building stone, a compact crystalline and often laminated deposit of calcite is formed, similar to the stalagmite of caverns in some respects, and for this variety the name travertine is employed by Merrill, in contradistinction to tufa or calcsinter, which he limits to the porous form. On the other hand, the Germans seem inclined to confine the term *kalk-sinter* to the compact laminated variety.

The laminated compact variety of deposit is largely used as a marble, and as such it passes under the name onyx marble and alabaster.

Calcareous deposits of a porous tufaceous character are sometimes formed in lakes and fresh-water pools. Some of these are called travertines; a good example is the Calcaire de Brie, or *travertin moyen* of the Lower Oligocene beds of the Paris Basin, which is quarried at Chateay, Landon, and Souppes, and has been used in the Arc de Triomphe and the basilica of Montmartre. A silicified form of this stone, known as *meulière*, is used for millstones and for building near La Brie and Ferte-sous-Jouaire. Other examples occur in the Oligocene and Eocene of France, the Hampshire Tertiary Basin, and in the Purbeck beds of the Isle of Portland and neighbouring parts of Dorset.

FOREIGN COUNTRIES.

France.—This country is well supplied with excellent limestones, especially in the Jurassic, Cretaceous, and Tertiary formations.

In Britain the best known of the French Jurassic limestones is the Caen Stone of Calvados. This is a fine-grained pale cream and yellowish homogeneous limestone, rarely with any obvious indications of oolitic structure. When first quarried it is very soft, but it hardens on exposure. The stone beds, 30 to 50 yards in thickness, occupy a horizon equivalent to the Fuller's Earth beds of the Great Oolite (Bathonian) of England. The stone is mined and quarried. The principal sources of Caen Stone proper are: Allemagne and Maladrerie; the beds mined at the former place are:

	Metres.
Gros banc	1·05
Banc crai	0·6
Pierre franche	0·8
Banc de 4 pieds	1·3
Banc de 30 ponces	0·8
Franc banc	1·2

Similar stone is worked in the Calvados district under the names Orival, Quilly, and Aubigny stone. The last

LIMESTONES

is more crystalline, denser, and harder than the typical Caen Stone; it is worked at the quarries of St. Pierre Canivet; it was used in the western doorway of St. Mary's Church, Stoke Newington (1858).

Caen Stone is unsuited for exteriors in this country; it may be used for interiors, and is capable of taking the most elaborate carving. It has been used in Canterbury Cathedral; Henry VII. Chapel, Westminster Abbey; 116, Piccadilly, at the corner of Down Street; and many other buildings.

Good limestones of Oxfordian age are worked in the departments of Ardenne and Meuse; the best known are those of Euville, Lérouville, and Savonnières. The Euville Stone is coarse in grain, composed entirely of crystalline fragments of crinoids; the first class, or 'ordinaire,' is pale buff; the second-class stone is rather warmer in colour. It is a very fine stone, and has been much used in Paris— in the Pont Neuf, Notre Dame, Austerlitz bridges, the Opera-House, Grand Hôtel, Hôtel du Louvre, etc.; and in many important buildings in Germany, Holland, and Belgium. The Lérouville Stone is similar. The Savonnières Stone is worked in many quarries in the Communes of Savonnières-en-Perthois and Brauvilliers. It is a pale grey to yellowish stone, shelly, oolitic, and of medium to fine grain. It has been extensively employed in Brussels, Antwerp, Amiens, Lille, etc. Another stone from this department is the Liais de Morley.

The Jurassic rocks of the Department of Yonne yield many important limestones—*e.g.*, the white, chalky oolite of Charentenay (Banc Royal and Banc Fin), used in the Louvre, Hôtel-de-Ville, Bibliothèque Nationale, etc., Paris; the Cravant Stone, a chalky grey or white oolite from the Palotte Quarries; the Roche du Larrys-Blanc, a milky white fine-grained oolite, much used in Paris and Brussels; the Andryes, Courson, Drayes, Fourneaux, Molesmes,

Anstrudes, and Auvigny stones; the Stone and Liais of Méreuil and Ravières, and the 'pierre blanche de Chassignelles.'

In the Department of Vienne are the Château-Gaillard or Planterie Stone (a white chalky oolite in the Banc Royal, shelly in the Banc Inférieur), used in the Palais de Justice, Brussels, and the Tuileries, Paris, etc.; the Chauvigny Stone, used in the restoration of Orleans Cathedral; and the stone of Tercé, from the quarries at Tercé and Normandoux, a white oolite of fine to very fine grain.

From the Department of Côte-d'Or come the stones of Comblanchien, compact, fine-grained, greyish-white; and Val-Rot and many smaller quarries.

In the Department of Corrèze the Lias yields the yellowish, fine-grained argillaceous limestone of St. Robert. From the same formation comes the Rencontre Stone of the Department of Cher.

Cretaceous rocks produce good limestones in the Department of Dordogne—*e.g.*, the milky white, fossiliferous stone of variable grain from Villars and Chancelade, and the yellowish granular stone of Columbier. These are used in Rouen, Louviers, Vernon, etc.

The Crazannes Stone, of the same age, a rather cellular, whitish-grey to reddish rock, of medium to fine grain, has been used in Bayonne Cathedral, restoration of Cologne Cathedral, in Antwerp, Ypres, etc.; and the similar stone of St. Savinien both come from the Department of Charente-inférieure.

Tertiary Limestones, of both marine and fresh-water origin, mostly pale yellow or cream-coloured, have been largely worked for building stones in the Paris Basin; in the great underground quarries beneath Paris, and in the neighbourhood of the city; also in Bordeaux, Marseilles, Montpelier, etc.

Carboniferous Limestone is quarried in the vicinity of Lille; and the *Triassic* Muschelkalk in Lorraine, Grasse, Toulon.

Austria.—Crystalline limestones of older Palæozoic age are quarried at Freiwaldau, Silesia (Reichstag buildings in Berlin; Hochschule, in Charlottenburg), and Meran, in Tyrol. The Kreswitz, or St. Theresa, marble of Chrzanow, Galicia, is of Devonian age. The Latschach marble of Villach, Kärnten, is a Carboniferous stone used for masonry and interior decoration. The Jurassic marbles of the Tyrol are used in masonry in that district. The Trias Muschelkalk, of Neumarkt, Galicia, is used both in engineering and architectural structures. In Istria and Southern Austria Cretaceous limestones are extensively worked. The 'Istrian' stone employed in London comes from one or other of the quarry districts, Capo d'Istria, Parenzo, Rovigno, etc. In colour it is white, cream, grey, or yellowish. Hard, medium, and soft varieties are found, some of which polish well. It is much used in architecture, statuary, and paving in Vienna, Trieste, Venice, Budapest, etc. Other Cretaceous stones are the Karstmarmor (Crocusstein) of Sisana—a white stone speckled with black, used for all kinds of work; the Untersbergermarmor of Salzburg, light red or yellowish, with white flecks and red spots, used in a like manner; and the Brunnlitz Stone of Polička, Bohemia. Tertiary limestones are the Hundsheim Stone, Zogersdorf Stone, and the nullipore or marine algæ limestones known as the Wöllersdorf and Mannersdorf Stones of Lower Austria; also the Breitenbrunn Stone of Eisenstadt, Hungary.

Belgium.—The most important limestones occur in the Carboniferous formation; they are quarried at Perlonjour, Soignies, Ecaussines, Tournay, Hainaut, Namur, and the Ourthe Valley. This stone is blue, dark grey to black, with small white crinoidal fragments (*petit-granit*), and white. It is a good sound stone, which works freely and may be polished. Large blocks are sometimes venty.

Jurassic limestones are worked at Montauban in Belgium, and Luxembourg and Eocene limestones at Brabant and in the neighbourhood of Brussels (used in the Hôtel de Ville, eleventh century).

Germany.—In this country limestones do not take a prominent place as building stones. Devonian limestone is used in Walheim, Rhine Province, and Helbron, Westphalia. A bright yellow oolite comes from the Brown Jura (Dogger beds) of Roncourt, near Metz. The Muschelkalk is used in Hanover and many other districts for local purposes. Red and yellow Jurassic and Cretaceous limestones are worked at Kapfelberg, and elsewhere in Upper Bavaria; and Jurassic stones are quarried also at Kelheim, Eichstadt, Offenstetten, etc.

In Italy Jurassic, Cretaceous, and Tertiary limestones are employed, including a considerable amount of calcareous tufa.

In Scandinavia the dark grey or red Orthoceros limestone of Ordovician age is used a great deal for buildings, steps, and paving-slabs.

United States.—The limestones of this country call for no special attention here. In New York limestones are quarried in the Calciferous, Chazy, Trenton, Niagara, Lower Helderberg, and Tully beds. The 'Bedford Oolite' of Indiana is a good stone of Sub-Carboniferous age. Silurian limestones are quarried in Vermont, Illinois, Indiana, Iowa, Pennsylvania, and Wisconsin. Cretaceous rocks are used in Nebraska and Texas; Permian limestone is used in Kansas. An interesting white, granular, crystalline dolomite, with much bitumen, is recorded by Merrill as being used in Chicago. The sticky bituminous matter causes the dust to adhere to its surface, giving it an antiquated appearance, which is much appreciated by some people.

CHAPTER VII

SLATES AND OTHER FISSILE ROCKS

The Nature of Slate.—Almost any kind of rock which is capable of being split into thin sheets or slabs has been called slate by quarrymen and others all the world over; but it is essential from the geological standpoint, and advisable from a practical point of view, to distinguish sharply and clearly between true slates and other rocks from which thin sheets may be obtained.

The essential characteristics of a true slate is a fissibility that is independent of original bedding. This is not merely a capricious or artificial distinction, for the fissibility of slates is the result of secondary or subsequent changes in the rock induced by pressure, and accompanied by mineralogical and internal structural changes which affect the strength and weathering properties of the stone in the most marked degree. By the thin original bedding laminæ of a shale, it may be split into flakes even thinner than many first-class slates, but the strength is infinitesimal; it crumbles at the slightest touch or with its own weight.

There need be no difficulty in distinguishing regular thin bedded shales from slates; the former are nearly always softer, their laminæ are more easily separated, and their relationship with the adjoining rocks in the field at once exposes their simple stratified nature. In hand samples, should there be any doubt, it may be at once removed by examination of a microscopic section, which will reveal the true character. A rock possessing true slaty cleavage may be split along the cleavage at any point, whereas in a shale the splitting can be effected only along the bedding planes.

We will examine first the characters of true slates. Slates are essentially altered—metamorphosed—rocks. They are formed from fine-grained sedimentary rocks, or from certain igneous rocks, by compression and shearing stresses set up tangentially in the earth's crust by forces which need not be considered here.

It will be clear that, if they are formed in this way, slates must be sought in those regions of the crust that have been subjected to tangential stresses acting over considerable areas. They are not to be found in areas of slight crustal displacement, as, for instance, the Jurassic and Tertiary districts of central and south-eastern England; they occur in the crumpled and broken strata of the older rock areas of Wales, Cornwall, the Lake District, and so on.

It will be evident, further, that slates may occur in strata of diverse geological periods, provided they have suffered the squeezing necessary for their formation; thus they appear in different parts of the earth in all the great geological systems, from the Archæan to the Tertiary, but in the younger systems only when the rocks have been uplifted by comparatively recent movements, as in the Alps. From the geological point of view, slates are common and widely-distributed rocks, but commercially good slates are not by any means so abundant.

Characters of Slate in the Mass.—The most obvious feature of masses of slate is their 'slaty cleavage'—that is, their property of splitting regularly in a definite direction. These directions of ready cleavage are, in the great majority of cases, disposed at a high angle with the horizon; they are often nearly vertical. Beds of rock in folded regions are often bent over into similar positions, but their ordinary characters of bedding are not thereby destroyed. In rocks possessing slaty cleavage, however, the original bedding is often completely obliterated. Traces of original bedding are not always entirely wiped

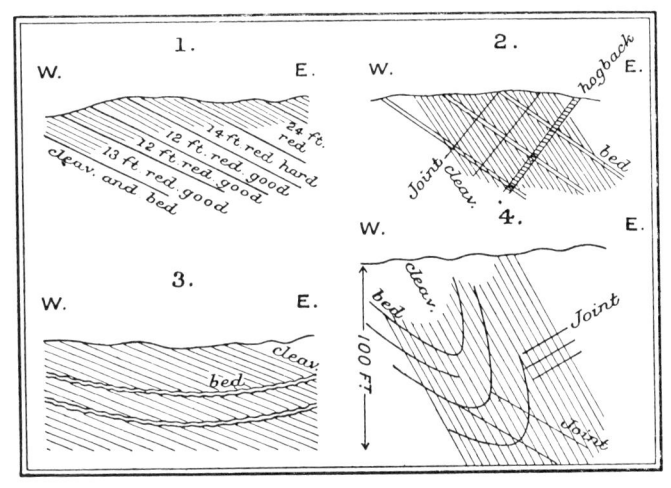

FIG. 16. — DIAGRAMS OF AMERICAN SLATE QUARRIES. (After Dale, Bull. 275, United States Geological Survey.)

1. Cleavage and bedding parallel. Empire Slate Quarry, Granville, New York.
2. Cleavage and bedding cut one another at a low angle. Sea green slate, Paulet, Vermont.
3. Crumpled bed. Fair Haven, Vermont.
4. Cleavage crossing folded beds. Sea-green slate, West Paulet, Vermont.

out; they may be indicated in various ways. Very frequently the beds are shown by what is called 'Stripe' in the slates—this is, an appearance of banding or striping, making itself evident by slight differences of colour or texture on the cleavage surface, and crossing it at any angle. Any structure that had been imposed on the bedding prior to cleavage will be evident to a greater or lesser extent in the cleaved state; thus the stripes may run regularly for long distances through the mass of slate, or they may be crumpled and folded or faulted.

Hard original beds of grit or quartzite are often found in masses of slate which have completely resisted the process of cleavage; they have frequently accommodated themselves to the pressure by shortening, either by crumpling up or fracturing, as shown in Fig. 16 (3). The independence of cleavage and bedding is often clearly exposed, as in Fig. 16, where it will be seen the cleavage is developed at various angles to the bedding, sometimes being coincident with it and sometimes normal to it.

The 'strike' of the cleavage planes, or the trend of their direction at right angles to their true dip, is usually constant, except for minor variations, over large areas. Hard masses of uncleavable rock will occasionally divert the strike of the cleavage from its regular course or produce variations in the dip. The notion that all slates have the strike of their cleavage orientated in the same manner, running, for example, always a few degrees off north and south, has been elaborated more than once in this country. The writer has even come across individuals who could not believe in the quality of an excellent slate because it did not conform to this rule. It is hardly necessary to point out that this idea is erroneous, for the direction taken by the cleavage planes is governed by the direction along which the compressive forces acted and the form and position of neighbouring rigid masses, and these differed in different parts of the crust; moreover, the pressure changed in direction at different times, even in the same area. Masses of slate are commonly broken up like other rocks by systems of joints, which facilitate the process of quarrying [Fig. 16 (2)]. It is a

SLATES AND OTHER FISSILE ROCKS

common occurrence to find the good slate of a certain locality passing within a short distance into useless rock through the multiplication of joints.

Microscopic Characters of Slate.—Slates are frequently classified as 'Clay Slates' and 'Mica Slates,' but it must be admitted that this classification is highly artificial. Examination of all kinds of slate under the microscope reveals a striking uniformity of mineral composition and mechanical structure, the differences between them being mainly those of degree and less those of kind. A so-called clay slate—that is, one in which there is no microscopic crystalline aggregate or very little of this structure—is practically non-existent, so far as commercial slate is concerned. Micaceous slates, on the other hand, show every gradation from an ultra-microscopic micaceous felt-work to coarse mica schists.

The minerals forming slate are these: *Muscovite*, perhaps most commonly in the form known as *Sericite*, *Chlorite*, *Quartz*. Then come the following, of lesser importance, and not invariably present: chalcedony, felspar, biotite, pyrite, magnetite, hæmatite, limonite, carbonates of lime, iron and magnesia, graphite or carbonaceous matter, epidote, andalusite, rutile, anatase, tourmaline, zircon, and a few other occasional minerals.

These minerals occur in slate either as original clastic grains belonging to the rock before cleavage, or as secondary products formed during the processes of cleavage and alteration. In some slates the only original mineral recognizable is zircon, in minute granules; in others a good deal of quartz, some mica and felspar, with tourmaline and rutile, are obviously in the condition in which they were deposited in the unaltered sediment. The coarser, rough-feeling slates usually contain more original mineral than the smoother and finer kinds.

Arrangement of the Minerals. — Characteristic of slates is the felted arrangement of mica and chloritic minerals. There is in most slates a finer-textured ground, or base, of very minute scales of sericite or chlorite; in this base are irregular streaks and layers of the same minerals in coarser flakes. The micaceous and chloritic flakes are irregular in form and have ragged edges; they may be so small as to be distinguishable only with great difficulty with the highest power of the microscope and the best illumination, or they may be visible, in parts of the rock at any rate, with the naked eye. By far the greater part of the felt-work is of secondary origin; it has been formed during the process of shearing and compression, and it is to this more than to any other feature that slates owe their useful properties.

Entangled within the felt are the other minerals; of special importance is the presence of minute granules of quartz, sometimes with chalcedony, both in the fine base and the coarser parts of the rock. In addition, we find sometimes larger clastic grains of quartz, felspar, and mica squeezed and drawn out into eye-like forms, round which the micaceous minerals closely wrap. These 'eyes,' or 'knots' resembling them, are also produced by other minerals, magnetite, pyrite, andalusite, and so on; they often give a lumpy surface to the cleavage face of slates.

Summarizing the more important points regarding the minerals of slate, we find:

1. Free silica occurs as original grains of quartz and as secondary quartz and chalcedony pervading the whole rock; the grains are often flattened and have ragged edges.

2. White mica—sericite—in many slates makes the bulk of the ground mass in elongated flakes, from 0·006 to 0·06 millimetre and more in length, and 0·00017 milli-

metre thick. Larger flakes of muscovite are sometimes original.

3. Chloritic scales are present in most slates; in some they form a considerable part of the ground mass. They are usually rather larger than the sericite flakes.

4. Felspar occurs in small original grains, and in some slates a good deal of secondary felspar is developed: most commonly a clear transparent albite, in small grains.

5. Biotite is less common; it occurs in small scales.

6. Calcite and carbonates of iron and magnesia occur in thin plates, rhombic crystals, and grains; also in the lenses, knots, and eyes.

7. Pyrites occur in small cubes and lenses; 300 have been counted in a square millimetre and in tiny spherules, 0·0017 to 0·027 millimetre, and as many as 600 have been observed in a square millimetre (Dale).

8. Magnetite occurs in distorted octohedra, usually of small size.

9. Hæmatite is common in red and purple slates in crystal flakes and small specks, 0·0004 to 0·01 millimetre in diameter.

10. Rutile is most commonly present in the form of microscopic needles, often extremely thin; as many as 2,400 have been estimated in a square millimetre of some slates. They are probably of secondary origin.

The appearance of thin slices of slate under the microscope will depend upon the direction taken by the section. In an average slate there are observable differences in the three principal directions of section, namely: (1) parallel to the dip of the cleavage and at right angles to the cleavage plane; (2) perpendicular to the direction of dip and the planes of cleavage; (3) parallel to the plane of cleavage.

Sections cut in the first direction exhibit the edges of the micaceous and chloritic flakes running through the slice in straight or slightly wavy lines, overlapping one another and closely adpressed, curving round the small lenses of quartz, magnetite, calcite, etc., if these are present, and in polarized light all extinguishing together as the slide or the nicols are rotated. Close examination of such a section may reveal the clastic grains of quartz, felspar, mica, or minute rhombs of calcite, dolomite, etc.

Sections cut in the second direction are very similar to the first, and sometimes are quite indistinguishable; but

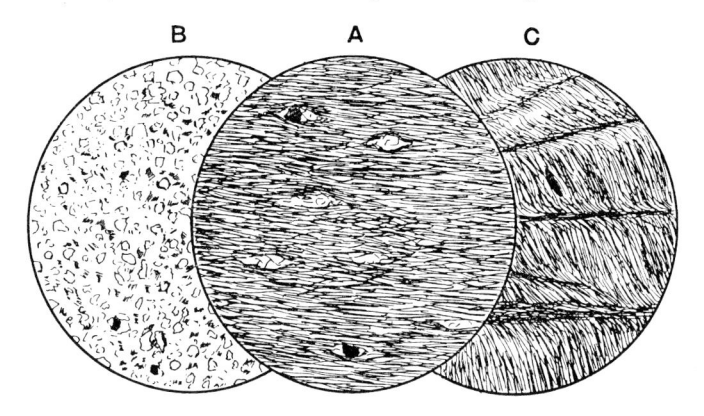

FIG. 17.—MICROSCOPIC SECTION OF SLATE.

A, at right angles to the cleavage.
B, parallel with the cleavage.
C, strain-slip cleavage.

in most cases they exhibit this important difference—the matted mica or chlorite flakes appear *shorter* in this direction than they do in the first.

Sections in the third direction—parallel to the principal cleavage—are strikingly different from either of the others; in these there is no trace of that fibrous, felty appearance so characteristic of the sections normal to the cleavage. Instead, we find a mosaic-like aggregate of ragged-edged, irregular flakes of mica and the other minerals, and the elongation of the flakes in one direction more than another is not always very clearly marked in the small area visible at one time.

The texture of slates may be coarse or fine, their appearance may be dull or shiny, they may feel rough or smooth to the touch; but no slate is worth consideration for use in exposed conditions and damp places unless it clearly shows under the microscope a thoroughly *crystalline* structure. To be sure that this is the case, the thin slices of the rock should be of the utmost possible tenuity, for many slates which appear dull and full of amorphous clayey matter in polarized light, if they are not thin enough, can be shown to be crystalline if they are examined in very thin slices, with high power and good illumination.

Structural Peculiarities of Slate.—Turning now to the examination of slates in blocks and slabs, there are one or two characters which require brief attention. On the cleavage surface of some slates, not in all, there appear rather irregular wavy lines, roughly parallel to one another, and traversing the slate in a fairly constant direction. These are the indications of what is called the *grain* of the slate. The grain is an obscure structural feature of many slates, which causes a tendency to fracture parallel to the grain; it is therefore a cause of weakness. Elongated slates should be so fashioned that their greatest length is in the direction of the grain, and not across it. In many cases it is nearly coincident in direction with the dip of the cleavage, and makes a low angle with the cleavage face; in the slate of Rimogne this angle is from 1 to 20 degrees, in the Fumay Slate it is 6 degrees.

It appears that grain is caused by some of the mineral flakes being arranged with their flat faces in the direction of pressure—that is, at a high angle to the cleavage. This has been observed in the case of flakes of chlorite and hæmatite, and in lenses and distorted (elongated) crystals of magnetite and pyrite.

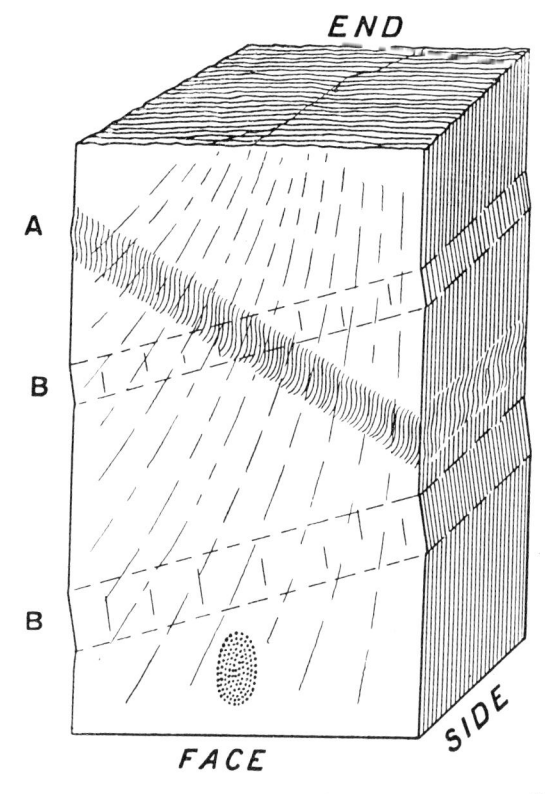

Fig. 18.—Diagram of Common Structures in a Block of Slate, showing a Cleavage Face crossed by Two *Stripes*, or Beds (B), and a Belt of Crumpling (A).

The fine-curved lines on the cleavage face are indications of the *grain* brought out by the act of splitting.

Another structure which is often a cause of loss and annoyance is *slip-cleavage* or *false cleavage*. This is a kind of subordinate, incipient, incomplete cleavage, developed in the rock at the same time as the primary or main cleavage or subsequently. Its effect is to cause the slate to break short in directions other than the main cleavage. Its presence is not always made evident in the rock mass, but it is readily detected by the microscope (Fig. 17). There may be one or more slip-cleavage directions in a rock, each set pursuing its own course and independent of the others.

It sometimes happens that a body of slate is rendered useless through the development of more than one strong cleavage; secondary and tertiary cleavages having been

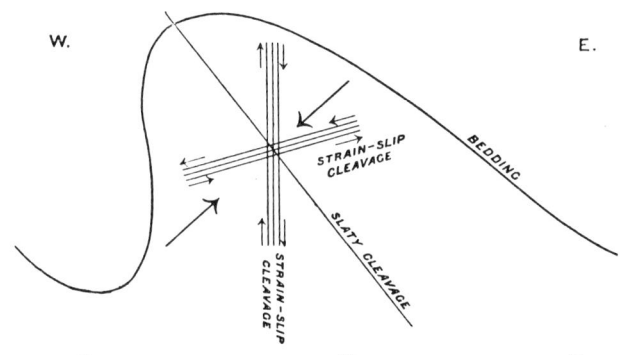

FIG. 19.—DIAGRAM SHOWING THE RELATIONS BETWEEN FOLDING CLEAVAGE AND STRAIN-SLIP CLEAVAGE IN THE SLATE OF ARGYLL.

(From Memoir of the Geological Survey 'The Geology of the Seaboard of Mid-Argyll,' by permission of the Controller of H.M. Stationery Office.)

superimposed on the original cleavage by repetition of the same processes at subsequent periods, by forces acting in different directions from that which produced the first cleavage. The effect of these additional cleavages is to cause the rock to break up into polygonal blocks.

Somewhat akin to slip-cleavage is incipient or partial shearing, which takes places as a secondary phenomenon in some districts. Its effects are best explained by reference to the figures. The first stage of the process is the production of minute bends across the cleavage, which pass over into fractures; along them more or less movement has taken place.

If these *shear zones,* as they are called, are of small amplitude, they give an irregular wavy surface to what would otherwise be a flat cleavage face; where they are more pronounced the effect on the surface resembles the rucking up of a tablecloth by pushing along the table from opposite ends. Devonian slates with these characters are used in parts of Cornwall for rough walling, but they are not suitable for any of the ordinary applications of slate.

This structure goes by different names—in Wales the sudden bends are known as 'crychs,' or 'curls'; in America they are 'hog-backs.'

The position of zones of shearing, as of the microscopic slip-cleavage, is often capable of being localized exactly in beds which have been folded. Thus it has been observed that the shearing tendency manifests itself in the neighbourhood of the apices of the folds.

When the bedding planes coincide with the cleavage faces, indications of fossils are quite common, as in the 'Delabole Butterflies,' fossil brachiopods of the Delabole Quarries, Cornwall; but where the bedding crosses the cleavage as stripes, the fossils are difficult to extract, although they may be present.

The **Colour** of slate is a matter of some importance, since, other conditions being equal, it is not seldom the principal factor in determining its selection. The colour ranges from pale to dark neutral grey, and from grey with a slight purple tinge to strong purplish-red, or from pale greenish-grey to bright green; many have a bluish tinge. Colour is no index of quality, and slates of several shades of colour are frequently obtained from different beds or 'veins' in the same quarry.

The colour is dependent upon the mineral contents of the slate. Dark greys, blues, and so-called blacks, owe their colour to iron and carbonaceous matter, and these varieties are usually more rich in pyrites than the slates of other tints. Purple or red slates are coloured by specks and plates of hæmatite (Fe_2O_3); the greater the amount of this mineral, the purer is the red colour. Green slates mostly owe their colour to chlorite, but this tint is produced, at least in part, by minute granules of epidote in certain slates. Ferrous oxide (FeO) is more abundant in the unfading green slates than in the red, but the colour is no index of the total iron content; some of the dark Welsh slates contain more iron than some of the reddest varieties from Vermont.

As in the case of building stones, the appearance is greatly influenced by the texture of the surface; it may be rough, dull, and dead in appearance, or with the development of much mica on the cleavage surfaces a bright and shimmering aspect may be imparted to slate of any colour. Messrs. Reade and Holland well describe this effect on the buildings in the Valley of the Meuse: 'The play of light and shade and the reflecting surface has a most charming effect, of which one becomes more conscious on an attempt to sketch any of the numerous interesting edifices with which the valley is adorned. At one time a portion of a roof will be in darkest shade, in another moment it will be resplendent as polished silver.'

For special purposes particoloured, stained, or faded slates are required. These are seldom to be obtained from the larger quarries producing ordinary goods, more because the management cannot be troubled with their selection than on account of their absence from the workings. A striking example of brick-red iron-stained dark blue or black slate is to be seen in the roof of Maentwrog Church, and one or two other structures in that neighbourhood.

Applied with moderation and judgment these slates are extremely effective, but their use requires considerable restraint.

Spots and blotches of different colour arise from several causes. Some brownish stains are only produced after exposure; these are generally due to oxidation and hydration of decomposing iron compounds, generally marcasite, or very fine granular pyrites, or ferrous carbonate. The Coryton slates of Devonshire exhibit this change to brown. Well-crystallized iron pyrites has usually no evil influence on the weathering of an otherwise sound slate.

Pale green ovoid spots are common in some dark grey and purple slates, and in some red ones; they have rarely any bad effect on the wear of the slate, though they may be objected to on the ground of appearance. In most cases these spots represent centres of reduction of the iron compounds in the beds prior to metamorphism. Spots originally spherical are frequently found to be elongated into ellipsoids, with their major axes in the direction of the cleavage planes (see Fig. 18).

Chemical Characters of Slate.—Dale (T. N. Dale, 'Slate Deposits and Slate Industry of the United States': United States Geological Survey, Bulletin 275, 1906) gives the range of composition in twenty-nine analyses of American slates as follows:

	Per Cent.
Silica	55 to 67
Alumina	11 to 25
Ferric oxide	0·52 to 7
Ferrous oxide	0·46 to 9
Potash	1·76 to 5·27
Soda	0·50 to 3·97
Magnesia	0·88 to 4·57
Lime	0·33 to 5·20
Water above 110° C.	0·33 to 5·20

Examination of a few selected analyses, however, will be more instructive. The twenty-one analyses of Table XI.

show how complex these rocks are chemically, and how little difference there really is between slates of very different appearance and character (see tables on pp. 288, 289, 290).

Chemical analysis throws no light whatever upon the differences found among slates in practice, except as regards fading. The researches of Hillebrand, Dale, Reade, and Holland clearly show that the content of lime in a slate—a point which is often emphasized in trade analyses—has little to do with the wear of the slate, much less than might be expected. A Buttermere green slate which was found by Reade and Holland to contain nearly 24 per cent. of lime was declared by them to be a splendid roofing slate, and the strong and durable Dinorwic slate appears to contain more lime than some of the other well-known and widely-used Welsh slates.

The purple and red slates contain more ferric and less ferrous iron than the green varieties.

Merriman has pointed out that the best and strongest slates have the highest percentage of silicates of iron and alumina.

The discoloration of slates is in a great measure due to the oxidation of the ferrous carbonate which occurs most commonly in isomorphous mixture with the carbonates of lime and magnesia. Dale states that the bleaching of the dark slates of Pennsylvania appears to proceed by these steps:

1. Removal of carbonates of lime and magnesia, accompanied by oxidation of the ferrous carbonate and deposition of limonite.

2. Oxidation and removal of carbon, also oxidation and removal of pyrites.

3. Removal of almost all limonite.

TABLE XI.

	1 Delabole, Grey.	2 Maenofferan, Dark Blue-Grey.	3 Alexandra Quarry, Reddish.	4 Ardennes, Devillian Violet.	5 Penrhyn, Purple.	6 Penrhyn, Purple.	7 Velenhelli, Blue.
SiO_2	56·75	53·92	60·17	60·56	57·75	63·01	63·30
Al_2O_3	21·81	24·09	18·89	19·72	16·44	16·28	14·99
Fe_2O_3	1·36	1·87	6·17	6·57	10·84	8·42	8·75
FeO	7·20	6·52	0·95	1·86	1·66	0·66	1·53
MnO	trace	0·19	0·20	0·11	0·25	0·12	0·25
TiO_2	0·71	0·80	1·15	0·55	1·27	0·58	1·04
CaO	0·84	0·24	1·75	trace	0·23	0·72	1·08
MgO	2·48	1·80	1·85	1·87	2·36	1·95	1·63
K_2O	3·72	3·64	2·76	4·64	4·23	2·98	2·84
Na_2O	0·97	0·74	1·39	0·55	1·30	2·14	1·92
CO_2	—	—	1·04	—	—	—	—
P_2O_5	trace	0·19	0·11	0·09	0·21	trace	0·07
Carbon	—	—	—	—	—	—	—
H_2O	4·23	5·87	3·70	3·85	3·35	2·95	2·88
BaO	—	—	0·04	—	—	—	—
SO_3	—	—	—	—	—	—	—
FeS_2	—	0·13	—	—	—	—	—
Loss on ignition	—	—	—	—	—	—	—
Alkalies, etc.	—	—	—	—	—	—	—
Moisture, etc.	—	—	—	—	—	—	—
	100·07	100·00	100·17	100·37	99·89	100·08	100·18

1. Delabole, Cornwall. Analysis by Reade and Holland.
2. Maenofferan, Wales. Trade analysis by Blount.
3. Alexandra Quarry (best slate), Moel Tryfaen, Wales. Analysis by Reade and Holland.
4. Ardennes. Analysis by Reade and Holland.
5. Penrhyn, Wales. Analysis by Reade and Holland.
6. Penrhyn, Wales. Analysis by Reade and Holland.
7. Dinorwic or Velenhelli, Wales. Analysis by Reade and Holland.

SLATES AND OTHER FISSILE ROCKS

TABLE XI—*Continued.*

	8	9	10	11	12	13	14
	Oakley, Old Vein, Dark Blue.	Oakley, New Vein, Dark Blue.	Lehigh, Co. Pa., U.S.A., Black.	Vermont, Black.	Vermont, Purple.	New York, Red.	Rochefort-en-Terre, Light Grey.
SiO_2	52·25	53·40	56·38	59·70	61·29	63·89	53·30
Al_2O_3	24·60	26·67	15·27	16·98	16·24	11·80	25·48
Fe_2O_3	10·40	9·53	1·67	0·52	4·63	4·56	—
FeO	—	—	3·23	4·88	2·62	1·33	7·80
MnO	—	—	—	0·16	—	—	—
TiO_2	—	—	0·78	0·79	0·77	0·52	—
CaO	1·00	0·90	4·23	1·27	0·60	2·25	2·90
MgO	2·09	1·85	2·84	3·23	2·99	4·57	2·40
K_2O	—	—	3·51	3·77	5·27	3·95	2·38
Na_2O	—	—	1·30	1·35	1·38	0·50	0·12
CO_2	—	—	3·67	1·50	0·54	3·15	0·12
P_2O_5	—	—	—	0·16	—	—	—
Carbon	—	—	0·59	0·46	—	—	—
FeS_2	—	—	1·72	1·18	0·04	0·02	—
Loss on ignition	4·06	4·47	1·11	0·30	0·56	0·77	5·20
Alkalies, etc.	2·04	3·18	—	—	—	—	—
H_2O	—	—	4·09	3·82	3·16	2·82	—
Moisture, etc.	—	—	—	—	—	—	—
BaO	—	—	—	0·08	—	—	—
SO_3	—	—	—	—	—	—	—
	—	—	100·39	100·05	100·09	100·13	—

8. Oakley Mines, Wales. Trade analysis.
9. Oakley Mines, Wales. Trade analysis.
10. Lower Franklin Bed, Old Franklin Quarry, Slatington, Lehigh, Pennsylvania. Analyses by Hillebrand, United States Geological Survey.
11. Vermont, U.S.A., Benson Village, Rutland County. Analysis by Hillebrand.
12. Vermont, U.S.A. Mean of two analyses by Hillebrand—one in South Poulteney, the other from near Hydeville, Castleton.
13. New York, U.S.A. Mean of four analyses by Hillebrand.
14. Rochefort-en-Terre, Morbihan, France. Trade analysis.

GEOLOGY OF BUILDING STONES

TABLE XI—*Continued.*

	15 Ardennes, Green.	16 Egryn, Green.	17 Velinheli, Second Green.	18 Elterwater, Green.	19 Tilberthwaite, Best Dark Green.	20 Vermont, Sea Green.	21 Vermont, Unfading Green.
SiO_2	57·89	59·16	63·06	52·67	50·16	63·33	59·37
Al_2O_3	20·40	18·42	18·03	12·63	17·85	14·86	18·51
Fe_2O_3	4·93	7·67	2·24	1·58	1·65	1·12	1·18
FeO	3·98	2·92	4·07	6·95	6·36	4·93	6·69
MnO	trace	0·14	0·30	0·19	0·48	—	—
TiO_2	0·89	0·17	0·73	1·18	1·03	1·73	1·00
CaO	—	0·04	0·81	5·78	3·67	1·20	0·49
MgO	2·42	2·54	2·21	6·37	6·35	2·98	2·36
K_2O	4·74	4·01	3·07	2·47	2·06	4·06	3·78
Na_2O	1·26	0·80	1·51	1·31	3·11	1·22	1·71
CO_2	—	—	—	5·41	2·45	1·41	0·30
P_2O_5	trace	0·07	0·06	0·16	0·16	—	—
Carbon	—	—	—	—	—	trace	—
FeS_2	—	—	—	—	—	0·11	0·14
Loss on ignition	—	—	—	—	—	0·69	0·51
Alkalies, etc.	—	—	—	—	—	—	—
H_2O	3·08	4·00	3·62	3·54	4·67	3·37	4·01
Moisture, etc.	—	—	—	—	—	—	—
BaO	—	—	—	0·02	0·03	—	—
SO_3	—	—	0·09	—	0·10	—	—
	99·59	99·94	99·80	100·26	100·13	100·01	100·05

15. Ardennes, France. 'Phyllade aimantifere.' Analysis by Reade and Holland.
16. Egryn, near Barmouth, Wales. Analysis by Reade and Holland.
17. Dinorwic or Velinheli, Wales. Analysis by Reade and Holland.
18. Elterwater (volcanic slate), Lake District, England. Analysis by Reade and Holland.
19. Tilberthwaite (volcanic slate), Lake District, England. Analysis by Reade and Holland (T. M. Reade and T. Holland, Proc. Liverpool Geol. Soc., 1897-1904).
20. Vermont, U.S.A. Mean of three analyses by Hillebrand.
21. Vermont, U.S.A. Mean of two analyses by Hillebrand.

Distribution of Slates.

Wales.—Slates in Wales are obtained from the Cambrian, Ordovician, and Silurian rock systems.

The *Cambrian* rocks are concentrated in two anticlinal tracts, one on the north-west side of the Snowdon district, passing south-westward from near Aber to near Llanllyfni, by Llanberis, and the other occupying the areas between Festiniog, Portmadoc, Harlech, Barmouth, and Dolgelly, and extending in a narrow strip down the coast to Aberdovey. The intervening Snowdon area is synclinal in structure, and is formed of Ordovician rocks.

The northern or Llanberis mass is by far the more important of the two Cambrian slate regions. It is in the Lower Cambrian that the slates are worked.

The strata composing these Cambrian rocks are thick masses of interbedded slates and grits, with occasional igneous intrusions. In descending order the general succession in the slate belt is as follows:

		Feet.
	Grits and slates.	
Workable Belt of Slate Rock	Green slates Blue and purple slates (principal group) Thin grit band Purple slates Thin calcareous grit Dark purple slates Grit ... Purple and green banded slates	500 to 1,000
	Green and grey grits. Crushed and useless slates. Grit. Conglomerate.	

This section varies, of course, in detail from point to point, but it holds good generally in the Llanberis and Nantlle districts. Most of the rocks are on the eastern side of the anticline, and both bedding and cleavage dip towards the south-east.

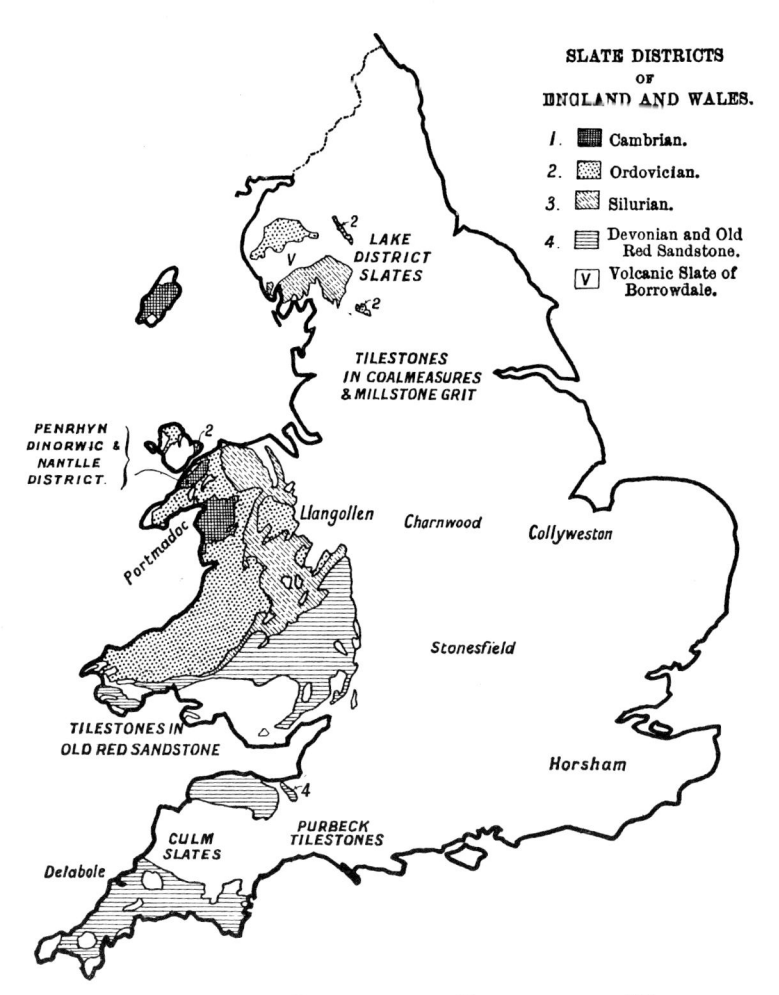

MAP VII.—SLATE DISTRICTS OF ENGLAND AND WALES.

SLATES AND OTHER FISSILE ROCKS

In the Penrhyn and Dinorwic Quarries the whole series is worked; but farther south, in which direction some of the beds increase in hardness, the whole set is not exposed in one quarry.

Davies gives the local names used at Llanberis and Nantlle as below:

LLANBERIS.		NANTLLE.
Green vein	Green	Green vein.
Goch galed, or silky red vein ...⎫ Glas galed, or hard spotted vein ⎭	Purple	⎧ Silky vein. ⎨ Blue mottled vein.
Glas Rhiwiog, or Royal Blue ...⎫ Goch Grychlyd, or curly red vein⎟ Hen las, or old blue vein ...⎟ Goch galed, or red hard vein ...⎭	Blue	⎧ Red and blue striped vein. ⎪ Red spotted vein. ⎨ Purple vein. ⎩ Purple striped vein.
Glynrhonwy vein		Faen Goch, or red vein.

As indicated above, these beds, or 'veins,' are not constant in character, as they are traced through the district; thus, the Faen Goch is a thick and valuable vein in the south, but is not much worked in the north; and in the southern area nearly all the veins tend to have a reddish tinge; the silk vein in the south is often too soft to make a good slate. The jointing is much more pronounced in the south of the district.

The two largest quarries, Penrhyn and Dinorwic, vast excavations on the mountain side, produce very similar material. The Dinorwic or Velinheli Quarry is situated on the face of Elidir Mountain, and from the base, in places below the level of the adjoining lakes, to the quarry top it has a working face of 1,800 feet. The work is carried on in step-like galleries, each cut 75 feet above the one below. The lower veins are obtained in the lower part, and the higher beds in the upper part of the quarry, both here and at Penrhyn.

The qualities produced at Dinorwic are given in Table XII.

TABLE XII.

Names of Dinorwic Slates.	Colour, etc.	Average Number per foot of Thickness.
Best Old Quarry	Bluish	70 (about)
Best New Quarry	Grey	62
Second New Quarry	Grey	42
Third New Quarry	Grey	26
Best Red Quarry	Reddish tinge	70
Second Red Quarry	Reddish tinge	45
Best Green and Wrinkled	Reddish tinge	70
Second Green and Wrinkled	Reddish tinge	40
Third Green and Wrinkled	Reddish tinge	28
Best Mottled	Grey, slightly spotted	62
Second Mottled	Grey, slightly spotted	42
Third Mottled	Grey, slightly spotted	28
Best Green	Sea green	62
Second Green	Sea green	42
Third Green	Sea green	28
Best Ton Slates	Various sizes from 42 to 26 inches long	30 sq. yds. per ton
Second Ton Slates	Various sizes from 42 to 26 inches long	24 to 25 sq. yds. per ton
Third Ton Slates	Various sizes from 42 to 26 inches long	18 to 19 sq. yds. per ton

The so-called red slate as sold is usually a reddish-purple, but there are veins of bright red spotted with green in this district.

In the Harlech anticline area there are plenty of good slates, but unfortunately there are usually too many associated grit beds to permit of economic quarrying on a large scale.

The slates of the Lower Cambrian of Wales are uniformly more gritty and siliceous than those of the higher beds; the difference will be noticed at once if they are scratched with a knife.

The darker slates of the Lingula Flags and Tremadoc

SLATES AND OTHER FISSILE ROCKS

Beds are quarried to some extent in Carnarvonshire and Merionethshire; their cleavage is irregular, and intervening beds make quarrying difficult.

Turning now to the next younger rock system, the *Ordovician*, we find the principal workings in the neighbourhood of Blaenau Festiniog. Here the slate is smoother and finer in grain than that of the Lower Cambrian, and it is rather darker in colour. There are open quarries in this district, but the most important excavations are mines. The dip of the bedding and cleavage about Festiniog is northerly at about 45 degrees.

There are five principal veins of slate, separated, as is usual, by beds of 'hards'—gritty bands—and penetrated by igneous dykes and veins of quartz. They are, in descending order, the Top Vein, Back Vein, Small or Stripy Vein, Old Vein, and New Vein. Most important is the Old Vein, 15 to 40 yards thick, of dark blue colour, free from spots, and with very perfect, even cleavage. In this district the following subdivisions are recognized in it from south to north:

1. Y Crystin, the crust.
2. Y Wythern Ddu, Black Vein.
3. Y Wythern Fawr, Great Vein.
4. Y Wythern Wen, White Vein.
5. Y Pump Wythern, Five (Stripes) Vein.
6. Y Crych Ddu, Black Curl.
7. Hard Slates.

Below (1) is a series of 'Hards,' and above (7) is another series, very constantly underlain by a thin seam of clay, which is a useful indication of the presence of the Old Vein. The Top, Northern, or Flag Vein, 100 yards thick, is most worked about Rhiwbach, where it is 230 yards thick. The Back Vein, 15 to 25 yards, sometimes called the 'Barred Vein,' on account of the presence of stripes and bands caused by sandy and soft courses, is variable in quality, but is very good at Maenofferen, where it is largely wrought. The New Vein, 10 to 50 yards thick, is a good dark blue slate, with regular, even cleavage; but it is more liable to 'crychs,' quartz veins, and pyrites than the Old Vein. It is, perhaps, in the best condition north of Rhiwbach. The small vein is 10 yards thick. All these veins are worked in the great underground chambers of the Oakley mines; at the Maenofferen mines the Old Vein and the Back Vein are obtained (Fig. 20).

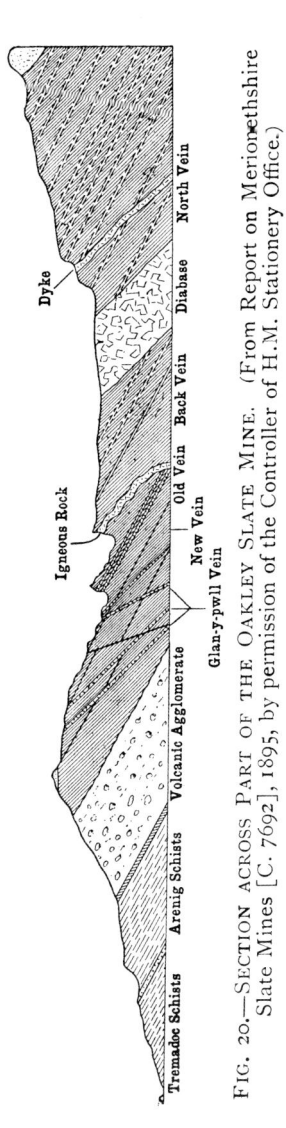

The other Ordovician slate quarries do not call for special notice; the principal quarries now working are included in the list (p. 430). To this a few brief notes may be added. The slates obtained from Cader Berwyn and Dolgelly are dark blue-grey, with fairly good cleavage; but they are somewhat softer than the ordinary Welsh Ordovician slate, and considerably softer than those from the Cambrian. In the Corris district there are two workable veins — the Narrow

FIG. 20.—SECTION ACROSS PART OF THE OAKLEY SLATE MINE. (From Report on Merionethshire Slate Mines [C. 7692], 1895, by permission of the Controller of H.M. Stationery Office.)

SLATES AND OTHER FISSILE ROCKS

and Broad Veins—separated by a considerable thickness of 'hards.' The former, blue-grey in colour, 15 to 30 yards thick, is more suited for slab than roofing stock.

The dark-blue slates of the Broad Vein mined at Minllyn, too hard for fine splitting, are worked for slabs alone; they are of excellent quality. Both veins are free from excessive jointing.

The Dolgelly slates are rather rich in pyrites. West of Snowdon, good slates are got at Bwlch Cwmlan, grey-blue in colour, free from pyrites, and of fairly good and tolerably smooth cleavage. There are two main veins, the Eastern and the Western. These slates dip south-westerly further west; similar beds dipping in the opposite direction have been worked in several quarries, mostly yielding grey-blue slate of fair cleavage, some rough and strong and some smooth (Dinas Lake, Cwm Trescol, Dolgarth, Hendre Ddu). The quarries of the Capel Curig and Bettws-y-Coed district produce dark blue slates, free from pyrites and of good, fairly even cleavage.

In South Wales, Pembrokeshire, and Carmarthenshire the Ordovician rocks (Llandeilo) yield rather pale blue, grey-green, or greenish-grey slates. Slates are quarried in the Bala Beds at Cader Berwyn and near Llangynog; the Rhewarth Quarries are very old. There are three main slate veins of moderate thickness, separated by beds of hard rock; they yield rather heavy slates, of dark blue colour and free from pyrites. They wear well and are strong. The Precelly green and rustic slates come from Clynderwen.

For list of chief Welsh quarries, see Appendix D.

Silurian slates are obtained from the Wenlock beds (Denbighshire Grits and Slates) in North Wales, near Corwen and Llangollen; they are dark blue-grey in colour and cleave well, but they are softer than most of the Welsh slates. Notwithstanding this, they appear to wear well, and are tolerably free from pyrites. The cleavage of these beds is generally found lying at low angles.

A cream-coloured slate, tinged with green, has been worked a little at Clegir Quarry, Bethws Gwerfyl Goch.

In the **Lake District** slates have been worked in three formations: (1) In the Ordovician Skiddaw Slates; (2) in the Borrowdale Volcanic Series (Green stones and Porphyries); and (3) in the Silurian Slates to the south of the latter (see Appendix D).

The grey Skiddaw Slates and soft greyish-blue Silurian Slates may be very briefly dismissed. There are numerous small quarries, but none is of importance, and when the building trade is in a bad condition many of these cease working altogether.

The well-known Green Slates of the Lake District are obtained from the great series of volcanic rocks of Upper Ordovician age (Llandeilo) (see Map VII.). In this great mass of material—8,000 to 9,000 feet thick, consisting of tuffs, breccias, and lava-flows—only about a dozen beds of ash or tuff have been cleaved in such a way as to make them of commercial value.

Most of the slates have a rather rough cleavage surface, and they are usually thicker than the average Welsh slate—$\frac{5}{16}$ inch is a common thickness. The colour is pale green (sea green), but in some of the quarries—Parrock, Honister, etc.—a darker olive green slate is obtained, which can generally be split rather thinner than the paler kinds. Most of the pale green slates fade a little on exposure to a very beautiful silvery greenish-grey.

The workings are either open quarries or mines; those on the high ground are in several cases open at the top, but the material is all removed from the bottom through an adit.

The Ashgill Slates are worked south-west of Coniston, and the Wenlock beds of Troutbeck and Kirkby Moor have been quarried for roofing slates.

Cornwall.—The slates of Delabole, in North Cornwall, are of Upper Devonian age. There are many small quarries in the same district, employing irregularly from one to three men, but there are only two of importance—the Old Delabole Quarry in St. Teath and Lamb's House Quarry at Tintagel, and of these the former is vastly the larger. The slates of Delabole and Tintagel are dark bluish-grey or neutral grey in colour, but those of Tintagel are somewhat rougher, and have a less regular cleavage. The outcrops of these two slates are quite separate. The strike of both the bedding and cleavage at Delabole is north-west to south-east, the dip of the cleavage being 24 degrees towards the south-west, and that of the bedding 30 degrees in the same direction. The quarry, which occupies the site of several older ones, is a large excavation of great depth (500 feet), worked in open galleries, the slate blocks and waste being drawn up by inclined tram and aerial rope-way.

The slate rock is divided into two good portions by a belt of fossiliferous poor rock (hards), 45 feet thick; above this is the Old Delabole Slate, worked to a depth of 450 feet, and below it is a thick mass of equally good, or even better, slate, worked to a depth of 100 feet, and proved a good deal deeper.

Most of the slate is free from spots, blotches, and striping; some pyrite is present in parts of the quarry, as well as veins of quartz and calcite in places. The smallness of the angle between the bedding and cleavage causes an irregular wavy striation of the split faces. The method of counting here is by dozens; three extra slates is the allowance for breakages in each lot of five dozen, or

60 per 1,000. The output comprises roofing slate, and sawn, planed, or sanded slabs for all kinds of work. More recently a 'green-grey,' or 'Abbey Grey,' slate and rougher 'rustic' slate has been produced at this quarry, for use when colour effects have to be considered. The dust has been tried for brickmaking.

These slates have been in use for several hundred years, and were largely exported to the Continent at an early date; their wearing qualities are undoubtedly very high indeed, and their lightness is a peculiar and important feature. Like most slates of this colour, they do not fade.

It has been employed in the Victoria and Albert Museum, South Kensington; New Patent Office; Royal Naval College, Dartmouth; Dockyard, Devonport; Truro Cathedral; Women's Hospital, Birmingham; numerous asylums, and many other buildings in England and abroad.

Slates of Devonian age in South Devon have been worked in the Torquay neighbourhood; and the Torcross slates of this age have been obtained near King's Bridge from several small quarries for local roofing. The largest quarries at the present moment are those about Diptford, near Totnes. The Devonian strata appear again in North Devon and Somerset, where they have afforded rough slates, strong and sound, but with rather wavy cleavage surfaces, and not capable of being split at all thin. They have been worked in a small way about Countisbury and Treborough, in the Brendon Hills. As a rule, these slaty rocks are much too crinkled and folded to make good slate.

In the Culm Measures (Carboniferous), which occupy a large portion of mid and western Devon, dark grey slates are constantly being worked in a small way, more often for walling than for roofing, as the cleavage is generally irregular and bad. The slates of Coryton weather very rapidly to a dull brown colour, and for this reason they are frequently employed when a 'rustic' effect is desired.

Scotland.—In Scotland slates are worked in the counties of Argyll, Perth, Dumbarton, and thin flags in Caithness. In the last-named county roofing flags have been quarried on a small scale near Pulteney Town from the Old Red Sandstone formation; this occurrence requires no further notice. In the other counties the slates are all obtained from the Highland Metamorphic Series. First in importance are the quarries of the west coast of Argyllshire; here a broad belt of rock, called the Black Slate, extends from the island of Scarba in a north-easterly direction up the Firth of Lorne and into the western side of Loch Linnhe, traversing in its course a good deal of the surface of the slands of Luing, Seil, Belnahua, and Easdale.

The slate rock is fine and even in grain, and bluish-black to black in colour. Stripes, ribbons, and spots are absent; but in many parts of the district much of the slate is unworkable, through the presence of strongly marked strain-slip. The slates are dull, owing to the absence of well-developed mica on the cleavage faces. They very often contain a considerable amount of pyrites. It occurs in small strings and in scattered cubes, $\frac{1}{16}$ to $\frac{1}{4}$ inch across; and it is worthy of note that this well-crystallized pyrites is most abundant in the hard blue-black slate, and is less evident in the more earthy varieties, which apparently contain *very finely divided* pyrite, or some marcasite, which causes them to discolour and wear badly. So well is this understood in the district that there is a saying: 'The best slate-rock contains pyrites.' These sparkling crystals of pyrites are called by the workmen 'diamonds,' and they have been observed still retaining their brilliant lustre in some slates that have been in place in roofs in Glasgow for over one hundred years. In some beds of slate the pyrites crystals interfere with the cleavage and cause some waste; and after the slate has been split they are occasionally rather inconvenient, as they interfere with the bedding of the slate on the roof; in other beds the crystals split through regularly with the slate.

In the greater part of this district the slate *beds* are much folded, and tend to tilt over towards the west-north-west; the cleavage dips at angles of 45 to 80 degrees towards the east-south-east. Mr. H. B. Maufe says that the workmen name the gently sloping limb of the folds the *sgriob* (pronounced 'skreep'); the flatter top is called the *bonn* (pronounced 'bown'); and the steep limb the *beal* (pronounced 'byel'). The largest and best slates come from the Sgriob, where it will be seen the dip of the bedding and of the cleavage most nearly coincide. The 'vein' of slate in this district is called a 'seam' or 'stone.' (See Fig. 19, p. 283.)

Thin beds of limestone occur in the slates, and close to them the quality is poor, but a short distance away it is very good. Dykes of porphyrite, lamprophyre, and basalt also traverse the slate rock, and in their neighbourhood the slate loses its tendency to cleave, as it does in similar cases elsewhere.

Most of the quarries lie on the low ground near the sea, and some of them are worked considerably below highwater mark. The veins, or seams, are followed along the strike in galleries, 12 to 20 feet deep, and worked towards the rise of the cleavage.

All the slates are used for roofing; no slabs are made. They are graded as 'full sizes'—average 115 square inches in area, but never less than 7 by 12 inches—and 'undersizes.' The thickness varies from $\frac{3}{16}$ to $\frac{3}{8}$ inch; average, $\frac{1}{4}$ inch. They are sold by the thousand of 1,200, plus

SLATES AND OTHER FISSILE ROCKS

forty to cover breakages. Most of them go to the Glasgow district.

The Ballachulish Slates are graded as 'Queens,' 'Duchesses,' 'Countesses,' 'Sizeables,' and 'Undersized.' There is a working face here of 43 feet.

The slate in Perthshire and Dumbartonshire occurs as a belt of slates and phyllites in the Metamorphic Series, running from south-west to north-east, near the boundary of the Old Red Sandstone. This belt is now principally worked in the quarries indicated in Appendix D, p. 432; the quarry at Luss is quite small. The slate is of excellent quality, and deep blue in colour.

Ireland.—There is a good deal of excellent slate in Ireland, and in the past there was a considerable export trade; but in the great majority of quarries the method of opening out the ground and conducting the business has been slovenly and erratic. At the present day there are only three important quarries—namely, those placed first in the list of Appendix D (Ireland); the others are only employing about half a dozen men each.

Irish slates come from the Ordovician, Cambrian, Devonian, Lower Carboniferous, and some, of low grade, from the Coal Measures. The slates of Tipperary are of Ordovician age. They are worked in the south-east at the Victoria Quarries; most of the material is good, sound, well-cleaved, pale grey slate, free from pyrites, and obtainable in large sizes. A green slate of good quality is also produced here. Near the south-west of the county, in the Gallymore Hills, a greenish slate has been quarried, but it does not split thin; and a similar slate occurs in the adjoining county of Limerick, in the Ahaphaca Valley. Both have been extensively used locally.

In the Killaloe Quarries, Portroe, there is a fine 'vein' of grey slate, some 400 feet wide, which is made up of subordinate veins, some with a ribbon and some without. Both roofing slates and slabs are produced.

The Madrenna slates of Co. Cork are of Carboniferous age; the vein is over 200 feet in width, and the colour is very dark grey. In parts this slate is subject to slip-cleavage, nodules, and patches and veins of pyrites. It is employed for slabs and flagging. Blue, purple, and reddish slates are found at many points, but they are not now worked, except occasionally, and on a very small scale.

In the Old Red Sandstone green and reddish slates occur, but they have been little exploited.

At the present moment the well-known slates of Valentia Island, Kerry, do not appear to be worked. The age of this slate is Old Red Sandstone. There are three main 'veins': the upper one, 9 feet thick, yields the best slates; the middle one, 16 feet thick, is softer, and more frequently jointed, but splits more readily; the lower vein, 14 feet thick, produces smaller-sized slabs and roofing slates from the waste. The slabs for which the quarry was famous were procurable in large sizes up to 14 feet by 5 or 6 feet regularly, and occasionally as much as 20 feet long. The slab here produced was somewhat harder than Welsh slab.

Ordovician slates have been found and worked more or less at the following and many other places: Broadford, Co. Clare; Glentown, Co. Donegal; the Ormonde Quarries, in Knockroe and Mealoughmore, Co. Kilkenny; in Glenpatrick (bright bluish-grey and pale grey), Waterford.

FOREIGN SLATES.

Slates are worked extensively in many parts of the world, and as the newer countries become better known and more populated, the number of productive centres is likely to increase. The slates which enter the English market most abundantly are those from France, Belgium, the United States, Portugal, and Italy.

France. — The principal slate-producing districts of France are situated in the Ardennes and in the Angers (Châteaulin) neighbourhood in Maine and Loire; also to a lesser extent in Morbihan, Savoie, Central France, and the Pyrenees.

First in importance are the slate-mines of the Anjou basin; there these are excavated in rocks of Ordovician age, which occupy a large area about Angers and to the north-west. The rocks are repeatedly folded, the strike of the folds being north-westerly and south-easterly. From north to south on the right bank of the Maine the following slate belts are recognized:

1. The belt of Avrillé, the slate proper of Angers.
2. The belt south of Avrillé.
3. The narrow belt of Beuriére; not exploited.
4. The belt of Saint-Nicolas.
5. The belt of Beaucouzé.
6. The belt of Bouchemaine.

On the left bank Belts 2, 3, and 4 have united, but they still retain the same characters. On the north the veins exploited for slate are the Veine du Nord, the Veine du Sud, and the Veine de l'extrême Sud. The principal quarries are in the more northern veins; they include those of Trélazé, the Grands Carreaux, Hermitage, Grand Maison, Petits-Carreaux, Montibert, Misengrain, Renazé, Pouéze, La Saulacé, Moulin-Carcé, and Fresnois, and farther west the Bel Air and La Forêt works of Combrée.

Slates of several grades are worked in the Maine et Loire district. The typical product of the Angers district, 'Best Angers,' is a high-quality slate, strong and tough, of a grey-blue colour. These slates have been used in very many public buildings in France, and a large export trade is done from the port of Nantes. Besides roofing slate, a good deal of slab is made, and some of this enamelled on the spot.

On the same outcrop of Ordovician rocks, farther west, near Rochefort-en-Terre, in Morbihan, the slate beds are worked in the Guenfol, Union, and other quarries. The quarries are open ones on account of the frequent jointing, which would make mining unsafe. The slates are strong and elastic and of a pale grey colour. (Coefficient of resistance to rupture by flexure 11·70, coefficient of elasticity, 15·50.)

In the Ardennes district the Cambrian rocks appear at the surface over a comparatively small compact area. These rocks have been classified in four groups, from north to south, as follows:

1. Phyllades and quartzites of Fumay (Fumacian).
2. ,, ,, ,, Revin (Revinien).
3. ,, ,, ,, Deville (Devillien).
4. ,, ,, ,, Bogny.

The beds are all inclined towards the south and are thrown into very numerous sharp folds; it is possible that Series 2 and 4 may be the same repeated by this folding.

The slates of Fumay are purple (so-called 'red') micaceous and chloritic slates, containing abundant crystals of ferrous carbonate; they frequently carry white spots, which are richer in silica than the surrounding coloured slate. There are two veins: the lower and more northern is the Veine Sainte-Anne (St. Anne's Slate), which is separated by about 120 metres of white quartzite from the Veine de la Renaissance. The cleavage dips 40 degrees E. 19° N., while the bedding dips 27 degrees E. 25° N.

SLATES AND OTHER FISSILE ROCKS

South of Fumay a vein of black slate, the Veine des Peureux, occurs in the Revinien group.

The Devillien group yields slates of grey-green and blue colour, also green slates with abundant crystals of magnetite. These magnetite crystals give a very characteristic appearance to the slates; they are readily observable in the cross-fractured ends of a slate block; on the cleavage faces they are invisible, but they produce the little elongated lumps which form a kind of coarse grain on the shearing surface of the slate, each crystal being covered by the mica and chlorite which wrap round it. This group furnishes the slates of Rimogne, Deville, and Monthermé. There are three veins of slates known from the quarries as the Echina, Saint-Barnabe, and Vanelles.

The Veine de l'Echina is made up of two parts *(ternes)*, separated by a bed of quartzite 30 metres thick: the northern, or *grand terne*, is 12 metres thick, and is compact; the southern, or *petit terne*, is formed of phacoidal masses and is only 3 metres thick. The *grand terne* passes under special names at different quarries—*e.g., Saint-Louis* at Terre-Rouge, *Echina* at Monthermé, and *Écaillette* near the same place; the *petit terne* is the *Sainte-Croix* at Terre-Rouge, *Rapparent* at Echina, *Saint-Honoré* at Monthermé, and so on.

The Veine Saint-Barnabé is divided in a similar way, and is worked at Baccarat (*grand terne*), and La Carbonière (*petit terne*), and at Malhanté.

Gosselet points out that when these *ternes* are thin they are subdivided into three zones. The two outer are green magnetite-bearing slates, while in the middle it has a beautiful grey-blue colour, and contains in place of magnetite reddish-brown grains of oxidized chalybite or ilmenite.

The quarries at Vanelles work only the magnetite-bearing green slate.

Devonian slates are worked here and there to the east of the Cambrian outcrop, and the Belgian slates of Herbeumont and Warmefontaine are from the same geological system; and again in Luxembourg, in the south-western continuation of the German Moselle district. These are often sold as 'Upper Moselle' slates, and are finer in quality than those of the lower reaches.

In **Austria** Cambrian slates are worked at Eisenbrod, etc., in Bohemia. They are fine-grained, tolerably hard, and split evenly; in colour they are green-grey, and bluish-grey. They are much used for roofing in Bohemia, and have been employed in the Votivkirche in Vienna.

Devonian slate in several shades of grey and greyish-blue is obtained in the Mähren district at Waltersdorf, Libau, Romerstadt, and Grosswasser, in the neighbourhood of Olmütz. In the same district very similar slates are worked in the Culm (Carboniferous) beds of Sternberg, and are extensively employed in the buildings of Olmütz and the vicinity. From the Culm rocks of Freudenthal and Troppau, Eckersdorf, Gersdorf, Dortschen, Freihermsdorf, etc., in Silesia, fine-grained, smooth-cleaving, dark blue-grey slates are obtained, and widely used in Vienna, Cracow, and other towns.

In **Germany** the principal slate-producing district is situated in the Lower Devonian rocks west of Coblentz, along the Valley of the Moselle, below Trier (Treves), and in the Hunsrück district; some quarries are worked on the east side of the Rhine, at Caub, Rüdesheim, and other places. In the northern part of the district there are quarries at Mayen (the largest in the district), Müllenbach, and Andernach; on the Moselle, south-west side, there are quarries at Thomm, Tries, Wintrich, and Zell. The quarries of St. Goar and Oberwesel are on the western side of the Rhine; while in the Hunsrück are the quarries of Bundenbach, Gemünden, Rhaunen, Grube, Herrenberg, Eschenbach, and others. The slates of this district are for the most part bluish in colour, but some of greenish-grey tints are also produced.

In Westphalia there are quarries in Devonian slates at Hörre-Raumland, Nuttlar, Fredeburg, etc.; and slates of the same age are worked at Langecke, Diez, Limburg, and in the Harz at Goslar.

In Saxony old palæozoic slates are worked at Lössnitz.

United States.—Slates are quarried on a large scale in Pennsylvania, Vermont, and Maine, and in a lesser degree in the States of Virginia, Maryland, New York, California, Arkansas, and several others.

The Pennsylvanian slate tract extends in a belt six to eight miles wide, in a west-south-west to east-north-east direction. The beds, which are of Ordovician age, have been strongly crumpled into a great number of minor anticlines and synclines, often closely squeezed together, and overturned slightly towards the north-west. The workable slates lie in two belts—an upper northern one, the 'Soft Vein,' and a lower southern one, the 'Hard Vein.' In both veins ribbons or stripes are well developed. In the former they are coarse and strong, and the good slate has to be obtained from between them; while in the latter, though they are closer together, they do not interfere with the splitting of the rock into roofing slate or slab. The principal centres are Bangor, in Northampton County, and Slatington, in Lehigh County. The Soft Vein is worked in the Bangor, Slatington, Pen Argyl, Danielsville, and Slatedale quarries, and the Hard Vein is worked at the Chapman and Belfast pits.

At Bangor there are several quarries; that known as 'Old Bangor' is the largest in America—1,000 feet along the strike, 500 feet across the strike, and 300 feet deep.

It lies near the top of the Soft Vein. The general structure is an overturned close syncline, striking N. 25° E.; the axial plane of the folds dips east-south-east at a low angle, and pitches 5°—10° S.S.W.; while the cleavage dips 5°—10° S. 32° E. The thickest bed here is 9 feet. This slate is very dark grey in colour, with a smooth non-lustrous cleavage face. It contains much pyrites in small spherules, and abundant carbonate in rhombic crystals and plates, so that it effervesces readily with acids; there is also much carbonaceous matter. The stripy beds are used for slab and the others for roofing slate. Other quarries in the Bangor area produce similar material.

South-west of the Bangor district is that of Pen Argyl; here the slate is worked in the lower part of the Soft Vein. The principal product is a very dark grey slate ('black'), but a certain amount of dark greenish-grey slate is quarried. Still farther in the same direction is the Slatington, or Lehigh, district; here the slate is dark grey, and very similar to the other Soft Vein slates. The beds range from 1 to 30 feet thick. From north to south the veins or beds are named as follows: Eureka and Snowdon, New Bangor, Mammoth, Big Franklin, Little Franklin, Washington, Trout Creek, Blue Mountain, and Williamstown.

The Heimbach Quarry, one and a half miles north-east of Slatington, contains two veins: the 'Heimbach Big Bed,' very dark grey, not so blue as Old Bangor; and the 'Heimbach Black Bed,' very dark grey to black.

Most of these slates are soft, easily quarried, and cheap. They are sonorous and pretty tough; they are well suited for slab and milled slate, and are much used for better-class roofing. After long exposure in their natural position they weather to pale creamy and brown tints. These dark slates are sold under various names—'No. 1 Pennsylvania Black,' and so on—but there is no fixed standard of quality. Of the Bangor slate, 'No 1 Bangor' is free from ribbon—*i.e.*, it is 'clear.' 'No. 1 Bangor Ribbon' is the same slate crossed by several narrow dark stripes at one end; for roofing the ribbon is usually covered by the lap.

Another type of slate, the *Peach Bottom Slate*, occurs in the counties of Lancaster, Susquehanna, and New York, in Pennsylvania, and extends into Harford County, of the adjacent State of Maryland. The slate belt is a low ridge, about ten miles long and a quarter to half a mile wide, formed by the outcrop of the south-eastern limit of a syncline in the Cambrian rocks, striking south-west to north-east, with the cleavage dipping to the south-east at a high angle or else vertical. The colour is very dark grey, with a faint tinge of blue—called black—and the split faces are distinctly lustrous; it is well crystallized, and contains much graphite and some andalusite, magnetite, and pyrite; it is sonorous, strong, and wears well, retaining its colour; it is used only for roofing, and the output is somewhat limited. The quarries in this slate are the Delta and the Cardiff, and several others.

The slates of Vermont are perhaps better known in this country than any other American slates. There are four slate districts, the most important being a belt of much folded and partly overthrust rock, lying between the Taconic Range and Lake Champlain; it runs for about twenty-six miles in Vermont, and extends south-westward into Washington County, New York, where, however, it is of less commercial value. Two geological systems yield slates in this belt: (1) Ordovician shales and grits, with red slates, and a few of bright green colour; and (2) Cambrian shales, limestones, and quartzites, with purple, greenish-grey—'sea green' and 'unfading green'— and 'variegated' slates, the last named being mixed greenish-grey and purple.

The 'sea green' slates are of various shades of greenish-grey when fresh, but after exposure they turn to all sorts of tints of brown, yellowish and grey-brown, and give the roof a mottled aspect. They are strong and split well, with a rather waxy-looking cleavage face. They are well crystallized, but contain a large amount of carbonate, here a mixture of carbonates of lime, magnesia, and iron in small rhombs, and it is the oxidation of these crystals which causes the discoloration. The slates with most carbonate appear to be softer than the others. They are used only for roofing.

The 'unfading green' slate is similar in colour, but it has much less carbonate, and, consequently, only changes colour to an insignificant extent on exposure. It is duller on the split face than the 'sea green,' and has also perfect cleavage; it is used both for roofing and for slab. The 'purple' and variegated slates are merely colour variations of the others, due to the addition of hæmatite; they are strong and sonorous, and are used for both roofing and slab. The colour is due to irregular mixtures of hæmatite and chlorite; the purple slates are dark purplish-brown, and are more even in colour than the 'variegated.'

The second slate area of Vermont is a tract of Devonian or Lower Trenton Ordovician rocks, striking north 10 degrees to 20 degrees east, and stretching northward into Canada. The cleavage dips westward at 75 to 80 degrees. This slate is quarried at Northfield; it is very dark grey-black, of fine grain, and with smooth, lustrous cleavage faces. It contains a fair amount of pyrites and graphite. Although it has the qualities of a good roofing slate, it is mainly employed for slabs. The two other black slate districts in Vermont are not of much importance.

In the State of Maine grey and very dark grey or black slates are quarried at Monson, Bronnville, Blanchard, and a few other places. They are of Early Palæozoic age, but the precise horizon has not been clearly made out. The slate veins are mostly small, and are regularly interbedded with quartzite. The texture is fine, and the cleavage face smooth but dull, except in the Bronnville quarries, when it is bright. All these slates are sonorous, and are employed both for roofing and slabs.

Dark grey and black slates of Ordovician age are worked in Virginia and West Virginia. Black, dark grey, red, and green slates of the same period are found in Arkansas.

The dark grey (black) slates of California, north of Glacerville, are from the Mariposa slates of Late Jurassic or Early Cretaceous age. The quarries are at Bangor, San Francisco, Chili Bar, and Eureka. In the last-named quarry is the very interesting green slate produced by the cleavage of a basic igneous rock (? gabbro); it is used for fancy trimming and lettering on black slate roofs.

In Georgia the grey Ordovician slate is used in the manufacture of Portland Cement.

The slates of Alabama, Minnesota, and Michigan are unimportant.

Other Countries.—In Norway slates are obtained from Östre Slidre, in Gudbrandensdalen; green slates from Etnadalen, in the Valders district; and several shades of grey and bluish-grey from the neighbourhood of Ose Fjord and Vossebygden.

Dark grey slates are imported from Valongo, in Portugal.

In Canada slates are well developed, but they are not much worked, partly owing, no doubt, to the proximity of the American slate district of Vermont. Cambrian slates are worked in Richmond County, Quebec, at the New Rockland Quarry, and a little has been obtained in Vancouver and New Westminster. In British Columbia and Nova Scotia various coloured slates are found. Good slates are now being worked in Newfoundland, principally in Smith Sound, Trinity Bay. Grey, blue, purple and blue tints, and a peculiar bluish-green, are procured. Some quarrying has also been done on the opposite side of the island, in the Bay of Islands.

Slates are obtained in India, in the Punjab, from the Kangra district, Rewari, Dalhousie, and Dharmsala, also from the Karahpur Hills; these are dark grey or black in colour, and generally of good quality.

A good strong slate, very suitable for all slab work, is worked at Monghyn, in Bengal. A little slate is quarried in Rajputana.

Slates are worked in mines in the neighbourhood of Genoa and Chiavari, in Italy. The most important district is the Valley of Fontenabuona, in Coreglio, Lorsica, Drero, and Rapallo. Next come the works of Cogarno and Lavagna, on the flanks of Monte San Giacomo, followed by those of Avegno, Tribogna, and Usica, in the Recco Valley, and there are small quarries in the Valley of Bargagli. These slates are of Eocene age; they occur associated with beds of sandstone, shale, and limestone. They are dark grey to black in colour, and are very soft when first quarried, but harden on exposure. The splitting has to be done while the slate is still wet. The best slates are called 'Lavagne.' Roofing slates, slabs, and building blocks are obtained; some of the slabs are imported to England. There are also slate quarries at Serravezza, in Tuscany.

The 'Farren' Slates of the Transvaal come from about ten miles south-east of Pretoria.

Slates are obtained in Australia and New Zealand.

On King Island, Tasmania, very promising samples of dark grey and warm red slates are being worked, most suitable, apparently, for slabs.

FISSILE STONE OTHER THAN SLATE.

Long before slates came into general use for roofing, thin-splitting, flaggy stones of diverse kinds and from various geological formations had been employed as roof covering.

Although these fissile stones, shingles, or stone tiles, are often called 'slates,' it is a misnomer, for they lack the essential qualities of true slates. In the first place, they possess no true cleavage; they cannot be split at any point, as the slate can; they yield only along certain planes at irregular intervals. Splitting can be effected only along the planes of bedding—that is, those of original deposition—and the stone between two adjacent planes is devoid of cleavability. In a word, the slaty *structure* is absent. The planes of easy parting are often determined by layers of clastic mica scales, sometimes by flattened fossils—oysters and other shells, or plant remains.

As regard materials, they may be sandstones, hardened shales, limestones, or sandy limestones.

Mention may be made of the schists, which differ from the aforesaid stones in that the tendency to split into thin slabs or plates is due, not to original bedding planes, but to others brought into existence by rearrangement of the original minerals and the production of new ones, all with a common orientation under the influence of pressure. Thus it would appear that the schists have somewhat in common with true slates; both result from metamorphic processes, and in some regions, where large igneous masses have formed zones of alteration in the rocks around them, slates may actually be found passing by insensible degrees into schists as they approach the intruded rock. The typical schist differs from slate in being more strongly and coarsely mineralized—if one may use the term in the sense that one or more minerals are developed with more individuality and on a grosser scale—than in the slates. Moreover, schists may be formed from any kind of rock, limestone, sandstone, and basic or acid igneous rock. Thus we have quartz-schists, hornblende-schists, chlorite-schists, and so on. The mineral quartz is nearly always present, and the other mineral—be it mica, chlorite, or what not—tends frequently to run in distinct layers, and the splitting of the rock takes place along the layers.*

The schists are naturally found amongst the older rocks, or in mountain regions that have suffered much compression. Mica-schists are the most common, and these are frequently found employed as paving slabs, as in many of the Norwegian towns, and occasionally they are used as roof covering of a very heavy kind.

Yet another kind of fissile rock occurs in the phonolite of France and elsewhere, which splits readily into thin flakes through the frequent parallel *jointing*. These phonolite plates have been used for roofing for a long period in Auvergne.

Returning to the fissile-bedded sediments, we may note the thin, highly micaceous sandstones found occasionally in the Silurian rock, as used on old buildings about Ammanford, in South Wales: these are small and heavy, but they have a pleasant mellow grey appearance on the roof. From the Old Red Sandstone marls of Stockholm, Pembrokeshire, thin sandy layers are obtained, and formerly were made into roof-tiles. These were very small and irregular in shape, but nothing could be more lively in effect than their mottled buff, red, and grey, with occasional micaceous sheen. No two stones are alike.†

In every district where Coal Measure Sandstones, Millstone Grit, or Lower Carboniferous beds yield thin, flaggy sandstones, the thinner slabs have been used on the roofs

* Iona Cathedral is being restored and roofed with spangled Moine schist from the Ross of Mull.
† The Caithness flags are used for roofing—*e.g.*, Kirkwall Cathedral.

SLATES AND OTHER FISSILE ROCKS

of older buildings—in Wales, Derbyshire, Yorkshire, Scotland, and Ireland. These Carboniferous stone-tiles are of fairly large size and of great weight; but the durability of well-chosen slabs is beyond question, and the colour, though sombre, is often relieved and brightened by the growth of lichens and moss. It need hardly be mentioned that their employment is almost a thing of the past. In some of the Yorkshire flag districts the thinner flags ($\frac{3}{4}$ inch to 1 inch) are used for roofing; they are called 'Grey Slates'; slightly thicker and larger slabs are known as 'Covers,' and are employed for lean-to out-houses, pigsties, etc. (Fig. 7.)

Jurassic rocks in England have yielded stone tiles at several horizons. The Lower Lias Limestones at Queen Camel, near Sparkford, Somerset, and Burley Down,

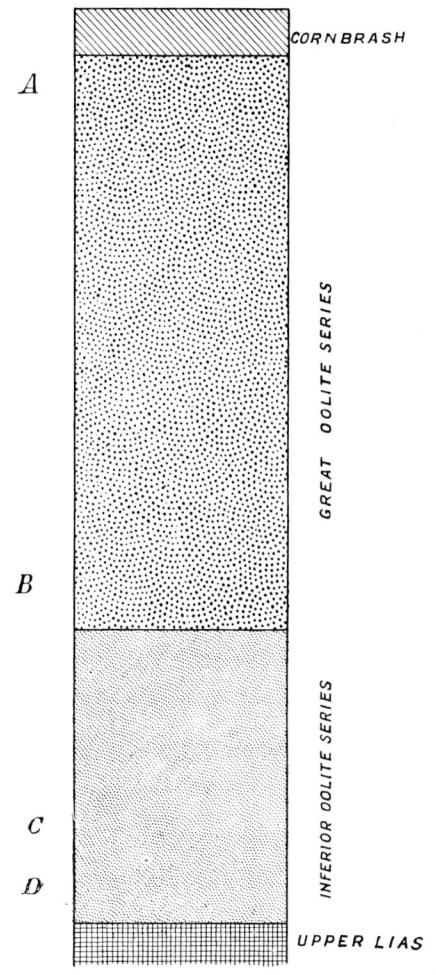

FIG. 21.—DIAGRAM TO SHOW THE RELATIVE POSITION OF THE PRINCIPAL TILE-STONES IN THE LOWER OOLITES.

A, Poulton slates; *B*, Stonesfield; *C*, Collyweston; *D*, Duston slates.

Shropshire; also flaggy beds in the Marlstone at Chacombe, Wilts, have been used for roofing. (Fig. 21.)

From the lower part of the Inferior Oolite, the Northampton beds of Duston, Northamptonshire, comes the White Pendle, or *Duston Slate*, obtained in a bed of fissile sandstone, with oolite grains, from 4 to 6 feet thick. (Fig. 21.)

Very thick and heavy stone-tiles were obtained from the Lower Freestone of the Inferior Oolite at Hyatt's Pits, south-east of Snowshill, and at Condicote and Lower Swell, in the northern part of the Cotteswold Hills. They may be seen in use in Snowshill and other villages, used in conjunction with 'Kyneton Slate' (see p. 321). From the same horizon a white oolitic siliceous freestone has been dug on the summit of Brailes Hill; after exposure to the weather this stone becomes distinctly fissile. Paving stones have been obtained from the lower part of the Inferior Oolite, near Bradford Abbas.

The most important of the Inferior Oolite stone-tiles come from Collyweston and Easton, south of Stamford, in Northamptonshire, and similar beds on the same horizon have been worked at Duddington, Medbourne, Kirkby, and Dene Park, all in the adjoining neighbourhood.

The Collyweston slates have been fully described in the Geological Survey Memoir, 'The Geology of Rutland, 1875,' by Professor Judd, and his account may be quoted here:

'The valuable fissile character of the beds is merely a local accident; and in some directions the bed of stone has been followed and found to become non-fissile, and in consequence worthless for roofing purposes. There is only a single bed of stone (the lowest of the series), which is used for making roofing slates. This varies greatly in thickness, being often not more than 6 inches thick, but sometimes

SLATES AND OTHER FISSILE ROCKS

swelling out to 18 inches, and in rare cases to 3 feet; while not unfrequently the bed is altogether absent, and its place represented by sand (or sandstone). Rounded mammillated surfaces, like the "pot-lids" of Stonesfield, abound in these beds.

'The slates are worked either in open quarries or by drifts (locally called "fox-holes"), carried for a great distance underground, in which the men work by the light of candles. The upper beds of rock are removed by means of blasting; but the slate-rock itself cannot be thus worked, for though the blocks of slate-rock when so removed appear to be quite uninjured, yet, when weathered, they are found to be completely shivered, and consequently rapidly fall into fragments. The slate-rock is therefore entirely quarried by means of wedges and picks, which, on account of the confined spaces in which they have to be used, are made single-sided. The quarrying of the rock is facilitated by the very marked jointing of the beds, a set of master-joints traversing the rocks with a strike forty degrees west of north (magnetic), while another set of joints, less pronounced, intersects the beds nearly at right angles.

'During the spring of the year the water in the pits rises so rapidly that it is impossible to get the slates out. The slates are usually dug during about six or eight weeks in December and January. The blocks of stone are laid out on the grass, preferably in a horizontal position. It is necessary that the water of the quarry shall not evaporate before the blocks are frosted, and they are constantly kept watered, if necessary, until as late as March. The weather most favourable to the production of slates is a rapid succession of sharp frosts and thaws. If the blocks are once allowed to become dry they lose their fissile qualities, and are said to be "stocked."

'The slates are cleaved at any time after they are frosted. Three kinds of tools are used by the Collyweston slaters: the "cliving hammer," a heavy hammer, with broad chisel-edge, for splitting up the frosted blocks; the "batting hammer," or "dressing hammer," a lighter tool for trimming the surfaces of the slates and chipping them to the required form and size; the "bill and helve," the former consisting of an old file sharpened, and inserted into the latter in a very primitive manner. This tool is used for making the holes in the slates for the passage of the wooden pegs, by means of which the slates are fastened to the rafters of the roof. These holes are made by resting the slate on the batting hammer and cutting the hole with the bill.

'The slates are sold by the "thousand," which is a stack usually containing about 700 slates of various sizes, the larger ones being usually placed on the outside of the stack. The slates when sold on the spot fetch from 23s. to 45s. per thousand. Many of the Collyweston slaters accept contracts for slating, and go to various parts of England for the purpose of executing their contracts.'

These slates vary in colour; those made from the unweathered heart of the stone are bluish-grey; the thinner layers and outer parts of the thicker ones are buff, yellowish, or mottled grey and buff. The paler slates are said to darken on exposure.

The average size of the larger slates is 18 inches long by 1 foot broad, and about $\frac{1}{4}$ inch thick. They are usually laid on the roof with 45 degrees pitch, in diminishing courses, with a 3-inch lap; the larger slates are fixed with oak or deal pegs, the smaller ones bedded in cement and pointed; the ridges and hips are of solid sawn stone, or of yellow or white tiles to match the stone. These quarries have been at work for several

SLATES AND OTHER FISSILE ROCKS

centuries, and still produce material for repairs and for special work.

'Wittering Pendle' is a less fissile, calcareous, sandy flagstone used for paving; it is from the same beds as the Collyweston Slate at Wittering, south-east of Stamford.

These slates come from below the Lincolnshire Limestone; a section of the strata is shown in Fig. 12.

The next deposit of slaty beds occurs at the base of the Great Oolite; it is known as the *Stonesfield Slate*. The method of working the fissile beds is very similar to that employed for the Collyweston Slates; they are quarried and mined principally at Stonesfield, near Woodstock, Oxfordshire, and similar stone has been obtained at Througham, near Bisley (Bisley Slate). Here the 'Slate Bed,' a fine micaceous sandstone, is 2 feet thick, but of this the top 4 inches will not split, and the lower part is of inferior quality, the middle 9 inches being all that yields good 'slate.' The slate bed underlies about 10 feet of calcareous sandstone. Stone-tiles have been obtained at Miscoden, Rendcombe, and Nettlecomb, near Birdlip; also at Oakridge Common and Battlescomb, east of Bisley. At Sevenhampton Common they were formerly worked. The Stonesfield Slate was formerly extracted in large quantities at Kyneton Thorns (Kyneton Slate), between Condicote and Naunton, and from this district north and north-east of the last-named place; and they have been worked to some extent near Aston Blank, Salperton, Harling, Pewsdown, Chedworth, Allington, near Bibury, and other places in Gloucestershire. On the same horizon fissile limestones have been obtained for roofing north of Bath, on Lansdown.

The slate beds are very variable, often altering in character rapidly within the space of a few yards; the following sequence, observed in one of the deeper mines at Stonesfield (66 feet deep), gives a fair idea of the beds:

Inches.

Stonesfield Beds.
- *Rag:* A hard, current-bedded oolite, forming the roof.
- *Top Soft:* A marly bed 5
- *Pot-lid* and *Overhead:* Blue-hearted, fissile, sandy limestone 9
- *Race:* Calcareous sandy bed, with some pot-lids ... ... ,. 8
- *Lower Head:* Similar to the Upper Head ... 9
- *Block:* Soft sandy bed.

The stone-tiles are made from the 'Overhead' and 'Lower Head' Beds, the former yielding the better 'slates.' The Overhead is yellowish-grey in colour where weathered, and contains thin seams of oolite; frequently it passes laterally into a 'pot-lid' bed, a soft, sandy calcareous layer, with irregularly disposed flat cake-like concretions—pot-lids. The Stonesfield Slates generally range in colour from brown to grey or blue, and in composition from calcareous sandstones, with little or no oolitic structure, to others with many oolitic grains. The larger sizes are about 1 foot 6 inches square, and are prepared in two qualities—'firsts' (thin) and 'seconds' (thicker). The more oolitic slates are usually the thicker.

The Stonesfield Slates are not satisfactory in larger towns, for after twenty years' exposure on Exeter College Chapel, Oxford, they had suffered so badly that they had to be removed.

Above the Great Oolite 'slates and planks' have been obtained from small quarries in the Forest Marble of Gloucestershire, Wilts, and Oxfordshire. No 'frosting' is necessary for the preparation of these stone-tiles, the laminæ of limestone or occasionally of calcareous sandstone occur separated by thin partings of shale or clay, and when they are quarried they only require trimming. They are very rough and heavy, but are said to be more

SLATES AND OTHER FISSILE ROCKS

durable than the Stonesfield Slate. They have been worked at Atford, in Wilts; Charlwood-by-Box, near Bath; Beverstone and Charlton, near Tetbury; Avening, Aldsworth, Chavenage, Cirencester, Burford, Fairford, and Poulton (see Fig. 22), where the 'Poulton Slates' are beds of flaggy, current-bedded 'bluestone,' or oolitic limestone alternating with grey clay beds.

At about the same horizon in Yorkshire, about Brands-

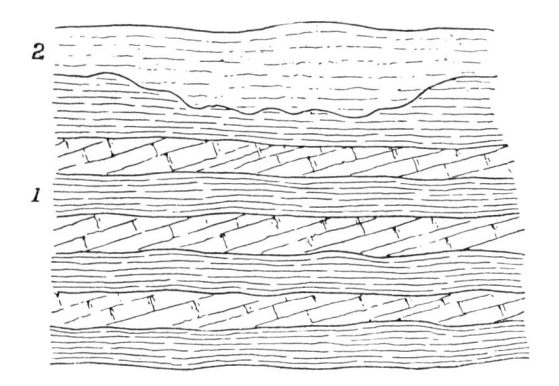

FIG. 22.—SECTION AT POULTON, NEAR FAIRFORD. (From Memoir of the Geological Survey, 'Jurassic Rocks,' vol. iv., by permission of the Controller of H.M. Stationery Office.)

1. Forest Marble, alternating layers of grey clay and current-bedded limestone = slate.
2. Soil and rubble.

by, in the Howardian Hills, the Grey Limestone Series yields a fissile, hard, siliceous limestone, from which in the past many large slabs and roofing tiles were obtained.

The Lower Purbeck Beds of Portland, in Dorset—including the 'Slatt Beds,' with the overlying fissile limestones—and in the neighbourhood of Weymouth and Portisham, have been used for roofing tiles, and examples may be seen on many old buildings. The same beds have produced tiles about Tilly Whim, Swanage, used in that town and at Corfe, and on farms in the district; also at Lumdon, used in Devizes and the vicinity. These stone tiles resemble those of Collyweston and Stonesfield in colour and appearance, but they are not so lasting.

The Horsham Stone of the Wealden Beds in Sussex is in places sufficiently fissile to be fit for paving slabs and roofing tiles; these are known locally as 'healing stones' (Saxon, *haelan*, to heal).

SLATE AS BUILDING STONE.

Wherever slaty rocks have been found they have been used for rough kinds of building in boundary walls, small dwellings, and in the older castles, mansions, and churches. For this purpose they possess many advantages, being easy to work, durable, and obtainable in convenient sizes. As a rule, the more gritty slate—material discarded by the roofing-slate quarries—makes the best building stone. Slate buildings of all kinds are found in the Welsh slate districts, and large blocks and scantlings are often introduced in the construction of quite small dwellings.

In the Lake District the rougher beds of the Skiddaw Slates, and much of the roughly-cleaved Borrowdale Rocks, and Bannisdale Slates, are used locally for building.

In Cornwall and Devon the Devonian Slate (killas) is extensively used all over the Devonian area, but particularly in the neighbourhood of the granite masses, which have hardened the slate and improved its quality as a building stone. Bodmin is perhaps one of the best examples of a town built almost wholly of local slate rock. The colour is dark grey, and rather sombre in effect, but variety is obtained by utilizing the red and brown iron-stained joint planes for facers. The stone is readily dressed with an axe, and the frequency of the

SLATES AND OTHER FISSILE ROCKS

clean-cut joints makes the preparation of convenient-sized blocks an easy matter.

In the Newquay district the grey killas is used in a similar way, to make neat coursed ashlar fronts, as well as for random rubble buildings, with dressings of granite or the local Elvan.

The Carboniferous Slates of Devonshire are put to similar uses, as about Okehampton, Coryton, etc.

Slates in the older palæozoic rocks of the Falmouth district are used for buildings of minor importance.

The Hornblende-epidote schists in the neighbourhood of Torquay have been largely used for dwellings about Salcombe, while the mica schists have been employed for rough walling.

Here may be mentioned the Schalstein of St. Germans, St. Budeaux, and the Shalstein tuffs of Weston Hill, not far from Plymouth; these are sheared and cleaved basic igneous rocks. They have been used in this district for many buildings. In the latter district the more gritty and coarse slates are used for building, and the same remark applies to the same areas of Scotland, the Isle of Man, and Ireland.

In Ireland slate is very widely employed for rough building in the villages, and in the past was used with good effect in castles, forts, and round towers. The round towers of Kilcullen, Co. Kildare, and Kinneigh, near Enniskeen, Co. Cork, for example, are constructed of slate, which in parts is regularly coursed and dressed to the curve; it has well withstood the wear of centuries in a most trying climate. Co. Cork provides the most numerous examples of the employment of slate for building. About Bantry a grey gritty variety is used for the main part of walls, a smoother, more tractable form being used for quoins, sills, and dressings generally. That used at Rushnacora is greenish; at Skillereen it is grey and reddish-brown. Co. Kerry slate is used at Tabert, Killarney, and elsewhere, and is preferred to the associated sandstone in this and other Old Red Sandstone areas. The ruins of Monasterboice, in Louth; Gosford Castle, in Co. Armagh; and many houses in Kells, Co. Meath, are built of Ordovician slate. Kylemore Castle, Co. Galway, is an example of a more modern large building in which slate has been largely used.

CHAPTER VIII

MISCELLANEOUS STONES

Flints. — For building purposes flint is obtained in England only from the Upper Chalk, in which it occurs in the form of irregular nodules or layers. The building flints are always made from nodules.

Flint is fairly uniform in composition; it consists almost wholly of silica, partly in the colloidal and partly in a crypto-crystalline condition. It is hard and compact, and breaks with a marked conchoidal fracture. In colour it ranges from black, through various shades of grey, brown, or greenish-brown, to white.

This stone is used in construction in a variety of ways, which have been aptly summarized by Mr. F. T. Baggallay, F.R.I.B.A., in a paper on 'The Use of Flint in Building, especially in the County of Suffolk,' Trans., R.I.B.A., vol. i., N.S. (1885):

1. Mixed with mortar or cement, and used as concrete in the body of many old walls.

2. Built up as gathered (surface or land flints), without preparation, into rough rubble masonry; as in buildings of all periods, from early Norman—or (?) Saxon round towers—to latest Perpendicular churches.

3. Picked in sizes and roughly coursed—much done in earlier Gothic periods—broken or unbroken, also as pebbles or mixed with other stones.

4. Broken flints alone, coursed or not.

5. For facing only, a further improvement is to split

the stone with greater care, so as to obtain a fairly even surface, and *to knock the white coat off*.

6. In later work the splinters were stuck into the mortar in the joints as thickly as possible (see p. 165); perhaps to wedge up the work and lessen the body of mortar. This is *galleting* or *garreting*.

The white coat referred to above is an outer layer of partially formed flint on those taken fresh from the chalk, or a decomposed layer sometimes seen in gravel flints.

The excellence of gauged flint cutting is probably due, as Mr. Baggallay points out, to gun-flint work, which has been in operation for many years, and is still carried on at Brandon, in Suffolk.

Building stones, or 'builders,' as they are called, are still made by the flink-knappers of Brandon. Mr. Skertchly ('On the Manufacture of Gun Flints,' Mem. Geöl. Survey, 1879) says that the cores from which flakes are made taper towards one end, and this is worked to a level face if possible; but if any projection exists on the face it is struck off, and the block is then called a 'chip-back,' because the face has been chipped back. These are not considered so good as those which have a smooth face, formed by a single blow of the quartering hammer.

'Builders' are known under the following designations: (1) Square block-faced builders; (2) square mixed-coloured builders; (3) round-black builders; (4) round mixed-coloured builders; (5) random faced builders; (6) rough builders; (7) land stones.

The best stones are obtained by primitive mines from the chalk; land stones, which are very hard and durable but cannot be dressed, are picked up from the surface.

The use of flint differs in different districts. In Suffolk mixed-coloured builders are preferred to black ones. An excellent example of black flintwork is the old Bridewell, by St. Andrew's Church, Norwich, A.D. 1400; in

MISCELLANEOUS STONES

Berkshire the small Decorated church of Shottesbrooke may be cited; but there are innumerable examples of the work in Norfolk, Suffolk, and Sussex.

The following notes are from Mr. Baggallay's paper:

'There are two distinct systems of using flush-panel and deeper work of split flint: one at the Thorpe Chapel, attached to the Church of St. Michael-in-Coslany, Norwich, and the churches of Lavenham and Long Melford, in which the dark flint is used as the ground to show up tracery of freestone; and the other at the large majority of Suffolk churches, in which the stone wall is divided into panels filled up with flint. The former can never, I think, be pleasant to an architect's eye, chiefly because it is . . . false construction; the small material, which we know to be a facing, showing as the body of the wall, while the decorative lines alone are of solid stone. For the latter I claim that it may be—would that it always were!—perfectly true as well as beautiful.'

Flint in Sussex is rarely used north of the Downs.

'The chalk district contains 162 parish churches, which have more or less retained their medieval character. With five exceptions, they are of twelfth or thirteenth century foundations. Probably in all flint forms the mass of the walling, but it is not used architecturally, except in a few later insertions. In the fifteenth-century churches knapped and squared flints are largely used, and chessboard panelling; but, with the exception of a few simple crosses, it is believed ornamental work, such as occurs in the eastern counties, is entirely absent'—*Note by Mr. Lacy W. Ridge (loc. cit.).*

Soapstone, or **Potstone**, is a soft grey stone, composed largely of talc and chlorite, formed by the metamorphosis of rock, originally rich in magnesia-bearing minerals, olivine, tremolite, etc., the talc being a pseudomorphous replacement of these minerals. It occurs usually among crystalline schists. A talcose chlorite-schist has been used in Inverary Castle.

The stone is practically unknown as a building material in Britain, but in some countries, notably in Scandinavia, it has been much employed for building and for situations subject to great heat. It has been quarried since prehistoric times in Shetland for potstones, tombstones, etc. Although very soft, and capable of being cut and carved with great freedom, it is very durable, as may be seen in Trondjhem Cathedral, the Citadel of Fredrikshald, and many other old structures. It is used in quite modern buildings in Christiania and elsewhere for enrichments and frontages. In Sweden it is quarried at Handöl, in Jämtland, and in Norway at Offa. It is obtained also in Piedmont and Switzerland; in India and China; at Potten, in Canada; in New York and North Carolina. The Santa Catalina 'serpentine' of California is a soapstone.

Serpentine Rock, mainly composed of hydrous silicates of magnesia, is frequently employed as a building stone, but it is difficult to procure sound blocks that will wear well. It is a common building stone in the Lizard, where it is used with granite quoins and dressings—*e.g.*, Landewednack Church. Many beautiful Serpentines are used as marbles, but its use as a polished decorative stone for exteriors is not to be commended. It is quarried on a large scale at Höle, near Stavanger in Norway, and has been used in the front of the Serpentingaarden, Christiania —the pale green variety for rock-faced blocks, relieved by polished work in the dark olive-green variety. In the United States the stone has been used largely in New York, Philadelphia, Washington, and Chicago, from the quarries in Pennsylvania. The Carthusian monastery of La Varne, in Var., is built entirely of this stone.

Laterite.—In tropical countries a common weathering

product, forming the surface deposit, is called 'laterite.' It is reddish-brown in colour and consists of varied mixtures of sand, clay and rock débris with abundant iron oxide cement. It is found in India, Burma, Ceylon, Africa, etc., and is frequently used for building and engineering works. St. Mary's Church, Madras, is built of this material, and many old temples in the coastal districts of India bear witness to its good wearing qualities. It is easily worked and hardens on exposure, but it is not attractive in appearance. In a more restricted sense the term 'laterite' is applied to weathering products rich in free aluminous hydroxide; these resemble commercial bauxite. (See *Bull. Imp. Inst.*, vol. vii.)

Artificial Stone.—The number of artificial stones that have been invented is already legion, and although every year adds to their number, many are short-lived. The briefest of notes upon the geological aspect of this important subject is all that may be attempted here.

Artificial stones may be classified roughly as follows:

1. Stones made of natural rock fragments, held together by cement—Portland cement, magnesian cement.

2. Similar stones subjected to a subsequent hardening process by treatment with water-glass—silicate of soda.

3. Stones in which granulated rock or sand is cemented with carbonate of lime.

4. Stones in which more or less of the carbonate of lime cement is replaced by silicate of lime.

5. Stones cemented by bituminous, asphaltic, or other organic substances.

In Class 1 come the ordinary cement blocks—'Pentuan Stone' and the like—and concrete, in which the quality of the product depends on that of the cement—Portland Cement.

In Class 2 the action of the cement is modified, and hydro-silicates of lime are formed, which harden the stone. Such are the 'Victorian Stone,' made from the granite of Groby; 'Atlas Stone,' 'Excelsior Stone,' made from similar material; 'Hard York Non-Slip Stone,' made from sandstone.

Under Class 3 would fall much of the stone made by Thom's patents—reconstructed stone. Many 'lime-sand' building blocks are of this character; so are the pumice- trass- and puzzolan-lime artificial stones.

The lime-silicate stones are not always marked off sharply from those with carbonate of lime only. Ford's stone is an example of a stone which purports to be cemented mainly with silicate of lime; in this sand and fat lime alone are used. Siebel's stone was made with the aid of calcium chloride solution. Ransome's stone was made from chalk, sand, and Farnham stone; the last-named stone was employed, as in the case of Victoria Stone, etc., for the production of the sodium silicate. Calcium chloride was added to assist in the hardening, silicifying process.

The nearest approach to natural limestones are those made by Mr. Thom from the quarry waste of Jurassic and Carboniferous Limestones; they can be made much denser and stronger that the natural stones after which they are named, if it should be desired.

Artificial sandstones, made from sand and lime, correspond to calcareous sandstones of Nature; those with a large proportion of lime silicate as a binding material have no natural counterpart.

Artificial stones of the better kinds yield satisfactory figures under all the ordinary stone tests, and they are being more and more employed in all countries. Only those which can be hand-worked by the carver or mason while soft are capable of giving the best artistic results at a cost reasonably below that of natural stone of the same class.

CHAPTER IX

THE DECAY OF BUILDING STONE

General Considerations.—It is the natural and proper desire of every architect that the creations of his hand and heart and brain should be touched as lightly as possible by the 'bawdy hand of Time'—a desire shared, we may presume, by his clients, and, indeed, by all men, if the work is good and true and fair to see. It is desirable, therefore, that the architect should have some knowledge, not only of the relative rates of decay in different stone, but of the causes of decay. Long accumulated experience, linked with that conservatism which hinders the introduction of new stones, has made the profession familiar with the behaviour of many building stones to a degree that far transcends anything that can be learnt from artificial tests upon new and untried material. It is none the less necessary that the processes and mechanism of decay should be understood, so that an examination of the characters of an untried stone may show wherein, and to what degree, it is likely to fail.

The student has often been urged to go into the country and see for himself the effects of the agencies of weathering in the quarries and escarpments, and wherever the rock is exposed. It is to be feared that this interesting and healthy exercise will teach little of the behaviour of *building stones*. Not only will he find, in nine cases out of ten, that the good weather beds in escarpments and cliffs and in many quarries are formed of material that the mason cannot or will not work, but the whole circumstances of natural weathering of the rock are different in kind and in degree from those which distinguish the stone when removed from its bed and fixed in the building.

We must differentiate, then, at the outset between the agencies effecting the degradation and decay of rocks in their natural position at or below the surface of the ground, and those which achieve similar ends acting upon the stones artificially removed from their original site. Full consideration of the former will be found in any good book on general geology; the latter may be briefly discussed here.

The agencies of decay in building stone are chemical and mechanical, the former assisting the operations of the latter, and *vice versa*. Since the range of chemical activity and the scope of mechanical disruption are influenced by the chemical, mineralogical, and structural character of the stone, it will be convenient to consider first the agencies which promote decay in stones of all classes, and then the nature of the decay set up in a few different kinds of stone.

Weathering Agencies—Causes of Decay.

Chemical.—Ordinary rain-water in country places acts as a solvent on all kinds of minerals. It produces oxidation and hydration of iron compounds, and attacks carbonate of lime and argillaceous matter in building stones; but since stones with clayey streaks and spots and those containing much mealy carbonate of lime ought never to be employed, its action on well-selected stone is in reality very slow. Its power as a solvent is increased by the presence of a small and nearly constant amount of *carbon dioxide* (the mean content is 3·34 volumes per 10,000), which enables it to remove alkalies from felspars, and to dissolve more calcium carbonate than it could if it were pure. Limestones and rocks with much calcite or dolomite suffer more than others from this solvent action; the influence on felspars and other silicates is very slight, so long as the fresh surfaces of the minerals are not exposed to the attack. It attacks the lime, potash, and soda of felspars, and the magnesia and iron of the ferro-magnesian silicates, forming carbonates of these substances, setting free silica and hydrated silicates of alumina (kaolin, etc.), and hydrated magnesian silicates (serpentine). If these minerals are in a fresh condition, the action of carbonated water is extremely small, but it is much more active when the minerals have already undergone weathering or alteration at an earlier period.

How slow the solvent action of rain-water may be, when other destructive agencies are excluded, is illustrated by the following figures, obtained by Hirschwald by careful experiments on thin slices exposed for a definite period. His results, scaled up to represent the loss of surface from rain action in an average year, give:

	Millimetres.
Calcite	0·007
Solenhofen limestone	0·0085
Magnesite	0·0098
Dolomite	0·0085
Chalybite	0·0105
Gypsum	0·2410

At this rate the superficial reduction of the Solenhofen Limestone (a compact, very fine-grained stone) would be no more than 0·85 millimetre in a century.

When, however, we consider the air of towns, we find that, through the continual combustion of coal, other and far more injurious substances are thrown into the air, which, together with the moisture, hasten the decay, especially of calcareous stones, in a marked degree.

By far the most injurious of these substances is sulphur, which in burning becomes sulphurous acid in the moist air, and is eventually oxidized to sulphuric acid. This attacks the carbonate of lime, forming calcium sulphate, which crystallizes as gypsum, and in the process of doing so assists in the disruption of the surface layers of the stone, thereby facilitating the farther attack of the acid moisture. The presence of sulphuric acid accelerates the action of moisture upon the other minerals, but to a much smaller extent.

In the neighbourhood of many manufactories other acids are liberated—nitric and hydrochloric acid, and in some cases ammonia. Their presence naturally accelerates stone decay in their immediate neighbourhood.

Angus Smith has estimated the amount by weight of these acids in 10,000 parts by weight of the air in three large towns as follows:

	H_2SO_4.	HCl.	HNO_3.
London	0·0241	0·0125	0·0084
Liverpool	0·3959	0·1016	0·0058
Manchester ...	0·4166	0·0579	0·0059

Another evil product of coal combustion without smoke consumption is the outpouring of particles of unburnt carbon and tarry particles. These, of course, are continually blackening the buildings in the vicinity, and to some extent they act as carriers and holders of injurious gases.

Much more injurious to stone than rain in towns are the foggy days, for the countless particles of condensed moisture carry the dissolved gases to the surface, and

THE DECAY OF BUILDING STONE

there they stay to do their work, and the stone face is not constantly washed, as it is in a good rain.

Snow carries a higher percentage of the injurious substances CO_2, H_2SO_4, HNO_3, and NH_3 than is found in rain-water; hence when snow is prevalent, its lodgment on ledges and in mouldings may act deleteriously upon the stone.

Many stones suffer more when exposed to sea air than they do in inland situations; this may be explained by the presence of sea salt borne by the wind. Thus the rain-water on the English coast has been found to carry an average of 0·22 parts per 100,000 of chlorine, and during a strong sea breeze this amount has been observed to rise as high as 21·80 parts per 100,000, which is equivalent to 35·91 parts of sodium chloride. The presence of sodium chloride in the carbon-dioxide-carrying rain-water increases its solvent action on the carbonate, sulphate, and phosphate of lime, and assists in the attack upon lime silicates. But this is not the only salt found in rainwater from the sea; magnesium chloride also facilitates the solution of $CaCO_3$, and attacks the lime, iron, and alumina silicates; so also does magnesium sulphate, while calcium sulphate may react on carbonate of magnesia in the stone, and, along with sodium sulphate, assist in the solution of $CaCO_3$. Thus it is clear that sea-water and the rain of maritime districts are more powerful agencies of chemical destruction than ordinary stream-water and inland rain in country places.

Particularly bad for stones is the continual deposition of dust in towns; some of this is inorganic, consisting largely of detritus from the roads. One can easily detect the mineral ingredients of macadam and setts on the nooks and corners of the highest parts of buildings. These, perhaps, do little harm chemically, but they are accompanied by much heterogeneous organic matter, and this, wherever it can lodge, acts as a powerful etching instrument. For, being repeatedly moistened and dried, cooled and warmed, its decay steadily proceeds, and one has only to lay a little moist decaying vegetable or animal matter on a slab of polished marble for a few days to see how rapidly it can act.

Organic.—The influence of the ubiquitous microbe in hastening the decay of stone has often been discussed, but the question as to how far these organisms are instrumental in causing the deterioration of building stones has never been fairly tackled. The activity of microbes in the decomposing elements of dust and dirt may undoubtedly assist indirectly in intensifying local destruction of stone; whether they are themselves primary agents of building-stone decay remains to be proved.

Microscopic algæ, lichens and mosses, wherever they can find lodgment on the surface of the stone, must assist the process of stone decay. The delicate hyphæ and root-hairs search out the minute crevices on the surface of the stone, and by the chemical action of the products of the activity of the living cells, together with the mechanical action attending their growth, each little patch of encrusting lichen or moss is the centre of a steady attack on the rock.

Urban conditions are not favourable to the growth of lichens; mosses, however, may be observed here and there in town buildings, but they do not flourish to the same extent that they do in the country.

Lichens may find a habitat upon stones of all kinds, but some show a marked preference for calcareous stones— *e.g., Verrucaria calcarea* and many other species, *Amphoridium dolomiticum*, and species of *Pyrenodesmia, Thelidium, Bilimbia, Rhinodina, Biatorina*, etc. Some prefer siliceous stones—*Rhinodina confragosa, Bilimbia coprodes*, and species of *Callopisma, Lecidea, Catocarpus*, etc. Some have a

liking for mortar—*Gyalolechia lutesalba*; others like tiles—
Buellia tegularum, Blastenia arenaria.

Calcareous rocks suffer most from the invasion of lichens. Their organization and influence on the rock surface have been investigated by E. Bachmann; he shows that it is the reproductive portion which penetrates the stone—often to the depth of a millimetre. The penetrating power of these delicate organisms is astonishing. It matters not whether the stone be fine-grained and compact or coarse in texture; and large crystals of cal-

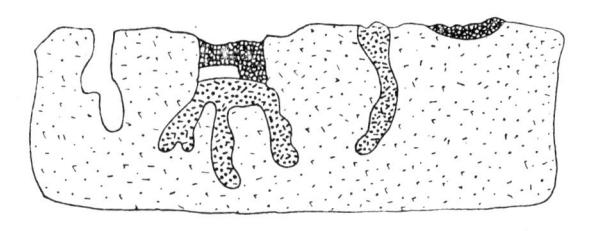

FIG. 23.—LICHEN GROWTH. (After Bachmann.)

cite are bored into without the least regard to the cleavage planes or other paths of lesser resistance. This action is clearly chemical, due to the excretion of special acids by the plant, aided doubtless by the evolution of CO_2, which is taken up by the moisture (Fig. 23).

Speaking of the mechanical influence Kerner von Marilaun says: 'A growing hypha penetrates wherever the merest particle of carbonate of lime has been dissolved, and accomplishes regular mining operations at the spot. Projecting particles of the carbonate not yet dissolved are separated by mechanical pressure from the main mass, and at the places in question where a lichen is in a state of energetic growth tiny loose rhombohedral fragments (of calcite) are to be seen, which are washed away by the next shower, or else carried off as dust by the wind.'

Marine algæ etch the surfaces of walls, piers, and embankments, in the same way as their air-dwelling relatives, and, like them, they are most destructive to calcareous stones.

James Hall ('Report on Building Stones,' 1868) stated that 'the lichen-covered rocks in Nature are usually those of great strength and durability. None of the softer or rapidly-decaying rocks produce this vegetation.' Of this it may be said that if a stone is so liable to superficial crumbling that lichens cannot ensconce themselves, it has no business in any building; and, setting aside rocks of this character, it may be observed that all other stones will support the growth, provided the conditions of moisture and temperature are favourable. Particularly liable to this cover of vegetation are those portions of buildings near the soil where ammoniacal and other salts creep up the surface of the stone, which is often found supporting a rich crop of the bright green *Pleurococcus*.

The rise of salts in solution in the ground water may be the cause of serious trouble. This is the case in the lower courses of stone buildings in Egypt, and was made the subject of a report by A. Lucas, who showed that the rapid decay was caused by the upward soakage of moisture carrying sodium chloride and smaller amounts of the carbonate and sulphate. The solution becomes concentrated within the surface layer by evaporation, and crystallized, causing disruption, with efflorescence. Various salts may be found in such situations—*e.g.*, the sulphates of sodium, potassium, magnesium and lime, and the so-called 'wall-saltpetre,' a double sulphate of lime and potash; carbonate of lime may also occur. Most of these salts act injuriously by increasing the solvent power of the water in the stone, as well as by causing disruption and an unsightly appearance where they crystallize.

THE DECAY OF BUILDING STONE

Mechanical Agencies.—Building stones are liable to suffer mechanical disintegration of the surface through the agency of wind, friction, simple changes of temperature, frost, crystallizing forces, and, in special cases, wave action.

Wind may be a powerful primary agent of stone destruction in certain situations, where it can pick up sand and coarse dust and drive this material upon the face of a building. It is not necessary to go to the well known examples in desert regions for illustrations of this action, but in countries like Britain the absence of large tracts of sand and the constant moisture of the air causes the destructive action of the wind sand-blast to be a rather localized phenomenon. None the less is it a force to be reckoned with in all coastal situations having sand tracts in the vicinity, and in the lower parts of buildings or monuments where dust may be blown. A walk with bare legs on the sands in a stiff breeze will convince anyone in a very short time. The glasses of lighthouses have been obscured and rendered useless in a single night by this sand-blast action. In a more extended sense the wind is always at work, moving minute loosened grains from the surface of stones like sandstones and weathered granites, and the effect of a strong wind in driving rain into the interior of porous stones is noteworthy.

Friction in other forms is mainly confined to the lower parts of buildings, by which people are passing frequently to and fro. 'Sheep polish' is familiar to the field geologist in pastoral districts where rocks abound, and some hard stones get polished in a similar manner where loafers congregate and sun themselves, or where there is considerable foot traffic; but the coarser and more friable stones, particularly those that are current-bedded, are often worn in a most unsightly way—a fact which architects often ignore until it is too late.

Change of temperature exerts a very important destructive influence on most kinds of stone. In typical desert regions, where there is no protecting mantle of vegetation, and cold nights sharply follow after hot days, the crackling of rocks is a familiar sound at nightfall, while the talus heaps of fragments afford abundant evidence of the intensity of this type of disintegration. In such regions the peeling of granite blocks is a very notable feature, and enormous masses of this rock are sometimes found split right through by this action. The change of temperature at nightfall is great and somewhat rapid in these latitudes, and a sudden heavy shower in West Africa has been observed to lower the surface temperature of a rock by 60° C.

In many American towns the effect of marked temperature variation has made itself felt, and a considerable difference may be noticed between the amount of scaling on the north and south facing fronts of east and west running streets; many buildings and monuments with a southern or south-western aspect have suffered considerably from the sun's rays.

The effect of temperature variation upon the internal structure of a stone will depend, amongst other things, upon its homogeneity or heterogeneity of mineral composition, its fineness or coarseness of grain, its colour or the colour of individual minerals, and the presence or absence of bedding, shear, or cleavage planes. Stones like the granites, consisting of different minerals in fairly large grains, appear to suffer most readily; the coefficient of expansion or the amount of deformation per unit length per degree of temperature is small in all the constituent minerals, but it differs in the several kinds. Thus, the mean coefficient of cubical expansion for Quartz is 0·00036, Orthoclase 0·00017, Hornblende 0·0000204, Calcite 0·00002, Dolomite 0·000035.

The tendency to produce minute intergranular thrusting

THE DECAY OF BUILDING STONE 343

during expansion is further enhanced by the fact that the expansion along one axis of a crystal is not the same as that along the others; thus in Quartz the mean linear coefficient of expansion parallel to the principal axis is 0·00000797I, while in the direction perpendicular to this axis the coefficient is 0·0000I337I; in Calcite the coefficients in these two directions are 0·0000263I5 and 0·0000054; in Hornblende the coefficients for the three axial directions are 0·000008I, 0·00000084, 0·0000095.

If we turn from the consideration of the changes of volume of the granular constituents of the stone to those produced in blocks of stone due to the summation of the minor expansions or contractions, we find that here, too, temperature changes may cause evil results.

The linear expansion of Granites has been determined by Totten to be 0·000004825 per inch for 1° F.; Adie found 0·00000438 per inch for 1° F., and Bartlett found 0·00000868 per inch for 1° F.

If we take Adie's figure and apply it to the case of a granite coping stone 5 feet long, we find that a rise of 50° F. will cause the whole block to expand 0·01314 inch. This expansion with contraction repeated many times is quite sufficient, although so small, to cause loosening of the cement·joints, with the concomitant facilitation of the ingress of water, with its accompanying bad effects.

Reliable figures for the many different kinds of granite, limestone, and sandstone are not obtainable. Totten gives 0·000005668 inch per degree F. for white marble; Adie gives 0·00000613. For sandstone, Totten gives 0·000009532; Adie 0·00000174. For slate, Adie gives 0·00001038. Experiments of the Ordnance Department of the United States Army, which consisted in heating bars of stone 20 inches long from 32° F. to 212° F. and then cooling them again to 32° F., showed that the heating produced a permanent change of dimension; in no case did the stone go back to its original length. The mean deformation found in granite and quartzites was 0·009 inch; in marbles 0·009; in limestones and dolomites 0·007; in sandstones 0·0047. This change in the stones was reflected in the lowering of the pressure resistance observed after heating.

Frost action is generally regarded, with good reason, as the building stone's most deadly foe in all latitudes where recurrent freezing and thawing take place. The intensity of frost action is very different, of course, in different regions, and in London and the larger manufacturing towns in Britain it is hard to say whether frost or acid-laden moisture is the more destructive.

Frost works through expansion of water in passing from the liquid to the solid state. The specific gravity of ice at 0° C. is 0·916—in other words, water in changing to ice increases its volume by nearly one-tenth; therefore, unless the pore space is more than nine-tenths full of water, the conversion to ice will exert no serious pressure upon the pore walls. The pressure due to the freezing in an enclosed space, full of water at the moment the change takes effect, is very great, being 138 tons per square foot. Fortunately, the whole of the pores in a building stone are seldom all filled with water under natural conditions. A few individuals or small groups may be full within the frozen zone; these will be local centres of stress, perhaps of permanent strain.

Some of Bauschinger's experimental results illustrate this point very clearly, as shown in the table on p. 345.

None but the most compact marbles can stand freezing after soaking *in vacuo,* and if the soaking is done under pressures up to 200 atmospheres, stones of all kinds will show the effects of freezing. It is obvious, therefore, that the degree to which the pore spaces of a stone can be filled under natural conditions will be a measure of the frost-resisting power of a stone.

From this a formula for the expression of the *Saturation-Coefficient* has been derived. If $w =$ the amount of water soaked into the stone naturally, and $w_1 =$ the maximum amount taken up under pressure, then the Saturation-Coefficient $S = \dfrac{w}{w_1} =$ the measure of pore-filling under normal conditions.

Stone.	Pressure Resistance.	Natural Soaking in Water.	Saturated under Air-Pump.
Limestone	685	No cracks after thirty-one freezings	Broke in two after freezing
Sandstone	583	No change after twenty-five freezings	Surface flaking after freezings
Coarse Granite	1,200	Unchanged	After eight freezings the micas split up and the polished faces became rough
Tuff	25	No flaws after twenty-five freezings	After fourteen freezings flaws were developed throughout the mass

As indicated above, the maximum value of S for frost-resisting stones will be 0·9. In practice the mean value 0·8 may be taken as the limit.

According to Hirschwald (*cf.* Table XVI., in Chapter X.), the principal criteria for estimating the frost-resistance of a stone are:

1. The saturation coefficient.
2. The special morphological arrangement of the pores, and the regular arrangement of such pores in parallel layers.
3. The weakening effect of water upon argillaceous, marly, earthy, calcareous, and fine clastic earthy material

in the stone; for these substances all tend to absorb the water.

It should be observed that the strength or pressure resistance of a stone is no index to its frost-resisting power.

The measure of weakening produced by soaking in water for a definite period (twenty-eight days; see Chapter X.) $= n = \dfrac{\text{Strength in wet state}}{\text{Strength in dry state}}$; and in all cases where this value falls below 0·6 the resistance to frost is said to be low.

It should be noted that open, porous stones—some of the tuffs, for example, or the very cellular travertines—do not by any means suffer the most from frost, the reason being that the pores are sufficiently open to permit of relief to the expansion of ice. On the other hand, very compact rocks, if they contain streaks or layers of softer material (see 3, above), and bedded rocks, with frequent well-marked planes of bedding, are very liable to suffer, either when surbedded or laid end on, or when they are laid on their proper bed in copings.

Crystallizing force, as indicated on p. 340, may cause serious damage to stone, much in the same way as ice. It may act where salts are produced by the decomposition of the stone through chemical agencies, as in the case of gypsum formed during the destruction of limestones, or in the efflorescing salts that sometimes come from mortar and cement, or from saline solutions rising from the ground, or those thrown on the building from the sea. Disruption of the stone surface is caused in these cases by the pressure exerted by the growing crystals.

Wave Action.—The mechanical effect of wave action in destroying the individual stones in sea-walls, etc., and in shattering the structures themselves, is due in part to the blows of sand-grains, pebbles, and large stones thrown by the waves against the wall, and in part to the air-pressure induced in all minute crevices on the surface of the stone and between the stones each time the wave strikes home.

Decay of Limestones. — These stones present the problems of decay in a simple form. All limestones fit for buildings are in practice almost identical from a chemical standpoint; their content of carbonate of lime should be so high that almost all other substances may be regarded as impurities of negligible influence. It is the carbonate of lime which sets the pace of decay.

We may leave aside for the moment the effects—such as they are—of the other constituents, and we shall see that the mode of decay of different limestones depends upon the state of aggregation of the carbonate of lime.

One of the most obvious effects of decay in limestones containing fragments of fossil shells, coral, crinoids, and the like, is the gradual removal of the exposed surface from between the shelly débris, leaving the latter standing up in relief, sometimes several inches above the matrix. Here it is evident that the fossils have resisted the weathering better than the matrix, and close examination of these fragments shows that they are composed of somewhat coarsely crystalline calcite, even where the original structure of the shell has disappeared and secondary calcite has taken its place. This effect may be seen very clearly in shelly Portland stone, in Purbeck marble, and in Carboniferous and Devonian limestones. So evenly does this process go on in the best limestones, as in good examples of Portland Whit Bed or Hopton Wood Stone, that, but for the presence of the upstanding fossils, it would be difficult to see that any decay had taken place. This evenness of wear is a peculiar quality of good limestones.

From the above observation one might be led to suppose that if the whole of the stone were in a more completely and strongly crystalline condition it would wear still better, and perhaps this is what has caused the textbooks to reiterate the dangerous generalization that the more perfectly crystalline the stone, the better will it wear. In a measure this statement is true, but in the case of limestones it must be taken with many reservations. The most perfectly crystalline of carbonate-of-lime stones is white marble, yet these same authorities tell us that it wears badly in towns, often worse than Portland stone.

An excellent example of the fallacy of such generalizations is afforded by the Ketton district. There we have the well-known Ketton Stone—an open-grained oolite, full of pores, and almost destitute of what is usually understood by a cementing matrix, a stone of splendid weathering quality; in the same quarries is the Ketton Rag, frequently quoted in lists of building stones. Now, this rag differs from the normal Ketton Stone only in the fact that it has a very perfect matrix of extremely well-crystallized calcite. The calcite grains are large; they fit closely together, and each individual crystal encloses a number of the oolitic grains. The result is that the rag is far more compact, dense, and impervious than the Ketton Stone; the weight of the latter is 128 pounds per cubic foot, while that of the former is 155, and it looks, when dressed, like a splendid building stone. Yet, notwithstanding the theoretical advantages of greater *density*, less *porosity*, and higher degree of crystallization, it is never used to-day as a building stone—not because of its greater cost to work, but because it will not stand frost.

It is of great importance to notice that it is not the solvent action of more or less acid-laden water that is the main cause of the disruption of the surface in the highly crystalline stone; for it is well known that these larger

THE DECAY OF BUILDING STONE

and more perfectly crystalline particles yield much more slowly to solvents than do the more finely granular or mealy particles.

Thus it would appear that the destruction of the surface of stones like white marble, which are composed entirely of rather coarse aggregates of crystalline calcite, is probably much influenced by the very perfect cleavages of the mineral, which, together with the intergranular spaces, provide a means of ingress for moisture, with the usual results after freezing or repeated wetting and drying and changes of temperature.

The Portland Whit Bed and Base Bed are usually contrasted strongly in books; but it is not so often realized that a good deal of the true Whit Bed is composed of material partly in the condition of the Base Bed. It is this mixture of the two kinds of texture which causes the mottled appearance on some blocks of Portland after exposure to the weather. The finer-grained, whiter parts of the fresh stone, with sparsely scattered oolitic grains, weather more rapidly and become black earlier than those parts of the fresh stone which have a buff tint and are composed of well-formed oolitic grains in close apposition. The buff parts of the stone will weather pure white if well exposed, and will stand up in relief above the other portions. (Plate VIII.)

These slight differences of texture may not be of serious consequence, and the various tones of grey in adjoining blocks in an ashlar face are not without their æsthetic value; but wherever great uniformity of tint is particularly desired, the blocks should be examined with a lens before they are dressed. This precaution is particularly necessary in the case of statuary built up of several blocks. The shelly patches and layers of Portland Stone generally weather well, and some architects like to have them because of the relief and variety they produce in the texture of the surface after exposure. Good illustrations of the variety of weathering surfaces in Portland Stone may be observed in St. Paul's Cathedral, Marylebone Church, the coping in front of Devonshire House, Piccadilly, and the arches at Hyde Park Corner.

In general, even-grained oolitic limestones weather fairly uniformly, and it must be admitted, in our larger towns, often rather too readily.

It is not possible to state any absolute rule, but it may be observed that stones possessing the best-formed grains united directly one with another, and having the least amount of mealy carbonate of lime, wear best.

The coarse-grained, shelly, and oolitic limestones of the Ham Hill type often make excellent and enduring stones for country wear, but they do not appear to be able to withstand the more searching ordeal of towns. Daly's Theatre was treated with the steam brush and patched up in 1908 after fifteen years, and now presents a very unhealthy appearance; the New Travellers' Club, in Piccadilly, was treated with some preservative in 1909, and in parts redressed, and at the present moment (1910) looks quite cheerful and fresh. Other minor examples of dressings of this type of stone could be cited which are in very poor condition after a short period of exposure. The fault of stones in this class lies not so much in the porous character of the mass of the material as in the prevalence of diagonal current-bedding, with occasional marly layers or 'clay-seams,' which present irregular lines of weakness; a kind of suppuration takes place along these lines long before the bulk of the stone is affected. So long as these stones are limited in use to flush ashlar they may wear well enough, but on every small moulding, cornice, or coping, and especially in carved work, they deteriorate quickly.

The somewhat similar Guiting Stone resembles in its

PLATE VIII

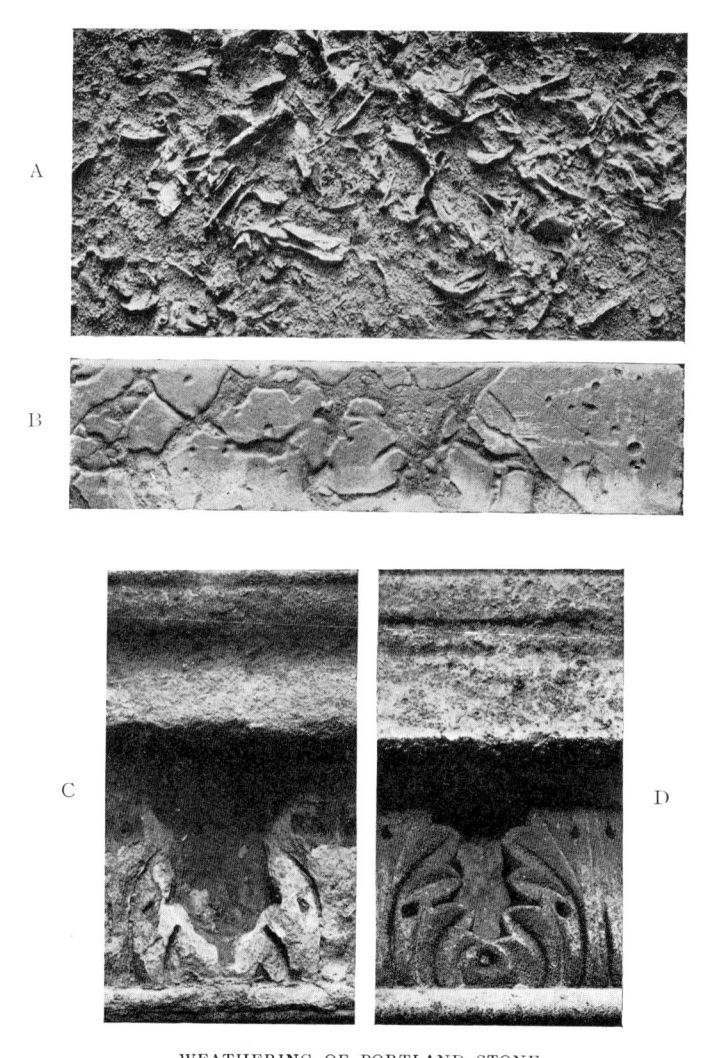

WEATHERING OF PORTLAND STONE.

A. Shelly Portland, magnified. B. Mottled Portland, reduced; both stones from the same building.
C. Portion of cornice from south side; D, from north side.

THE DECAY OF BUILDING STONE

mode of weathering the now obsolete Barnack Rag, which has stood for centuries in country situations; it seems to behave well as dressings in suburban houses, but it is often treated very unfairly by architects.

It may safely be asserted that any thoroughly calcareous limestone that is sufficiently crystalline, and neither too mealy, like some chalks, nor too argillaceous, like some marls and Lias limestones, will behave well in country places, even in a moist and variable atmosphere like our own, and in dry climates will be even more permanent.

In towns, however, all kinds of limestones are much more fiercely attacked, mainly through the influence of the sulphuric acid resulting from the combustion of coal. This acid readily attacks the carbonate of lime with the formation of the soluble sulphate gypsum. On well-exposed surfaces and on vertical and highly inclined surfaces the corrosive action is slower, because the rain removes the solvent and its results rapidly. On the other hand, beneath cornices and in all nooks and crevices it has time to operate on the stone, and the resulting gypsum tends to accumulate and form a vantage-ground for still further solution.

Professor Church cites the following items in the composition of the dark stalagmitic incrustation from beneath the cornice, above the colonnade, and below the dome of St. Paul's Cathedral:

	Per Cent.
Gypsum	73.80
Carbon	1·01
Ammonium sulphate	0·93
Tar	0·60
Calcium phosphate	2·22
Calcium carbonate	none

How serious this action of sulphuric acid may be can be gathered from the figures recently calculated by Dr.

Rideal for London. Taking the average quantities of coal, gas, and mineral oils used, as determined by the Board of Trade, he finds that *every day* there is sent into the atmosphere 981,792 pounds of *sulphur* from coal, 893 from gas, and 743 from mineral oils.

Professor Church puts the case in another way: 'I will direct attention to the amount of calcium carbonate which the minimum quantity of *sulphuric acid* yearly produced from coal-burning in London might transform into gypsum: 500,000 tons of this acid might destroy 510,200 tons of $CaCO_3$, evolves 204,080 tons of CO_2, and produce 877,544 tons of gypsum.'

Of course, much of the sulphuric acid produced in the air is washed or blown away, but sufficient remains to make its mark in an unmistakable manner on all calcareous stones. Wherever moisture can lodge, even for a short time, the solvent action is at work, and foggy days are in this way particularly destructive.

The destructive action is not limited to cornices, string courses, and mouldings, though these suffer most rapidly, but plain vertical surfaces are often made to scale. This is caused by the acid moisture soaking some little way into the stone, and remaining there to attack the material, while the outer surface is less decomposed, owing to the removal of the more acid moisture by the less acid rain. Now, once the destruction of the inner layer has commenced, it proceeds until the outer skin is pushed or falls away; for besides destroying the cohesion of the stone by solution, the operation is intensified by the forces of expansion and crystallization of the gypsum—100 volumes of carbonate of lime (calcite) become 120 volumes of gypsum.

It is not always clear why a skin is formed in one case and not in another, but it is evident that its formation is most common on plain vertical surfaces, while crumbling from the surface inwards takes place usually on narrow edges and moulded work.

Lichens and small algæ and moss can rarely grow on limestone, except in country places, because the surface of the stone is too rapidly removed to permit of their lodgment. When lichens do grow they have been supposed to act as preservatives, but this can only be apparent. It is true they always present the same front to the eye, but the products of their vitality must be attacking the stone all the time.

Crystals of pyrites are not common in good limestone building stones; where they occur, however, as in some palæozoic limestones and in certain argillaceous limestones, they facilitate the decay of the stone, for they decompose, yielding sulphuric acid with moisture, and this attacks the surrounding carbonate of lime, forming small pits round the pyrites grains.

Crystalline limestone with even a small amount of bituminous matter—often in patches—seems to be more resistant than the non-bituminous portion of the same stone. Much of the substance is harmful.

The Decay of Magnesian Limestone and Dolomite.— These stones are attacked by weathering agents in the same manner as the limestones, but with rather different results in certain cases.

In country districts the principal solvent is rain-water bearing carbon dioxide in solution; this attacks the carbonate of lime more readily than the magnesian carbonate, with the result that some stones, in which dolomite rhombs lie in a matrix of mixed carbonates, weather just like sandstones. The matrix is weakened, and the dolomite rhombs are liberated. In some natural exposures the 'sand' composed of these rhombs may be scooped up in handfuls.

In towns a similar stone suffers much more severely;

sulphate of magnesium and calcium are formed, and the former is more soluble than the latter, so that decay is comparatively rapid. See W. Pollard, Summary 'Geological Survey,' 1901 (1902).

Reference to p. 194 will show that magnesian limestones and dolomites vary very much in texture and in composition. In country situations those with the nearest approach to the composition of the *mineral* dolomite and those that are at the same time most perfectly crystalline will wear the best. In towns these will prove the best of their kind, too, but the rate of decay will, of course, be greater. English dolomitic rocks need special care in their selection, because the stone is liable to rapid variation in the bed.

The bad state of the dolomitic limestone in the Houses of Parliament does not prove that stones of this class are worse than other limestones for town use, but that slovenly work will produce the results to be expected of it. Nevertheless, it must be admitted that in mouldings, cornices, and carvings, the rapid solution of the magnesian sulphate, formed during the attack of the sulphuric-acid-bearing moisture, leads to serious mechanical disintegration.

The Decay of Sandstones.—The weakness of sandstones lies in the cement; the grains of sand, by which we mean quartz grains, are for all practical purposes indestructible. Decay sets in by the breaking down of the cement, and this may take place either by chemical processes or by purely physical actions.

A moment's consideration of the different types of sandstone structure, described on p. 119, will show in what way the destroying agencies may attack the stone. In sandstones of the highly siliceous quartzite type chemical action is negligible; these stones give way only after changes of temperature have produced a superficial loosening of the outer layers of grains. In the minute intergranular fissures so formed moisture can penetrate more readily, and the activity of the frost is brought into play. These stones, too, are often subject to cracking—small, almost invisible, cracks—on a small scale before they are exposed in the building.

The durability of sandstones depends mainly upon the contact cement—that which unites grain to grain; to a much smaller degree upon the cement which merely fills up the pores. For this reason the sandstone with a large amount of so-called 'binding material' may not be a worse weather stone than another with much less, provided that the contact cement is of a good kind, and gives as good a hold in each case.

Here we are speaking of sandstones in which the grains are virtually in contact with one another at several points; if the sand-grains are not in contact, but lie in a matrix or ground mass of other materials, they are taking no part in the mechanical structure of the stone, and the weathering of such stones is determined solely by that of the matrix. Thus, unless there is a good deal of silicification, sandstones with an argillaceous matrix will weather like shales or mudstones; those with a calcareous matrix like limestones, and so on. Thus the Kentish Rag, which often contains a large percentage of sand, weathers like a limestone, and the silica is here a cause of weakness, for the loosening of the sand-grains facilitates the crumbling and peeling of the surface. Similarly the Red and White Mansfield Stones are often classed as sandstones, but they usually weather like a dolomite, and the presence of the sand is probably beneficial only in the case of paving stones, steps, and landings, where it increases the hardness and security of foothold.

The best kind of contact cement is silica; ferric oxides are not bad, but calcite, dolomite, and argillaceous (unsilicified) matter are not satisfactory. The significance of the pore cement is by no means easy of estimation; its influence on the wearing of the stone will be determined by the kind, the amount, the texture, and the distribution of the cement.

Calcite and mealy carbonate of lime are attacked by the weathering agents precisely as in limestones. Felspar grains behave as they do in granites, but if they are already reduced to a kaolinaceous powdery condition, the material is easily washed from the surface of the stone by rain. If pyrites is present, it will sooner or later be converted to coloured hydrated sesquioxides of iron, with the liberation of sulphuric acid; hence, in the presence of carbonate of lime, especially the less perfectly crystallized kinds, this mineral is injurious. If carbonate of lime is absent, pyrites is only objectionable if it is sufficiently abundant to cause much staining, and when well crystallized its decomposition is very slow. On the other hand, no sandstone containing marcasite should ever be employed, for it will inevitably effloresce, and cause disruption.

Mica in a sandstone, if it is uniformly scattered throughout the stone, tends to make the dressed surface slightly more friable; but unless there is so much of it that the general cohesion of the stone is weakened, its presence is not seriously detrimental. If, however, the mica is very abundant on certain layers, as in some flaggy sandstones, it may do much harm by providing too free a passage for water, with the attendant danger of splitting. This is particularly true where the stone is not laid on its proper bed.

It has been pointed out earlier that the pore cement of many sandstones is of mixed constitution, and in these cases the chemical weathering is complicated. The general result of destructive agencies is to make the pore cement more porous, and so prepare the way for frost action in thrusting apart the quartz grains.

Sandstones about the ground line in damp situations are liable to be damaged by the absorption and upward soaking of salt solutions from the soil, leading to crystallization of the salts and disruption of the stone in dry periods. Sandstones are also subject to disintegration at the surface from the impact of wind-borne dust, which bores out the softer pore cement and the more tender minerals, thus loosening the grains of sand.

Some sandstones exhibit current-bedding in a marked degree. If the contact cement is good throughout stones of this kind, no ill effects need be expected. Nowhere, perhaps, are more current-bedded sandstones employed than in Nuremberg, where the most fantastic bedding patterns may be encountered at every turn, and rarely is it seen to lead to undue decay. Should there be any marked tendency to split readily on any kind of bedding plane, the stone should not be used for mouldings, carving, or turned work of small diameter. Corsehill Stone, with very distinct fine bedding, may be seen in many parts of London and other cities in perfect condition; but numerous examples could be pointed out where small balusters have degenerated into a shocking state, and where it has been used on end as facing it has scaled badly.

The Decay of Granites and other Igneous Rocks.— In spite of the fact that the crystalline igneous rocks exist in such great variety of texture, structure, and mineral composition, the manner of their decay, *where they are employed in buildings*, is so uniform, that nothing is to be gained by considering the different kinds separately.

Much time has been wasted and wrong notions have been propagated through confusion of the circumstances which accompany the decay of these rocks in their natural situation with the very different conditions promoting destruction in the stones in the building.

It has long been recognized that in the field the felspars of igneous rocks are subject to decomposition, whereby their alkalies are dissolved and removed in solution, silica is set free, and incoherent hydrated silicate of alumina is left behind. In this process rain and superficial waters have no doubt taken an important part. By similar processes the more basic silicates—the amphiboles, pyroxenes, and dark micas—are disintegrated and decomposed; their iron is oxidized and hydrated; the magnesia and lime are converted to carbonates; their alkalies are removed in solution; and the silica and alumina are treated more or less as in the case of the felspars. Where, however, these rocks are set up in buildings, the chemical alterations so obvious in the surface of rock masses hardly come into operation at all; and even in a moist climate, like that of England, provided the rocks are *fresh*, the amount of decay through chemical agencies is practically negligible.

When the rocks are not fresh—that is, when they have undergone some process of *alteration* or remineralization prior to their extraction from the quarry—portions of the rock may or may not be rendered more liable to the chemical attack of atmospheric moisture. In granites and other acid rocks, for example, the felspars may have been 'micacized'—converted into an aggregate of minute flakes of white mica and quartz—and though none of the original felspar crystals may remain, if the silicification is satisfactory the rock will be little, if any, the worse as regards its capacity for wear.

Another form of alteration consists in the production of an aggregate of finely granular or crystalline hydrated silicate of alumina (kaolinization), with more or less silicification. This tends to make the altered crystals of felspar softer; but since in a usable stone it has never gone so far as to destroy the cohesion of the rock, its effect in reducing the resistance of the stone to weathering is not really serious. In Cornwall very considerably kaolinized granites have been employed in local buildings without showing signs of marked deterioration—for instance, in the tower of the Church of St. Stephen's-in-Brannel. Such stones may show a lower figure under crushing stress, but it does not follow that they will weather badly.

Alteration of the lime-bearing plagioclase felspars is often accompanied by the separation and crystallization of calcite in the aggregate of new minerals; and when this is largely developed in a stone, it will be attacked just as carbonate of lime is attacked in other stones, and decay will be hastened. It is, however, by no means established that fresh plagioclastic felspars regularly decay in a building stone more rapidly than the orthoclastic varieties, although statements supporting this contention —based on laboratory experiments—will be found in many places.

The white mica in building stone is practically unalterable. Turning to the ferro-magnesian minerals, which are sparsely represented in most granites and syenites, but are important constituents of the more basic rocks, we find that where they are fresh and unaltered they are scarcely at all influenced by chemical weathering. Where these minerals have been altered and their place taken by aggregates of crystals of the carbonates of calcium, magnesium and iron, chlorites, hydromicas, and serpentine, and various oxides of iron, and by certain zeolitic minerals, the decay of the stone by chemical attack will be facilitated; but even in stones of this class the alteration may include the deposition of secondary silica or the growth of fresh secondary felspar, which will improve the quality.

In any of the igneous rocks the presence of an appreciable amount of carbonate, especially if associated with pyrites, is detrimental.

Well-crystallized pyrites in a fresh granite is not a

source of much danger, but it, like all other iron compounds, may produce, by oxidation and hydration, either local spots or a uniform tinge of brown stain, in accordance with its distribution in isolated crystals or in uniformly scattered granules. The basic rocks usually turn brown after prolonged exposure, while the acid ones do not as a rule become more than slightly tinged.

The principal cause of decay in crystalline igneous rocks is variation of temperature. In hot, dry countries the effects of the diurnal range of temperature are quite sufficient in course of time to produce very striking results. During the day the heated surface of the stone is caused to expand, but owing to its poor conductivity the layer in which expansion takes place is a shallow one; and when night sets in, with its rapid lowering of temperature, powerful stresses are set up between the now rapidly contracting outer layer and the comparatively unaffected inner parts of the stone. This produces the rapid exfoliation especially noticeable in granites in such climates.

Whether or not flaking of the surface takes place during a marked rise of temperature, stresses are set up between the adjoining mineral grains in a granite, syenite, diabase, etc., on account of the variable masses of these grains and their unequal conductivity and coefficients of expansion. The conductivity and expansion coefficients are not even constant in value in different directions within the same crystal; and the consequence of these oft-repeated stresses, though they may be small in amount, is a permanent deformation of the surface of the stone. The felspars, amphiboles, pyroxenes, and micas open out little by little along their cleavage and parting planes; quartz develops hair-like cracks along planes of inclusions or along directions of earlier strain; and furthermore, each grain tends to separate from its neighbour. Thus even in temperate climates the surface of the stone is prepared for the reten-

THE DECAY OF BUILDING STONE

tion of moisture and the action of frost; and the pressure of wind and the fall of rain dislodge grain after grain by purely mechanical means.

How hot a stone may become in this country is within the experience of all who *in puris naturalibus* have attempted to recline on a summer sun-bathed rock. Granite masonry embankments having a southern aspect and subjected to tidal wetting are prone to flaking through the alternation of hot dry and cool wet conditions. Cleopatra's Needle, on the Thames Embankment, and similar monuments removed to America, although they had retained the sharpness of their incised inscriptions for 2,000 years in the dry climate of Egypt have had to be treated with preservatives in their new situations to stop the very rapid deterioration that commenced, owing to moisture and frost taking advantage of the numberless minute cracks produced during the long exposure to dry changes of temperature.

The amount of movement set up by the daily march of the sun is well illustrated by measurements made with the hollow granite obelisk of Bunker's Hill. On a sunny day this structure, which is 221 feet high and 30 feet square at the base, swings in such a way that a pendulum hung from the centre of the top describes an irregular ellipse having a maximum diameter of nearly half an inch.

CHAPTER X

THE TESTING OF BUILDING STONES

Object of Testing.—Before we consider the methods and results of testing building stones, it may be well to inquire, What is the objective of the tests ? In brief it is to establish the relative fitness of stones for use under the conditions in which it is desired to place them. In other words, it is to discover whether a stone is strong enough to withstand the particular stresses to which its position in the structure render it liable, and whether its other qualities are such as will insure a long life in the climatic circumstances which will surround it.

Determination of the quality of *strength* is a comparatively simple matter ; it is carried out by investigating the behaviour of the stone subjected to stresses of pressure, tension, bending, shearing, and the like, when under specified conditions.

The estimation of the *wearing quality* of a stone is a far more complicated and difficult business, involving the investigation of the chemical and physical influence of water, the effect of frost, the effect of changes of temperature, and of acid-laden air. To these may be added the influence of wind, aspect, and general climate, the position in the building, and the effect of form, in hastening or retarding the processes of decay ; and, in certain cases, the operation of abrading agencies and of plant life.

No series of tests, however elaborate, however accurate,

THE TESTING OF BUILDING STONES

would enable anyone totally ignorant of the behaviour of stones in existing buildings to form a correct conception of the relative value of a number of individual stones of different kinds. In estimating the value of a new stone, we draw consciously or unconsciously upon our experience of the quality of stones already in use; and the surest way of securing a sound knowledge of the attributes of stones in general is only to be obtained by the complete investigation of *all* the properties of stone, whose qualities of strength and resisting power can actually be observed in buildings of known age. The knowledge so gained can then be applied to the interpretation of the results of examination and testing of untried stones in terms of strength and durability.

Compression — Crushing Strength. — The crushing strength of a stone is determined by applying gradually increasing pressure to a block placed between two steel plates, which may be vertical or horizontal, according to the type of testing machine employed. Pressure is applied until the stone breaks down, but it is usual to observe and note the formation of cracks that are sometimes produced before the sample gives way. The results are expressed in tons per square foot, pounds per square inch, or kilogrammes per square centimetre.

To secure fair and comparable results, the following points should receive attention:

(*a*) *Form.*—The samples are usually made into cubes, not because it is theoretically the best form, but because of its convenience and the influence of custom. Blocks of square section and height one and a half times the length of the base would probably give better results. Two-inch cubes are very commonly used, but for weaker stones cubes of 4-inch and 6-inch sides have been employed. The actual size is of small moment, but it should be observed that it places a coarse-grained rock, like certain granites, at a considerable disadvantage if small test-blocks are taken.

(b) *Bedding.*—The highest and fairest results are obtained with blocks that are sawn true and then rubbed fine and smooth. Hammer-dressed or chiselled surfaces should not be employed. No bedding material should be applied between the stone and the steel plates. All sorts of material have been tried, with the object of neutralizing inequalities of bedding and distributing the pressure evenly; but it was long ago shown that such substances as wood, lead, millboard, blotting-paper, and the like, tend to lower the result by exerting a lateral thrusting action on the stone under pressure. Many tests have been made with a thin layer of plaster of Paris and some with cement on the two opposed faces in contact with the steel plates. All such bedding materials act unequally on different stones, and are better omitted. Hudson Beare, who used plaster in his tests (see p. 372), showed that the loss of strength caused by using sheets of lead for bedding ranged from 36 to 55 per cent.

(c) *Drying.*—The samples should all be dried in the same way and to the same degree—preferably by prolonged repose on racks at a moderate temperature of 30° to 35° C. Higher temperatures are most commonly employed (110° C.), but it is well known that this is sufficient to cause modifications of the strength of the stone—small in amount, it may be, but it varies with the kind of stone, and will thus introduce small but undesirable errors.

The United States War Department tested a series of 20-inch bars by accurately measuring them, and placing them in a water-bath at 32° F., then in a bath at 212° F., and then back to the cold bath again. A permanent expansion of the stone was observed, which amounted to a mean

THE TESTING OF BUILDING STONES

swelling of 0·009 inch in granites, quartz, quartzite, and marbles; 0·007 inch in limestones and dolomites; and 0·0047 inch in sandstones. The crushing strength was reduced after this process, but the result is unfortunately complicated by the use of water.

(d) *Number of Test Pieces.*—It is very desirable that a good number of samples of each kind should be tested; this, of course, applies to all kinds of tests in a greater or less degree.

The range of variability under pressure tests may be illustrated by the following results, obtained in the Charlottenburg laboratories:

Granite.—In sixty different kinds the test pieces from the same quarry showed an average individual difference of 31·6 per cent.

2 kinds	showed	differences of	2 to 10	per cent.
6	,,	,,	,,	11 to 20	,,
23	,,	,,	,,	21 to 30	,,
21	,,	,,	,,	31 to 40	,,
*6	,,	,,	,,	41 to 60	,,

Porphyry.—In three different kinds the mean variation in the test pieces was 18 per cent., with a range from 11 to 27 per cent.

Basalt.—In six different kinds the mean variation in the test pieces was 26·8 per cent., with a range from 14 to 43 per cent.

Limestone.—In eleven kinds the mean variation was 38 per cent., with a range from 11 to 56 per cent.

Sandstone.—In fifty-six kinds the mean variation was 35 per cent.

11 kinds	showed	differences of	11 to 20	per cent.
15	,,	,,	,,	21 to 30	,,
21	,,	,,	,,	31 to 40	,,
11	,,	,,	,,	41 to 50	,,
*4	,,	,,	,,	51 to 60	,,

* Some gave even higher differences.

Even if two or three test pieces are sawn from the same block, appreciable differences are always observed in the pressure results. It is obvious, therefore, that tests recorded from one or two samples are hardly worth the trouble taken to perform them.

Until quite recently the crushing strength, determined by some variant of the method outlined above, has been the principal, often the only, test applied, not only in this country, but in Europe and America. The test studied in relation with other experimental data may yield information of value, but the mere statement of numerical results is of extremely small value, and throws no light whatever on the wearing power of the stone.

It is of far greater importance to know something of the last-named quality than to have a measure of the resistance to crushing. 'It has been computed that the stone at the base of the Washington Monument—the highest structure in the world—sustains a maximum pressure of 22·658 tons per square foot, or 314·6 pounds per square inch. Certain contractors require a stone to withstand twenty times the pressure to which it will be subjected in the wall, while others require resistance to only ten times that pressure. Even if requiring a factor of safety of twenty, the strength required for a stone at the base of the monument would be only 6,292 pounds per square inch. The pressure at the base of our tallest building can scarcely exceed one-half that at the base of the monument, or 157·3 pounds per square inch. According to the above estimate, stone used in the tallest buildings does not require a compression strength above 3,146 pounds per square inch (using a large factor of safety). There is scarcely a building stone of importance in the country that does not give a higher test than this. Ordinary building stone has from two to ten times the maximum required crushing strength. A stone having a crushing

THE TESTING OF BUILDING STONES

strength of 5,000 pounds per square inch is sufficiently strong for any ordinary building' (Buckley, 'Building Stones of Wisconsin').

Since it is desirable to employ the stronger kind of stone for those situations in the structure which will be subjected to the greatest pressures, the simple compression test may be of use in making the selection; but here due attention should be paid to the environment of the stone, and if it is to be in a wet situation, the tests to be compared should be those made on the stone in a saturated state (see p. 392).

Shearing.—Although stones in buildings succumb to shearing stress much more frequently than to pressure, comparatively few results of shearing tests are available for comparison.

The United States Army shearing tests were made upon blocks approximately 6 inches long, 4 inches broad, and 2 inches deep. These prisms were faced up and supported on end blocks 6 inches apart, and pressure was applied to one side in the ordinary machine by means of a 'plunger' the same width as the test piece and 5 inches long, so that there was a clearance of $\frac{1}{2}$ inch on either side of the plunger between the ends of the underlying supports. A few of the results are given below:

	Shearing Strength.	
	Pounds.	Pounds per Square Inch.
Granite	138,800 =	2,872
Granite	84,400 =	1,742
Granite	50,600 =	1,052
Sandstone	66,400 =	1,382
Sandstone	47,600 =	992
Marble	75,100 =	1,554

Tensile Strength.—It is very generally stated—with perfect truth—that stones in masonry are rarely called upon to withstand stresses of tension, and it is no doubt on this account that the tensile strength of stones is seldom determined and only occasionally requested. Now, although it is correct to assume that individual blocks of stone in a structure are not often subjected to tension stresses *in the block*, they are none the less liable to such stresses *internally* whenever they are exposed to the action of frost or considerable changes of temperature.

The method of testing universally employed for cement might well be introduced as the principal strength test for building stone. It is generally admitted that most stones have more than enough strength to withstand any simple pressure that may be put upon them in a building; in most cases, then, it seems to be waste of time to test the crushing strength unless the result is used for the estimation of some other property. The man who actually performs the crushing

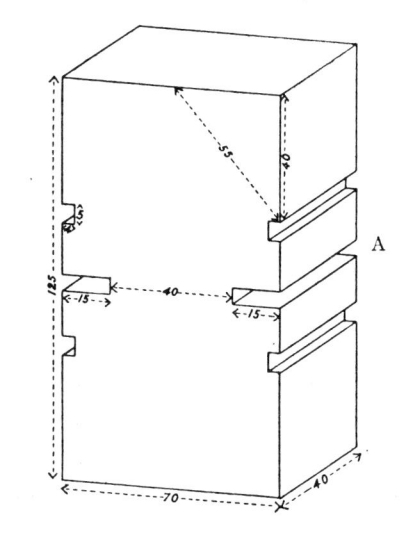

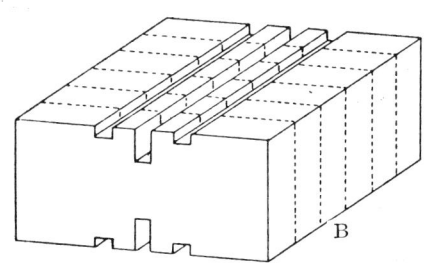

FIG. 24.—BLOCKS OF STONE PREPARED FOR THE TENSION TEST. (After Professor J. Hirschwald.)

test may learn a good deal about the stones from their respective behaviour during the test and their form after testing; but just as much and more can be learnt from their behaviour under tension, with the additional advantage that the fracture surfaces in the latter case are in a much better condition for examination with a lens than they are after crushing. The character of the fracture faces will often throw much light upon the quality of the stone.

The test for tensile strength may be carried out exactly as for cement, and some tests have been made on blocks of stone in shape and size like the ordinary cement briquette.

Preparation of test pieces of this form out of stone is more difficult and less satisfactory, for, unlike cement, the sample cannot be moulded: it must be cut.

The dimensions and form of test pieces employed by Hirschwald are indicated in Fig. 24, A. A number of samples are cut from single blocks, as in Fig. 24, B, by a lapidary's saw, and the grooves are cut very carefully

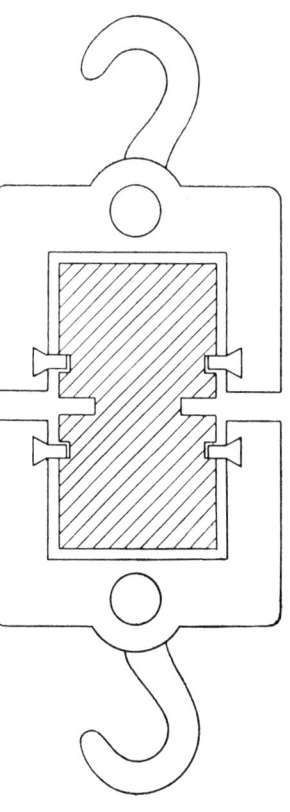

FIG. 25. — BLOCK OF STONE IN SUPPORT FOR TENSION TEST. (After Professor J. Hirschwald.)

with truly parallel sides by the same means. The test piece is supported in the testing machine by nuts, as in Fig. 25; this part of the apparatus should be of hardened steel.

Examples of tension results will be found in Tables XIII. and XX.

One advantage of the tensile strength test is that smaller stresses are required than for the pressure test. The examples of Table XIII., taken from Bauschinger's results (*loc. cit.*, p. 433), will serve to illustrate this point.

TABLE XIII.

Name of Stone.	Resistance to Pressure.	Resistance to Tension.
Granite, Hauzenburg	1,020	210
Gneiss, St. Gothard Tunnel	790	132
Gneiss, St. Gothard Tunnel	1,100	195
Granite, Fürstenstein	850	92
'Greenstone,' Ottendorf	1,070	300
Trachyte, Sohndorf	575	118
Mica schist, St. Gothard Tunnel	910	256
Muschelkalk (limestone), Randersacker	440	69
Bunter sandstone, Bettingen	775	115
Molasse sandstone, Algäu	510	24
Mean	804	151

Relation, 1 : 0·187.

Elasticity.—The modulus or coefficient of Elasticity—

$$E = \frac{\text{Stress per unit area}}{\text{Strain per unit length}}$$

—is a property difficult to determine in stone, owing to the extreme smallness of the quantities to be measured; consequently, there are few results available.

The methods usually employed are compression and bending. In the former case blocks are treated in the pressure machine as for the crushing test, and the amount of deformation is measured indirectly by means of a mag-

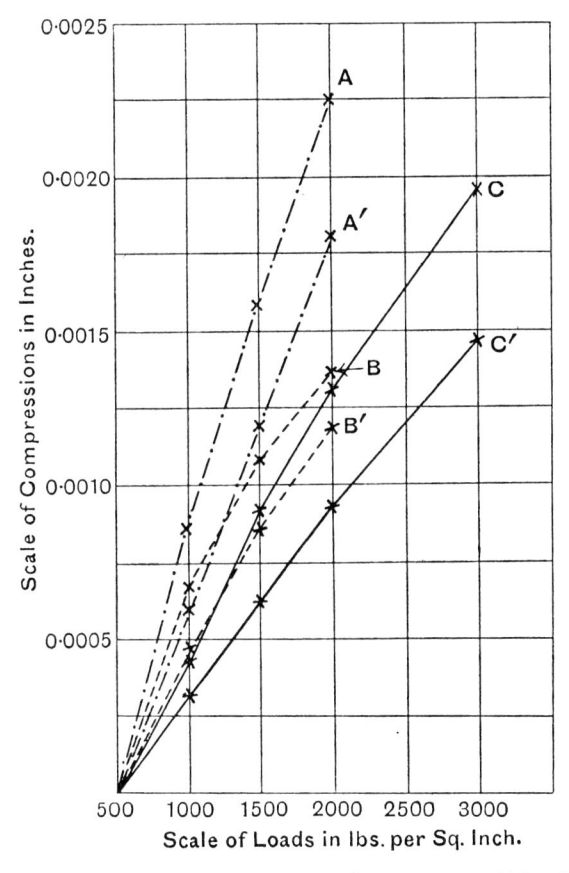

FIG. 26.—STRESS-STRAIN CURVES IN SANDSTONE. (After Professor Hudson Beare, by permission of the Council of the Institution of Civil Engineers.)

    A and A'. First and second tests of Darley Top Stone.
    B and B'. First and second tests of Bramley Fall Stone.
    C and C'. First and second tests of Aspatria Stone.
    (Compressions measured on a length of 1½ inches.)

nifying lever arrangement, or directly by the vernier microscope, for each increment of stress.

TABLE XIV.—CRUSHING STRENGTH, DENSITY, WEIGHT, AND ABSORPTION OF BUILDING STONES. (HUDSON BEARE.)

Name of Stone or Quarry.	Crushing Load (Mean) in Tons per Square Foot.	Density Average.	Weight per Cubic Foot in Pounds.	Absorption per Cent. of Dry Weight absorbed.
GRANITES:				
West of England, Penryn	1060·2	2·65	165·4	0·12
Cornish Grey	959·9	2·59	161·7	—
Corennie, Aberdeenshire	1234·3	2·61	162·9	0·38
Cove, Aberdeenshire	987·1	2·71	169·1	0·55
Kemnay, Aberdeenshire	1008·5	2·63	164·1	0·42
Kemnay, Aberdeenshire	1211·1	2·58	161·0	0·21
Craighton, Aberdeenshire	1282·0	—	—	—
Peterhead, Aberdeenshire	1207·7	2·54	158·5	0·29
Dyce, Aberdeenshire	1105·8	2·65	165·4	0·19
Hill of Fare, Aberdeenshire	1360·3	2·55	157·9	0·40
Sclattie, Aberdeenshire	850·5	2·58	161·0	0·10
Persley Grey, Aberdeenshire	942·8	2·60	162·3	0·19
Rubislaw, Aberdeenshire	1098·8	2·64	163·7	0·09
Rubislaw, Aberdeenshire	1289·7	2·61	163·7	0·09*
Ben Cruachan	876·9	2·75	171·6	0·29
SANDSTONES:				
Prudham, Four Stones Quarry	455·3	2·28	142·5	4·16
Corncockle	383·8	2·12	132·6	4·57
Gunnerton	354·6	2·09	130·8	5·16
Cragg	573·8	2·18	136·1	4·13
Corsehill	444·9	3·09	130·4	7·94
Polmaise	551·5	2·27	141·7	4·58
White Plean	612·8	2·21	138·2	4·25
Arbroath	558·1	2·42	151·0	2·32
Auchinlee	203·6	2·07	128·9	6·90
Craighleith	861·9	2·22	138·6	3·61
White Hailes	662·0	2·30	143·8	3·71
Dean Forest	530·0	2·42	151·4	2·71
Gatelaw Bridge	495·7	2·07	129·5	5·84
Blue Hailes	459·7	2·29	143·2	4·70
Binnie	569·1	2·16	135·1	5·22
Hermand	457·4	2·28	142·6	4·70
Howley Park	466·7	2·25	140·3	4·90

* 177 feet below surface.

THE TESTING OF BUILDING STONES

CRUSHING STRENGTH, ETC., OF BUILDING STONES—*Continued*.

Name of Stone or Quarry.	Crushing Load (Mean) in Tons per Square Foot.	Density Average.	Weight per Cubic Foot in Pounds.	Absorption per Cent. of Dry Weight absorbed.
SANDSTONES—(*continued*):				
White Grinshill	209·3	1·96	122·5	7·80
Darley Top	516·7	2·23	139·0	3·40
Hercules Ridge	335·7	2·21	138·0	3·60
Bramley Fall	238·4	2·11	132·2	3·70
Ackworth	389·1	2·25	140·7	5·00
Robinhood	574·0	2·31	144·6	3·90
Aspatria	239·9	1·97	123·2	8·50
Lightcliffe, Bed 3	1025·1	2·39	149·6	2·30
Lightcliffe, Bed 4	1015·9	2·40	149·6	2·30
DOLOMITES:				
White Mansfield	461·7	2·24	140·1	5·01
Red Mansfield	591·9	2·29	143·2	4·58
Yellow Magnesian Limestone	577·4	2·33	145·4	4·62
Anston	301·9	2·11	132·2	7·50
LIMESTONES:				
Ancaster, Brown Weather Bed	552·6	2·50	156·3	2·42
Portland Base Bed, P Quarry	287·0	2·20	137·6	6·84
Portland Base Bed, R Quarry	204·7	2·12	132·3	7·51
Ketton	101·7	2·07	127·9	8·10
Limestone, Crystalline	956·2	2·80	174·7	—
Corsham Down	94·5	2·07	129·0	11·17
Farleigh Down	62·5	1·93	120·5	12·88
Monks Park	139·6	2·19	136·7	8·02
Box Ground	97·5	3·05	127·9	8·29
Coombe Down	117·7	2·06	128·6	5·99
Corn Grit	134·5	2·14	133·6	9·88
Stoke Ground	90·0	2·02	126·3	10·84
Winsley Ground	100·7	2·13	132·9	7·74
Westwood Ground	110·2	2·08	130·3	8·85
Westwood Ground (fluated)	106·9	2·12	132·3	8·03
Doulting, Fine	103·9	2·00	125·0	10·87
Doulting, Chelynch	180·8	2·41	150·4	3·36
Ham Hill	165·3	2·18	136·0	—
Ancaster Freestone	184·0	2·22	140·4	6·27

For British stones the best available set of results is that obtained by Beare (Table XIV.). In these experi-

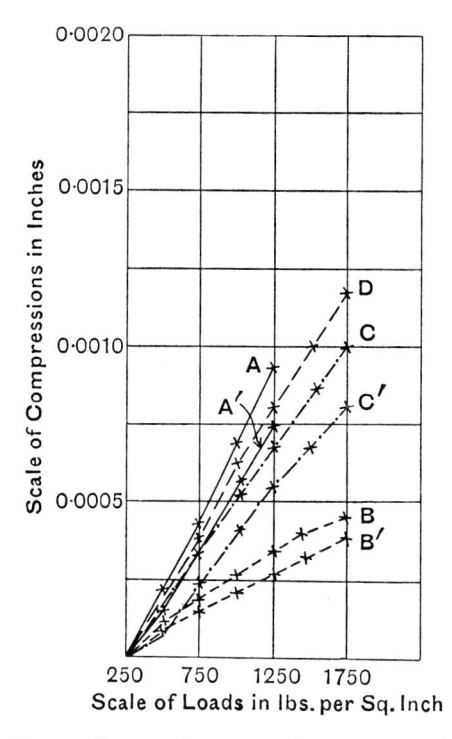

FIG. 27.—STRESS-STRAIN CURVES IN LIMESTONE. (After Professor Hudson Beare, by permission of the Council of the Institution of Civil Engineers.)

> A and A'. First and second tests of Doulting Fine Bed.
> B and B'. First and third tests of Doulting Chelynch Bed.
> C and C'. First and second tests of Ham Hill Stone.
> D. Ancaster Freestone.
> (Compressions measured on a length of $1\frac{1}{2}$ inches.)

ments he employed test-cubes of $2\frac{1}{4}$ inches in the Greenwood and Battley machine used for the pressure tests, with a specially sensitive measuring gear. A load was

THE TESTING OF BUILDING STONES

first applied sufficient to keep the test piece, with the gear attached, in its position between the dies of the machine, and the corresponding scale reading was taken as zero; this load varied from 250 to 1,000 pounds per square inch. 'The loads were then increased gradually by 250 pounds to 500 pounds, or 1,000 pounds, per square inch at a time, according to the strength of the stone, and each time the readings of the position of the pointer on the scale taken, till the highest load it was desired to apply was reached; the load was then let back to the starting load, and the permanent set measured. A fresh set of observations was then made with this new zero, the same method of procedure being followed' (*loc. cit.*).

The stress-strain curves for three sandstones and four limestones are shown in Figs. 26 and 27.

The mean value of the coefficient for all the sandstones tested (except one) was for the first test 108,040 tons per square foot, and for the second test 132,280 — or the stone is from fifteen to twenty times as compressible as steel.

The corresponding mean values for the magnesian limestones was 254,500 and 321,000; for the limestones, 133,530 and 150,750; and for the granites, 479,000 and 522,100 tons per square foot.

The United States Engineer Corps tests were made by subjecting the stone to compression and to bending. In the former test prisms 24 inches by 6 inches by 4 inches were used; deformation was measured by means of a micrometer. Some of the results (abbreviated) are given below:

Granite, Milford, Mass.:
    E = 5,128,000 lbs. between 1,000 and 2,000 lbs. per sq. in.
       = 7,272,700 ,,   ,,   600 ,, 8,000   ,,   ,,
Granite, Rockport, Mass.:
    E = 6,666,700 ,,   ,,   1,000 ,, 2,000   ,,   ,,
       = 9,523,800 ,,   ,,   10,000 ,, 12,000   ,,   ,,

Marble, Pickens Co., Ga.:
  E = 6,896,500 lbs. between 1,000 and 3,000 lbs. per sq. in.
    = 9,090,900 „   „   5,000 „ 7,000   „   „
Dolomite, Tuckahoe, N.Y.:
  E = 13,333,000 „   „   1,000 „ 3,000   „   „
    = 15,384,600 „   „   8,000 „ 10,000 „   „
Oolitic limestone, Ky.:
  E = 9,290,000 „   „   1,000 „ 2,000   „   „
Sandstone, Cromwell, Conn.:
  E = 7,711,000 „   „   1,000 „ 5,000   „   „
Sandstone, Mass.:
  E = 1,780,000 „   „   1,000 „ 2,000   „   „
Slate, Maine:
  E = 12,250,000 „   „   1,000 „ 5,000   „   „

For the bending tests the prism was placed on end supports and pressure applied in the middle, the distance between the supports being 19 inches.

Chatley says ('Stresses in Masonry'): 'Taking 144,000 tons (per square foot) as a convenient and fair average (for the value of E), we have

$$\text{Strain} = \frac{\text{Stress}}{144,000},$$

or for a stress of 144 tons per square foot (1 ton per square inch) the strain is $\frac{1}{1000}$ . . . In tension no stone can bear this stress, although in compression it will do so apparently. It will not actually bear it, for the concurrent shearing stress along any diagonal plane inclined 45 degrees to the horizontal is upwards of $144 \div 2 = 72$ tons per square foot, which is above the figure given (for shearing). As a practical rule it may therefore be said that scarcely any stone will bear an extension or compression amounting to $\frac{1}{1000}$ of its linear dimension. Since for ordinary blocks of stone this elongation or shortening is imperceptible, we arrive at the common result that stone is not visibly elastic.'

It may be noted here that the maximum compression observed in the American tests cited above was 0·384 in the case of one of the sandstones and 0·0411 in one of the granites. The permanent set was greatest in the granite (maximum 0·0067); in the slates it was least.

Tests made in the University of Wisconsin gave the following values of E in pounds per square inch:

	Minimum.	Maximum.
Granite	156,000	2,070,000
Sandstone	32,000	400,800
Limestone	31,500	1,835,700

It should be observed that the value of E for one kind of sandstone — the 'brownstone' of Lake Superior — ranged from 56,000 to 387,900 pounds per square inch; and generally the sandstones gave a much lower coefficient than either the limestones or granites.

Transverse Strength, Cross-breaking Test.—For this test rectangular blocks of stone are prepared, and in the experiment they are supported at the ends and the load is applied in the middle. The stress at rupture in pounds per square inch for a prism loaded and supported in this way is calculated from the formula

$$R = \frac{3Wl}{2bd^2}.$$

For British stones the tests made by Baldwin-Wiseman may be quoted ('The Effect of Fire on Building Stones,' Transactions, Surveyors' Institute, xxxviii., 1906). For test pieces he employed dressed prisms, 6 inches by 1 inch by 1 inch, or thereabouts; these were supported on two steel-knife edges, 4 inches apart, with a thin sheet of rubber between the support and the stone. The load was applied

TABLE XV.—CERTAIN PHYSICAL PROPERTIES OF STONE. (BALDWIN-WISEMAN.)

Stone.	Weight per Cubic Foot in Pounds.	Specific Gravity.	Percentage Volume of Pores.	Crushing Stress of the Stone (Dry) in— Pounds per Square Inch.	Tons per Square Foot.	Crushing Stress of the Stone in various Conditions, soaked in Water.	Breaking Stress of the Stone (Dry) in— Pounds per Square Inch.	Tons per Square Foot.	Breaking Stress of the Stone in various Conditions, soaked in Water.
SANDSTONES AND QUARTZITES:									
York	135·1	2·162	11·99	8,480	546·6	0·977	1,124	72·3	0·606
Red Mansfield	136·2	2·179	14·98	2,520	162·1	1·007	666	42·8	0·754
Aspatria	125·4	2·006	7·61	2,430	156·3	1·008	498	32·0	0·526
Quartzite	184·0	2·944	2·88	3,680	236·6	1·005	2,354	151·4	0·986
Soft Daresbury	111·9	1·790	16·58	900	57·9	0·445	126	8·1	0·769
Hard Daresbury	115·7	1·852	17·75	2,020	129·2	0·866	282	18·1	0·332
LIMESTONE AND CALCAREOUS FREESTONES:									
Doulting	129·1	2·065	14·46	1,190	76·6	0·992	320	20·6	0·606
Portland Base Bed ...	136·2	2·180	12·80	4,080	262·4	0·725	1,448	93·1	0·708
Monks Park	125·8	2·013	20·51	1,570	101·0	0·380	630	40·5	0·664
Box Ground	119·2	1·907	22·33	940	60·3	0·171	404	26·0	0·695

THE TESTING OF BUILDING STONES 379

LIMESTONE AND CALCAREOUS FREESTONES—(continued):									
Bradford Oolite ...	109·9	1·758	22·24	860	55·3	0·691	401	25·8	0·493
Bath Oolite	120·4	1·926	17·81	1,140	73·3	0·675	609	39·2	0·604
Hopton Wood ...	141·4	2·263	8·36	2,920	187·8	1·003	1,003	64·5	0·948
Micheldever Chalk ...	96·6	1·545	46·79	390	25·0	0·310	162	10·4	0·551
Mottisfont Chalk ...	100·7	1·611	45·86	290	18·7	0·366	215	13·8	0·612
MARBLES:									
Carrara	146·2	2·340	0·78	4,450	286·1	1·007	823	52·9	0·924
Rouge Royal	138·2	2·212	0·93	6,420	412·8	0·701	1,933	124·3	1·008
St. Anne's	157·9	2·527	0·74	9,250	594·8	0·921	2,243	144·2	0·954
Dove	156·9	2·510	0·87	5,940	381·9	0·914	2,235	143·7	0·800
Black	158·9	2·542	0·73	6,710	431·5	0·999	2,955	190·0	0·788
Belgian Granite ...	194·3	3·109	1·06	5,460	351·1	1·004	2,108	135·5	0·776
GRANITES AND IGNEOUS ROCKS:									
Red Granite	157·5	2·520	0·28	5,480	352·4	0·931	1,437	92·4	1·009
Grey Granite	168·3	2·693	0·12	5,950	382·6	1·008	2,712	174·4	0·938
Diabase	163·1	2·610	0·24	6,070	390·3	1·004	1,820	117·0	0·874

to the centre of the slab by means of a steel saddle, with rounded bearing, carrying a hook and scale-pan for the weights, which were added in increments of 5 pounds at a time.

The results are given in Table XV.

For the Wisconsin tests prisms of from 4 to 7 inches in length and a cross-section of 1 inch were tested in an Olsen machine.

The United States Engineer Corps used prisms 24 inches by 6 inches by 4 inches, with an effective length of 19 inches and depth of 6 inches.

Some of the values for the modulus of rupture are given below:

Granite:
  R = 3,909 lb. per sq. in. } 4 in. long
    = 2,324   ,,   ,,
Limestone:
  R = 1,609   ,,   ,,   7   ,,
      2,042   ,,   ,,   6   ,,
      3,923   ,,   ,,   5   ,,
      4,659   ,,   ,,   5   ,,   } Wisconsin.
Sandstone:
  R =   391   ,,   ,,   7   ,,
        498   ,,   ,,   7   ,,
        655   ,,   ,,   3·5 ,,
        845   ,,   ,,   4   ,,
        992   ,,   ,,   4   ,,
      1,324   ,,   ,,   4   ,,
Granite:
  R = 1,745   ,,   ,,   19   ,,   , 4 in. broad, 6 in. deep } U.S.E.C.
  R = 2,416   ,,   ,,   19   ,,   4   ,,   6   ,,

Weight and Specific Gravity.—The weight of a stone per cubic foot is presumably a constant of some value, and between the heavier and the lighter stones there is a considerable difference, which may be of importance in comparing freightages or the load to be borne at certain points in a masonry structure. When, however, we look at the weights recorded by different authorities for the same stone, we find discrepancies amounting sometimes to as much as 15 per cent.—a very serious matter when it is necessary to calculate the weight of large quantities of stone.

Most of the published weights are said to be determined upon the stone in its 'normal' condition, or as it is supplied by the contractor for a building. This 'normal' condition is a very variable one.

The actual weight of a stone varies from day to day, and the same stone from the copings or basement portions of a building will almost always differ in weight from that in flush ashlar. The weight depends upon the specific gravity of the *material* of the stone, on its porosity, and on the amount of water it contains at the time of weighing. In order to have tolerably comparable determinations without going into unnecessary refinements of manipulation, it would be well to record the weight in a thoroughly dry and in a thoroughly saturated state (p. 389).

The specific gravity is a property useful in determining several of the characteristics of a stone. It is commonly measured by weighing the stone in its ordinary state in air and then in water; then

$$\text{Sp. gr.} = \frac{W}{W - W_1},$$

where W = weight in air and W_1 = weight in water. The sample for this purpose should not be less than a 2-inch cube, and the result may then be near enough for some purposes. It may be used, for example, to estimate the weight of a stone per cubic foot; thus, if the specific gravity is 2·7, the weight of a cubic foot of water being approximately 62·4 pounds, the weight of the same volume of the stone will be 2·7 × 62·4 = 168·5 pounds approximately.

Conversely, the specific gravity may be determined by directly measuring the volume of the stone weighed, as was done by Beare for the results in Table XIV.

The above-mentioned methods may not accurately represent the specific gravity, because they neglect the factors of porosity and rate of absorption. To obtain the true specific-gravity of the *stone material* the small cube may be thoroughly dried and weighed in air, and then completely saturated and weighed in water, but the pyknometer (or specific gravity bottle) method is quicker and better. The specific gravity by this method is calculated from the formula

$$\text{Sp. gr.} = \frac{W_p}{W_w + W_p - W_{pw}},$$

where W_p = the weight of the powdered stone, W_w = the weight of the water alone, W_{pw} = the weight of the powder and water. About 20 grammes of the powder should be used, and it should be moderately coarse. When the powder is in the water, the pyknometer should be placed under an aspirator or air-pump until some time after bubbling has ceased.

Tetmajer (*Methoden und Resultate der Prüfung Kunstlicher und Natürlicher Bausteine*) found the density or weight of unit volume of stone, *inclusive of pores*, by a volumometer method. He dried small blocks of the stone 15 to 20 c.cm. at 50° C., and when quite dry brushed the surfaces free from loose dust; then warmed them to 54° C., and dipped them into molten paraffin until a layer 1 to 2 millimetres thick was deposited all over.

The density of the stone $\rho = \dfrac{W}{V} = \dfrac{W}{V_{sp} - V_p}$.

W = Weight of stone.
V = Volume of stone.
V_p = Volume of paraffin = $\dfrac{\text{Weight of paraffin by difference}}{\text{Sp. gr. of paraffin at } 15° \text{ C.}}$.
V_{sp} = Volume of paraffin + Stone.
ρ = Density of stone.

Porosity and Absorption.—By the methods mentioned in the preceding section we are provided with a means of estimating the porosity of a stone.

If P = the volume of pores in percentage of the volume of the stone, Sp = its specific gravity; ρ = its density, inclusive of pores, then

$$P = \frac{(Sp - \rho)\,100}{Sp}.$$

Or, if d = the weight of the stone when dry,
    w = the weight of the stone when thoroughly saturated,

the percentage porosity can be determined from the formula

$$\frac{(d-w)\,Sp}{(d-w)\,Sp + d}\,100, \quad \text{or} \quad \frac{d-w}{w-S_w}\,100,$$

where S_w = the weight of the saturated stone when suspended in water, as for the specific gravity determination. Hirschwald used the former method; Buckley and others have employed the latter.

By porosity is understood the percentage of pore-space in a stone, or the ratio of the volume of pores to that of the stone. The difference between absolute porosity and the ratio of absorption, or relative porosity, must be borne in mind. It has been a very common practice to express the porosity as

$$\frac{\text{Weight of water absorbed}}{\text{Dry weight of stone}},$$

but this is incorrect, and will always give a result that is too low.

Again, the experimental method of obtaining the result has in many cases been unsatisfactory as regards the soaking of the stone. Many results have been obtained by taking the dry weight, and then placing the sample in water until all air-bubbles had ceased to form on the

GEOLOGY OF BUILDING STONES

TABLE XVI.—SATURATION UNDER DIFFERENT CONDITIONS: COEFFICIENT OF SATURATION—RELATIVE POROSITY (HIRSCHWALD).

	Saturation in Percentage of Weight of Stone.			S = Saturation Co-efficient. $\dfrac{W_2}{W_p}$	Saturation in Percentage of Total Volume of Pores.			P = Relative Porosity Co-efficient.
	W_1	W_2	W_v	W_p		W_1	W_2	W_v	
SANDSTONES:									
Mürlenbach ...	4·89	5·66	7·89	9·23	0·613	52·97	61·30	85·46	20·16
Neukirche ...	5·57	5·89	9·48	10·25	0·575	54·41	57·56	92·57	20·80
Gerstungen ...	4·92	5·02	5·75	6·67	0·753	73·25	74·78	85·55	15·04
Obernkirchen ...	5·72	6·35	11·14	11·45	0·555	50·02	55·54	97·33	23·26
Grüssau ...	2·47	2·49	2·61	2·75	0·905	90·13	90·57	95·10	6·77
Löwenberg ...	7·76	7·81	12·58	12·85	0·608	60·50	60·86	97·92	25·38
LIMESTONES:									
Kauffüngen ...	0·35	0·49	0·55	0·59	0·831	59·47	84·27	94·67	1·57
Limburg ...	0·66	0·85	0·89	0·90	0·944	72·99	92·29	92·30	2·51
Kammerforstberg ...	2·20	2·41	3·71	4·58	0·526	48·15	54·03	81·10	11·50
Hottensen ...	7·51	7·88	19·08	21·19	0·372	35·46	37·20	90·04	36·47

SLATES:								
Nordenau	0·52	0·58	0·86	0·86	0·674	60·35	67·50	100·00	2·33
Simmerath	0·92	0·98	1·18	1·41	0·695	65·00	69·58	83·33	3·81
Caub	0·51	0·55	0·70	0·70	0·786	72·92	79·16	100·00	1·91
Ruppachtal	1·04	1·10	1·80	2·16	0·509	48·19	51·11	83·33	5·66
GRANITES:								
Strehlen	0·74	0·74	0·98	1·31	0·565	56·46	56·46	74·53	3·34
Strehlen	2·65	2·91	3·88	3·88	0·750	68·44	75·06	100·00	9·32
Jannowitz	0·40	0·57	0·66	0·84	0·680	47·89	68·00	78·27	2·17
Trusenthal	0·28	0·59	0·59	0·69	0·855	41·79	85·58	85·58	1·82
OTHER IGNEOUS ROCKS:								
Siebengebirge (Basalt)	0·27	0·27	0·39	0·39	0·693	68·46	68·46	100·00	1·14
Naumburg (Basalt)	1·07	1·07	1·17	1·17	0·915	91·14	91·43	100·00	3·37
Burgberg, Ilfeld (Porphyry)	2·28	2·90	3·44	3·90	0·744	61·76	78·46	90·23	9·15
Nahetal (Porphyry)	1·24	1·24	1·24	1·40	0·886	88·75	88·75	88·75	3·48
Weibern, Rhine (Tuff)	22·11	23·41	30·25	33·75	0·694	65·51	69·37	89·64	45·14
Brohltal, Rhine (Tuff)	13·48	13·48	15·00	15·54	0·867	86·71	86·71	96·52	29·35

surface, and then reweighing, after wiping, to find the water absorbed. Results so obtained are not comparable for different stones.

The method of soaking is of great importance, as the results indicated in Table XVI. will sufficiently demonstrate.

Buckley's method was to obtain the porosity from the data furnished by his specific-gravity measurements. For saturating the stone he used the apparatus shown in Fig. 28. The samples, 2-inch cubes, after being dried at 110° C. and weighed, were placed in the bottle a, which was itself placed in a water-bath kept at 100° C. The air was then partly exhausted from the bottle by the aspirator b, until the pressure was about half an atmosphere, as registered by the manometer c. Boiling distilled water was then admitted slowly through the tube d, thus allowing the ascending water to drive out the air from the stone as it rose. The stones were kept in the hot water at the reduced pressure for thirty-six hours, and then taken out and quickly weighed.

In many respects this method is better than the old one, but even in this case the pores are not so completely filled as is possible by causing the soaking to take place under pressure greater than one atmosphere.

The effect of different conditions in determining the amount of saturation is illustrated in Table XVI., the examples being selected from a much larger number given by Hirschwald.

The first column (W_1) gives the result of what he calls the quick-soaking method. For this purpose he dried the samples (about 30 grammes in the case of dense stones, and 70 grammes for more porous stones) at 50° C. for three hours, cooled them in a desiccator, weighed, and then placed them in distilled air-free water until they maintained a constant weight; this was reached in a few hours, or only after several days, according to the character of the stone. (It may be noted that the variations in the manner of wiping the

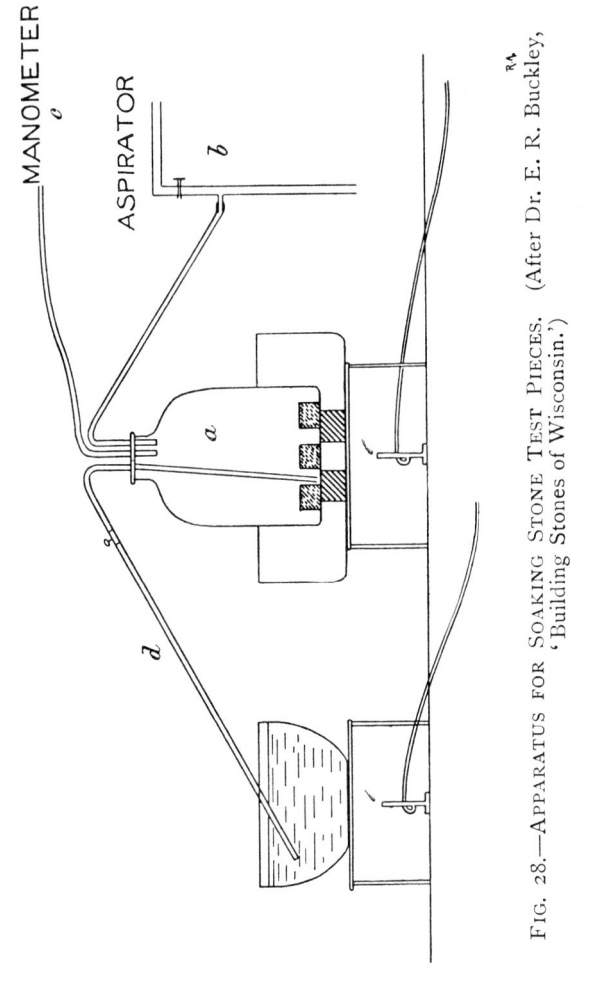

FIG. 28.—APPARATUS FOR SOAKING STONE TEST PIECES. (After Dr. E. R. Buckley, 'Building Stones of Wisconsin.')

wet stone before weighing made a difference of as much as 2 per cent. in the result in the case of rough porous stones.)

The second column (W_2) gives the results of the slow-soaking process carried out in the apparatus shown in Fig. 29. The water was allowed to drop into the large

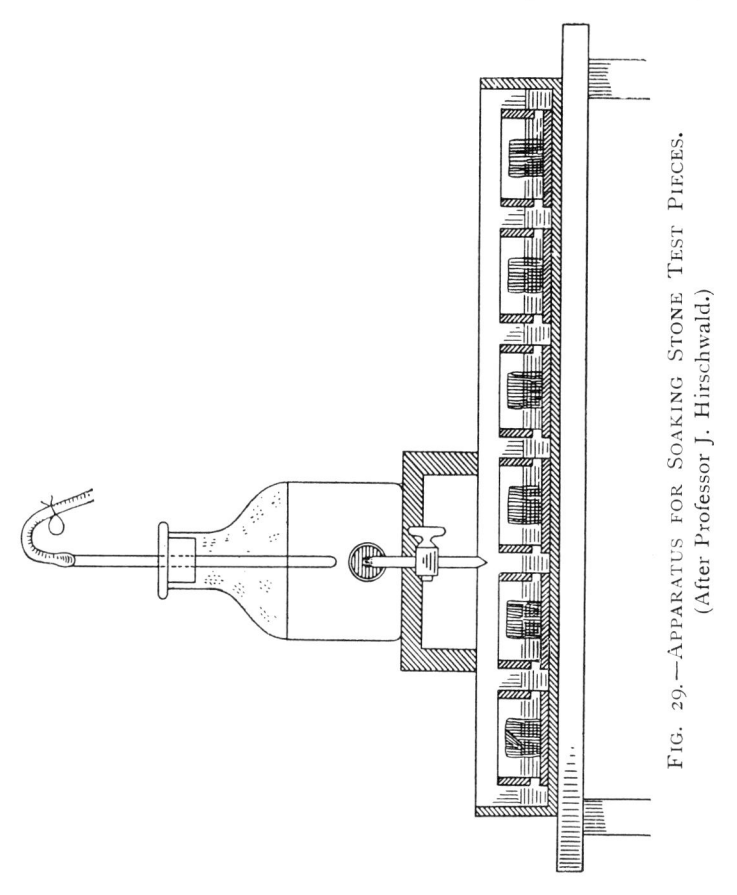

Fig. 29.—Apparatus for Soaking Stone Test Pieces. (After Professor J. Hirschwald.)

trough at such a rate that the test pieces were covered after four hours; they remained in the water until they had attained a constant weight, as in the first method. The advantage of the small separate trays is that any small fragments that may drop off can be conveniently collected and weighed.

The third column (W_v) gives the results of soaking under reduced pressure produced by a water aspirator, or in the case of dense stones, by a mercury pump.

The fourth column (W_p) gives the result of soaking under pressures of 50 to 250 atmospheres until constant weight was obtained. The air was first exhausted as far as possible.

The fifth column gives the so-called Saturation Co-efficient S, or the ratio of the water absorbed under normal conditions, by the better method (W_2) to that absorbed under pressure (W_p).

The old dicta that 'The more porous the stone the greater the danger from frost,' and 'The value of a stone for building purposes is inversely as its porosity,' can no longer be maintained. Much more depends upon the type of porosity than upon its amount. A stone which has open pores and absorbs water quickly will, under suitable conditions, part with it quickly, and thus it will have more chance of being safe when frost comes.

J. C. Jones (Transactions, American Ceramic Society, vol. ix., p. 528) has laid stress on the relative effective diameter of the pores and the rate of flow in them, or that of soakage and drying. He points out that in a wall there are three broadly defined zones: (1) A zone of drainage, (2) a zone of capillary soaking (a little above the ground), and (3) a zone of immersion (in and near the ground). In the first of these the most important function to determine is the relative rate of absorption, and he measured this by the ratio of water absorbed in a partly immersed brick (or stone) in fifteen minutes to that absorbed after a definite period of prolonged soaking.

Table XVII. gives the effect of different types of porosity

GEOLOGY OF BUILDING STONES

TABLE XVII.

Source of Stone.	Structure and Other Qualities.	Porosity Type.	W_1	W_2	W_v	W_p	$S = \dfrac{W_2}{W_p}$
Granites:							
Weinberg	Fine grain, very strong	II.*a*	0·36	0·39	0·71	0·72	0·544
Jannowitz	Medium grain, very strong, slightly cavernous	III.*b*	0·39	0·49	0·50	0·71	0·690
Jannowitz	Medium to coarse grain, cavernous	III.*a*	0·40	0·57	0·66	0·84	0·679
Strehlen	Fine grain, friable (from old building)	II.	1·70	1·80	2·32	2·38	0·756
Limestones:							
Beckum	Strong, micro-crystalline	III.*b*	0·26	0·40	0·50	0·74	0·541
Schönecken	Fine grain, crystalline, with obscure bedding	I.	0·55	0·71	0·94	1·00	0·710
Ellrich	Hard, cavernous	III.*a*	2·20	2·41	3·71	4·58	0·526
Rüdersdorf	Schaumkalk	III.*a*	3·92	4·43	5·89	8·96	0·494
Near Schöppenstedt	Shelly, cellular, tolerably weak	VIII.	7·81	8·11	11·85	13·16	0·616
Sandstones:							
Dortmund	Very strong, siliceous, with felspar and carbon particles	III.*b*	1·47	1·63	1·74	2·21	0·738
Langendorf	Well bedded, with kaolinized felspar	V.	3·32	3·58	6·10	6·35	0·564
Mürlenbach	Fairly strong, coarse grain	VIII.	4·89	5·66	7·89	9·23	0·613
Aschaffenburg	Bedded, with dusty Fe_2O_3	V.-VI.	4·32	4·75	7·36	7·44	0·614
Röhrtal	Friable, with strongly decomposed felspar	V.	6·45	6·80	9·39	9·68	0·702
Nebra	Fairly strong, with pyrites	V.	8·05	8·48	12·83	12·90	0·657
Near Börssum	With carbonaceous layers	V.	11·35	11·39	14·65	14·75	0·772
Slates:							
Goslar	Very thin lamination, ridged surface, little mica	II.	0·56	0·58	0·78	0·80	0·725
Clotten	Fairly thin lamination, wavy surface, little mica	II.	0·61	0·64	0·93	1·00	0·587
Ruppachtal	Fairly thin lamination, ridged surface, fair amount of mica	II.	1·04	1·10	1·80	2·16	0·509

W_1, By rapid soaking. W_2, by slow soaking. W_v, by soaking *in vacuo*. W_p, by soaking under 75 atmospheres' pressure. S = saturation coefficient. Porosity Type: The figures in this column refer to the kinds of porosity described by Hirschwald (*Die Prüfung der Natürlichen Bausteine*, p. 45).

TABLE XVIII.

Sandstones.		Condition in Old Buildings.		Saturation in Weight Percentage of Stone.				Quotient of Absorption ⊥ and ∥ to Bedding.				$S =$ Normal.
Source of Stone and General Structure.				W_1	W_2	W_v	W_p	$\dfrac{W_1 \wedge}{W_1 \vee}$	$\dfrac{W_2 \wedge}{W_2 \vee}$	$\dfrac{W_v \wedge}{W_v \vee}$	$\dfrac{W_p \wedge}{W_p \vee}$	
HORN, LIPPE: Bedding hardly perceivable		Completely frost-resisting	{ ⊥ ∥	4·48 4·63	4·55 4·69	8·94 8·75	9·13 9·04	} 1·03	1·03	1·022	1·01	0·528
KIRN: Very obscurely bedded		Not completely frost-resisting	{ ⊥ ∥	3·47 4·38	5·63 5·36	6·33 6·72	7·47 7·57	} 1·26	1·05	1·01	1·01	0·814
SOLLINGEN: Marked, thin bedding		Tolerably resistant to badly flaking	{ ⊥ ∥	4·69 3·48	4·77 4·07	6·26 5·70	6·63 6·38	} 1·35	1·17	1·098	1·039	{ 0·670 to 0·830
WÜNSCHELBURG: Obscurely bedded		Completely frost-resisting	{ ⊥ ∥	4·47 3·90	4·64 4·46	9·02 8·81	9·11 9·01	} 1·15	1·04	1·024	1·019	0·467
FROM ANOTHER BED: Fairly well-marked coarse bedding		Frost-resisting	{ ⊥ ∥	3·51 3·74	3·54 3·80	7·08 7·60	7·16 7·65	} 1·07	1·07	1·07	1·07	0·563

on the amount of absorption and the saturation coefficient S.

Table XVIII. shows the effect of bedding in stratified rocks, when soaking was permitted to take place in one direction only, either parallel or at right angles to the bedding planes. It will be observed that the absorption

TABLE XIX.—COEFFICIENT OF WEAKENING BY SATURATION.

		Mean Resistance to Pressure in Kilogrammes to the Square Centimetre.		Coefficient of Weakening (n).	Saturation in Weight Percentage.
		Dry.	Saturated.		
Granites (Hanisch's Tests)	Säusenstein, Austria	1,923	1,728	0·93	0·5
	Gmünd, Austria	1,070	908	0·85	0·94
	Mean of 11 stones	1,482	1,321	0·88	0·77
Granites (Gary's Tests)	Hälleforschult, Sweden	3,519	3,351	0·95	0·06
	Lysekil, Sweden	2,362	2,109	0·89	0·11
	Istaheda, Norway	1,163	1,106	0·95	0·40
	Mean of 25 stones	2,440	2,296	0·94	0·20
Crystalline Limestones (Hanisch's Tests)	Koholz, Austria	1,417	1,196	0·84	0·39
	Sterzing, Austria	618	547	0·88	0·42
	Mean of 9 stones	981	880	0·90	0·51

is not always greatest in the direction of the bedding plane when the soaking is done under normal conditions, but the values approximate closely when it takes place under pressure.

Weakening in Water.—Many experiments have been made to show the loss of strength of stones in the wet condition. Tables XIX. and XX. illustrate this feature.

THE TESTING OF BUILDING STONES

Hirschwald gives for the coefficient of weakening by saturation

$$n = \frac{\text{Crushing strength wet}}{\text{Crushing strength dry}},$$

after twenty-eight days' 'slow' soaking (p. 388). Stones containing the most argillaceous, earthy, calcareous, or fine dusty material, exhibit the greatest amount of

TABLE XX.—COMPARISON OF THE BORING AND TENSION METHODS OF DETERMINING THE COEFFICIENT OF WEAKENING (HIRSCHWALD).

Sandstone from—	Kilogrammes Weight on Borer.	Bd.	Bw.	$n = \frac{Bw}{Bd}$.	$n = \frac{Tw}{Td}$.
Bretzenheim, Rhine	4·5	6·45	0·68	0·11	0·18
Nebra, Saxony ...	6·5	3·36	2·13	0·63	0·60
Löwenberg, Silesia	9·5	4·86	2·33	0·48	0·62
Near Obernkirchen	12·5	7·0	6·67	0·95	0·94

Bd = Number of revolutions made by the borer in sinking 10 millimetres (Dry Stone).
Bw = Number of revolutions made by the borer in sinking 10 millimetres (Wet Stone).
$n = \frac{Bw}{Bd} =$ { The coefficient of weakening determined from the boring test.
$n = \frac{Tw}{Td} =$ { The same coefficient determined by the tension test for comparison.
The above are average results from numerous tests.

weakening; in these cases n is likely to be less than 0·5. Any stone with a value for n less than 0·6 may be considered a bad one for damp or frost-exposed situations.

The weakening in compact rocks with fresh mineral ingredients—such, for instance, as a fresh granite, or a sound marble or quartzite—will be probably less than 0·1.

The reduction of strength in these dense stones may be due in part to the transmission of the pressure through the water within them.

The Weakening Coefficient of Slates, etc.: Sclerometer Tests.—The hardness of single minerals is conveniently estimated by comparison with Mohs' scale (see p. 411).

Refined determinations of the hardness of minerals in definite directions are of no value whatever for the study of building stones. These are made with some form of sclerometer—an instrument for making a scratch on the object to be tested under measurable conditions. The instrument usually consists of a scratching point—an octahedral diamond crystal—which can be gradually weighted, and a sliding platform for the test piece. The least weight required to make a visible scratch is the measure of relative hardness.

This method can give good results only in the case of even-grained stones; its range of utility is therefore limited; but it is a convenient test for slates which are difficult to treat fairly by the pressure or tension test.

In Table XXI. a few results of tests on dry and wet slates, with the calculated coefficient of weakening, have been selected from Hirschwald's experiments.

Abrasion or Attrition Tests.—The power to resist abrasion is a property comparatively rarely required in building stones. The fact that certain portions of a building will be subjected to more attrition than others can seldom be taken into account in selecting the stone; æsthetic and other considerations will usually have greater weight. For paving stones, steps, landings, and for all kinds of road construction, the ability to resist wear is of the first importance.

Abrasion tests have been employed to measure the relative *hardness* of stones, but it must be observed that hardness is only one of the factors that influence the results.

The principal methods used for testing resistance to abrasion are (1) the Boring Test; (2) the Grinding Test; (3) the Sand-blast Test; and (4) the Rumbler Test.

Boring Test.—The comparative hardness of stones has been measured by a boring test; the apparatus consisted

TABLE XXI.—TESTS ON ROOFING SLATES TO DETERMINE THE COEFFICIENT OF WEAKENING BY THE SCLEROMETER METHOD (HIRSCHWALD).

Roofing Slates from—	Minimum Weight on the Scratching Point in Grammes.		Coefficient of Weakening.
	Dry. Hd	Wet. Hw	$n = \dfrac{Hw}{Hd}$
Sauerland (quarry)	2	2	1
Nuttlar, Westphalia (quarry)	4	3	0·75
Clotten, Mosel (old roof)	2	1	0·5
Rhine District (old roof)	3	1	0·33
Hüttenrode (quarry)	5	1	0·2
Caub, Rhine (old roof)	4·5	0·75	0·17

of a weighted chisel-shaped borer, which was caused to rotate eccentrically on the test piece.

For comparing the hardness of different stones, either the weight may be kept constant and the number of revolutions counted in which the borer sinks to a fixed depth, or the number of revolutions may be constant, and the least weight necessary to sink the borer to the required depth may be taken.

In Table XX. the coefficient of weakening in water was determined by noting the number of revolutions required under the same weight for the dry and wet test.

This test cannot be employed satisfactorily for stones with a very strong siliceous binding material, and in obtaining the coefficient of weakening of coarser-grained stones a saw-shaped cutting edge was used.

Grinding Test. — In the Charlottenburg Institute Bauschinger's method is used. The abrading machine consists of a cast-iron disc, made to rotate horizontally, with a velocity of twenty-two revolutions per minute. On this is placed the stone to be tested, in the form of a cube, with a smooth face parallel to the bedding, offering a surface of 50 square centimetres to the disc. The cube is so placed that the centre of the face is 22 centimetres from the middle of the disc, and by means of a lever it is kept under a constant load of 30 kilogrammes. At the commencement of the test, and once a minute afterwards, the plate is fed with 20 grammes of No. 3 Naxos Emery. The loss of weight of the cube is determined after five minutes, or 110 revolutions. This is repeated four times with each cube, and the sum of the losses taken as the result.

The United States Officer of Public Roads employs a core 1 inch in diameter, cut from the solid rock, and faced off at the ends. The loss of weight after the stone has been held with a standard pressure against a steel disc fed with sand during 1,000 revolutions is the measure of 'hardness.' The result so obtained is scaled down by subtracting one-third of the loss of weight in grammes from 20. Below 14, rocks are called soft; from 14 to 17, medium; above 17, hard.

The Sand-blast Test has been used by Professor Gary in testing roadstones, woods, tiles, linoleum, and other floor coverings ('Mitt. a. d. K. Materialprüfungsamt,' xxii.,

THE TESTING OF BUILDING STONES

Heft 3; Berlin, 1904). He employed flat pieces of stone, with smoothed surfaces, at a distance of 6 centimetres from the nozzle of the machine, which delivered 'standard sand' under a steam pressure of 3 atmospheres for two minutes. Experiments were made parallel and at right angles to the bedding, and the loss of weight and of volume were recorded. The results show very fair agreement with the figures determined by the grinding disc; but the test is a more useful one, for it shows at once any irregularities of texture, and picks out in a most graphic manner the hard and soft minerals. Care is required in eliminating all coarse grains from the sand.

The Rumbler or Rattler Test has been used in France, America, and England for the comparison of roadstones, and occasionally for building stones. Various forms of machine have been used, the essential features being a rotating cylinder, horizontal or inclined, provided with longitudinal, steel, removable internal ribs, and sometimes with loose balls of steel or lumps of flint; but these constitute an undesirable addition. A considerable charge of stone is placed in the machine, and the weight of abraded chips and dust is determined after a separate dry and wet test.

This test is of little value for the comparison of building stones, but it is undoubtedly the most satisfactory one for macadam. A large number of British stones of all kinds have been tested in this way by Mr. E. J. Lovegrove, and the tested samples, with full particulars, have been deposited in the Museum of Practical Geology (see also 'Attrition Tests of Roadmaking Stones': E. J. Lovegrove, J. S. Flett, and J. Allen Howe; London, 1903).

Tests to Estimate Comparative Toughness have been made usually by dropping weights. The United States Office of Public Roads (Form 28) drops a standard weight upon a specially-prepared test piece until it breaks. 'The height in centimetres of the blow which causes the rupture of the test piece is used to represent the toughness of the specimen.'

Corrosion Tests.—Tests have frequently been made upon building stones with the object of ascertaining their relative power of resistance to the chemical action of town atmosphere.

Rivington recommends soaking a stone for some days in a 1 per cent. solution of sulphuric acid and of hydrochloric acid. Merrill dried and weighed samples, and suspended them in water, through which he passed a stream of CO_2 for six weeks. Buckley dried 1 inch to $1\frac{1}{2}$ inch cubes at 110° C., weighed and placed them in wide-mouthed bottles containing vessels of water to keep the air moist. The bottles were provided with tubes for the inlet of gas, and the cork was sealed. Washed CO_2 was passed into one set and washed SO_2 was passed into the other, until the bottles were respectively filled, and the charge was renewed twice a week for six weeks. The samples were then removed, washed in distilled water, and dried at 110° C. and reweighed.

These and similar tests give quite inconclusive results, and they are not worth repeating.

Staining Test.—A very handy test—one which often throws much light on the structural characteristics of a stone—is the staining method described by Hirschwald.

Small test pieces should be roughly squared with a hammer; stones with a splintery fracture are better sawn. A convenient size for compact rocks was found to be 6 by 4 by 3 centimetres, and for porous rocks 7 by 5 by 4 centimetres. These are dried at about 40° C., and soaked for forty-eight hours in a saturated solution of nigrosin in absolute alcohol. The samples are dry again in about half an hour.

Fresh crystalline minerals—felspar, calcite, quartz—show only a slight staining along fractures and cleavage planes; but decomposed and weakened felspars and other altered silicates take the stain more readily. Powdery minerals, such as earthy carbonates, and to a greater degree argillaceous matter, the iron oxides and hydrated oxides, take the stain greedily, and tend to filter the solution as it enters the stone; so that sediments with much of this powdery material stain darker on the outer surface, forming a narrow dark band there.

The more compact the stone and the smaller the pores, the broader is the outer band of colour. If the powdery material has been extensively silicified, or filled with crystalline calcite, it reduces the amount of stain taken up. Highly porous stones are stained completely through and strongly, and all cracks and large pores show up clearly.

Hirschwald goes so far as to recognize by this method thirty-five types of structure, with four degrees of intensity of staining in each. A few of these are indicated in Fig. 30.

Tests to Ascertain the Resistance to Freezing.—Tests with this objective have been made in the following ways:

1. By imitating the action of freezing through the crystallizing force of salts dissolved in water, and noting the loss of weight caused by the action.

2. By alternately freezing in the open air and thawing, and noting the loss of weight.

3. By alternately freezing in the open air and thawing, and noting the loss of strength.

4. By conducting the experiments (2) and (3) in a refrigerator.

400 GEOLOGY OF BUILDING STONES

By the first method the stone was dried, weighed, and soaked in a hot solution of sodium sulphate or of sodium

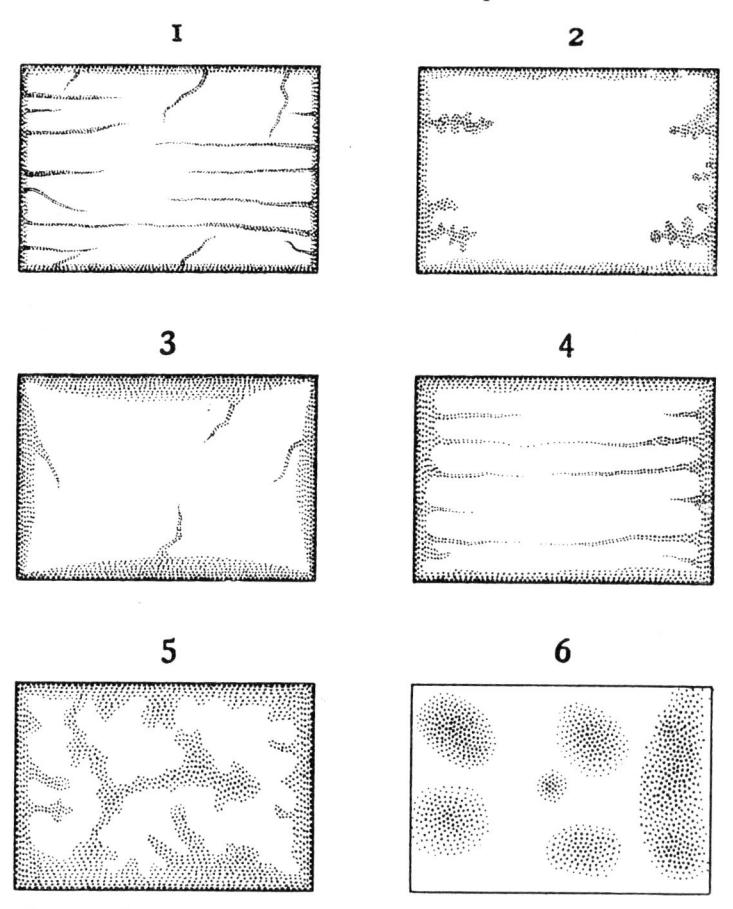

FIG. 30.—ILLUSTRATION OF THE EFFECT OF THE STAINING TEST ON DIFFERENT TYPES OF STONE. (After Professor J. Hirschwald.)

thiosulphate (hypo.); it was then allowed to cool and dry, and the loss by flaking from the surface was determined by weighing. This method is now generally abandoned, as the results are not at all satisfactory.

Whether the freezing is done in the open naturally or in a refrigerator is a small matter—the latter is the more convenient—but in any case the method of estimating frost action by weighing before and after a certain number of freezings is unsatisfactory, and it would be better not to employ it.

TABLE XXII.—TENSION TESTS TO ILLUSTRATE LOSS OF STRENGTH DUE TO FREEZING (HIRSCHWALD).

	Td.	Tw.	Twd.	Tf.	Tfr.
Sandstones:					
Aschaffenburg ...	89·12	61·26	106·49	105·88	88·51
Soest	62·55	57·66	75·33	72·14	59·36
Albendorf	42·17	35·85	53·81	53·36	41·72
Bevergern	19·66	9·15	21·40	20·37	18·99
Anröchte (glauconitic limestone)...	153·58	132·09	186·06	132·09	99·81

Td = Breaking stress of the stone in a natural state, dried at 30° C.
Tw = Breaking stress of the stone after fourteen days' soaking.
Twd = Breaking stress after fourteen days' soaking, followed by drying at 30° to 35° C.
Tf = Breaking stress after fourteen days' soaking, followed by thirty times freezing, and finally drying at 30° to 35° C. before testing.
$Tfr = Tf - (Twd - Td)$ = The loss of strength due to frost action alone.

The above results are the means of numerous tests, expressed in kilogrammes per square centimetre.

Repeated soaking and drying was found not to affect the value of Tw.

For determining the relative amount of frost action, the crushing strength—or, better, the tensile strength—measured before and after freezing, gives more concordant results.

Table XXII. brings out several points that demand careful attention when the comparative merits of freezing-test results are being considered.

It will be observed under column Twd that the effect

of continued soaking, followed by drying, is accompanied in each of these cases by an increase of strength. This is not invariably the effect of wetting and drying, but it has been observed in many stones by numerous experimenters. Further, in the examples given in the table the result of freezing, after the repeated wetting and drying, is to show an apparent increase of strength due to freezing as compared with the dry strength. The effect has been a source of trouble to many authorities, and they have been at some pains to explain away the results; but if correction is made for the real increase in strength due to the drying, it will be found in all cases that there is a real decrease in strength due to the frost action.

The relation between the resistance of a stone to frost and its coefficient of weakening in water is illustrated in Table XXIII.

TABLE XXIII.—RESISTANCE TO FROST AND THE WEAKENING COEFFICIENT. (HANISCH,* BAUSCHINGER.†)

Source.	Resistance to Pressure. Dry.	Resistance to Pressure. Wet.	Coefficient of Weakening. n.	
Sandstone:				
Weidling *	498	226	0·45	} Did not resist frost.
Kritzendorf *	689	401	0·58	
Ribnik *	1,101	734	0·64	
Ebelsbach (Keuper) †	519	351	0·70	} Resisted frost.
Ebelsbach (Molasse)†	999	756	0·75	
Mömmungen †	810	662	0·81	
Miltenberg †	748	706	0·94	
Kleinwallstadt †	460	385	0·83	} Did not resist frost.
Limestone:				
Epfenhofen	1,073	923	0·86	

The two last stones failed under frost action through the influence of qualities other than the weakening in water.

In carrying out the freezing tests, Hirschwald employs test pieces formed as indicated on p. 368 for the tension test, and the stones are put through this test before and after freezing. The experiment is conducted in two parts: in one the stone is soaked for thirty days; in the other it is soaked, according to its compactness, from two to thirteen hours. The object of the first is to test the fitness of the sample for damp situations, and that of the second is to show its suitability for the 'drainage zone' of walls. In the first test the stone after freezing is thawed out in about two hours in water at 20° C. In the second test the stone is thawed in a vessel over water which is kept at about 45° C., in order to prevent evaporation. The freezing is in each case repeated twenty-five times in a refrigerator of the Bauschinger or Linde type.

In order to avoid the correction for the drying effect (see p. 401), the tension test for the determination of the weakening effect of frost should be made only on the soaked sample before and after freezing.

Stones which show distinct indications of bedding should be tested, as in the case of the weakening in water, in pieces cut both parallel and at right angles to the bedding planes.

Weathering Tests under Natural Conditions.—The tests intended to give information upon the relative resistance to weathering agencies, by abbreviating the process artificially, as by the freezing, soaking, corrosion, and other experimental methods, are all far more drastic in their action than the processes of ordinary weathering. Thus in the tests for freezing the test pieces contain from twenty to thirty times the water they would hold when in their place in a wall—except when this was on or very near the ground line in damp situations—while the freezing test would be the equivalent of from one to five years of frost action in a climate like that of England. Evaporation between periods of rain is continually operating to prevent complete saturation of the stone in the drainage zone of buildings; the mean value for evaporation, taken from a number of experiments, has been given at 18·9 per cent. of the absorbed water in saturated samples, after forty-five hours' exposure to the still air of a chamber maintained at a temperature of 17° C.

Moreover, the action of the weather on stones in a building is applied, as a rule, on two faces at the most, and in the majority of stones on one face only; and it is the sum of the effects of a number of distinct forms of destructive activity, some of which, as alternate wetting and drying, may actually tend to strengthen the stone.

More than forty years ago Pfaff (*Zeitschrift d. Deutsch. Geol. Gesellschaft*, 1872, Bd. 24) exposed polished slabs of stone for periods of two and three years to the atmosphere, and determined the loss by weighing and measuring, then calculated the loss in a century. He found a polished slab of granite would be reduced 0·85 millimetre, and an unpolished slab of the same stone 0·76 millimetre in a century; while the compact Solenhofen limestone, on the measurements for two years, would lose 0·86 millimetre; on the measurements for three years, 1·33 millimetres in the same period. Thus it is more than probable that his results are too low.

The more complete experiments of Biszinger (*Ueber Verwitterungsvorgänge bei Krystallinischen und Sedimentärgesteinen*, Erlangen Dissertation, 1894) show in the clearest manner that the most prominent result of weathering is a mechanical disruption of the stone, and that the attendant chemical decomposition is exceedingly small.

Biszinger employed about 20 kilogrammes of each stone, broken to fragments of about 15 millimetres diameter, thus providing an attackable surface of about 3 square metres. The material was exposed in zinc boxes, having an opening for the admission of rain of 1,000 square centimetres. Once a year, or at longer intervals, the disrupted matter was collected and graded in sieves, and the several grades of residue were analyzed and compared with that of the

TABLE XXIV.—RAIN-ABSORPTION TESTS ON STONES FROM A BUILDING.

SEE FIG. 31.	1.	2.	3.	4.	5.	6.	7.
(a) Sandstone from Postelwitz	98·780	98·573	0·207	0·21	7·288	7·307	7·380
(b) Sandstone from Nebra	104·122	103·921	0·201	0·19	5·782	5·893	5·902
(c) Sandstone from Nebra	124·85	124·55	0·3	0·24	8·836	8·870	9·020
(d) Granite from Camenz	88·811	88·61	0·2	0·22	0·622	0·641	0·702
(e) Granite from Camenz	54·785	54·498	0·287	0·52	0·502	0·545	0·545

1. Weight of rain-soaked stone.
2. Weight of dried stone.
3. Weight of rain-water absorbed.
4. Weight of rain-water absorbed in percentage.
5. Weight of water absorbed in percentage of the dry weight of a control sample: ordinary method of soaking—after one hour.
6. After two hours.
7. After three hours.

The letters (a), (b), (c), (d), (e), refer to the position on the wall from which the samples were taken, as indicated in Fig. 31.

original sample. Further, the water which ran off the stones was analyzed and compared with the rain which fell freely during the same period. These tests showed that, while the mechanical disruption was very evident, the amount of stone substance removed by rain-water was not more than 0·003 to 0·01 per cent. of the original mass. These experiments were carried on for seventeen years.

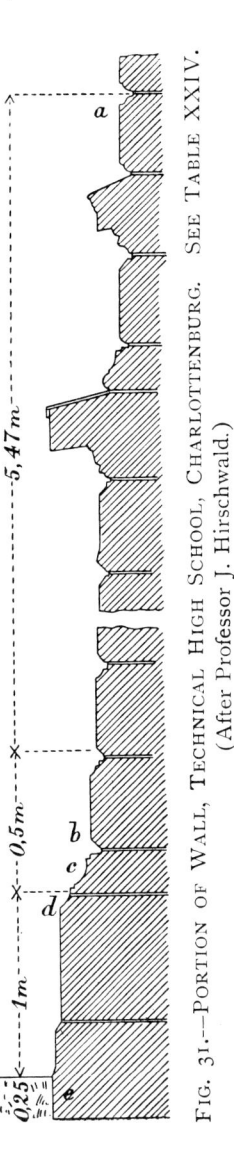

The amount of water absorbed by stones in a wall is well illustrated by the following investigation, described by Hirschwald: After a day's heavy downpour (6·4 millimetres) in December, portions of stone were taken from the wall, as indicated in Fig. 31 and Table XXIV. The absorption under these conditions is extremely small; and in general, it is found that the absorption by stones in buildings is far smaller than that by the same stones in the ordinary test pieces during testing. There is less difference between the absorption results, artificial and natural, in the case of compact stones than in the case of the more porous kinds.

Chemical Analysis. — For practical purposes the chemical analysis of building stones is of very little value, except to add a certain dignity to trade circulars. Stones of extremely divergent structure and wearing power may appear almost identical in the results of their analyses, and the chemical composition of the stone cannot be correlated with any of the important properties it possesses.

It has already been observed (p. 334) that the destruction of building stone through chemical agencies — except in the case of stones with much

FIG. 31.—PORTION OF WALL, TECHNICAL HIGH SCHOOL, CHARLOTTENBURG. SEE TABLE XXIV. (After Professor J. Hirschwald.)

calcareous matter, in towns—is exceedingly slight, and the presence or absence of the more readily attackable minerals may be determined with sufficient accuracy by easier and quicker methods than the complete chemical analysis.

Microscopic Examination.—Almost all authorities are agreed that the microscope affords a most convenient and reliable means for the investigation of the character of a building stone.

For its successful employment a certain amount of training in petrological methods is necessary; but when these are mastered, it is a comparatively easy and rapid way of comparing the mineral composition and structural type of stones. While the mineral composition is being observed, a fairly good idea of the chemical composition is being attained. Indeed, by employing micrometer measurements with the microscope, it may be made to yield information as to the chemical composition that will be sufficiently accurate for all practical purposes.

Some modification of the method of Rosiwal (*Verhandl. der k. k. geol. Reichsanstalt*, Vienna, xxxii., 1898, p. 143) is usually employed. This consists in measuring the diameters of each of the mineral grains along a large number of different lines in the thin rock slice by means of an ocular micrometer. The several measurements are summed for each mineral, and the relative proportions of the diameters correspond to their respective volume in the rock. Many modifications of the micrometer eyepiece are in use; sometimes a square field is employed, divided into 100 equal squares, by which the percentage of the minerals present may be determined directly.

The following examples (Table XXV.) will serve to illustrate the employment of the method for estimating the chemical composition of stones. For building stones it is rarely worth the trouble of working out.

TABLE XXV.—COMPARISON OF MICROSCOPIC AND CHEMICAL ANALYSES.

Red Granite, Trojak.

Minerals.	A.	B.
Quartz	31·1	31·4
Orthoclase and Microcline	31·8	31·5
Oligoclase	33·1	32·9
Biotite	4·0	4·6
Total	100·0	100·4

A = Results of micrometer measurements by Rosiwal's method in percentage volume.
B = Percentage weights calculated from the figures in A.

	C.	D.	E.
SiO_2	73·7	76·08	−2·4
Al_2O_3	14·5	13·30	+1·2
Fe_2O_3	1·1 ⎫ 1·6	1·11 ⎫ 2·09	0·0 ⎫
FeO + MnO	0·5 ⎭	0·98 ⎭	−0·5 ⎭
CaO	1·3 ⎫ 1·7	0·96 ⎫ 1·12	+0·3 ⎫
MgO	0·4 ⎭	0·16 ⎭	+0·2 ⎭
K_2O	4·9 ⎫ 8·5	4·33 ⎫ 7·70	+0·6 ⎫
Na_2O	3·6 ⎭	3·37 ⎭	+0·2 ⎭
P_2O_5	—	0·20	—
Loss on ignition	0·2	0·30	−0·1
Total	100·2	100·79	—

C = Chemical composition calculated from B.
D = Chemical analysis by John.
E = Difference between the micro-measurement estimate and the chemical analysis,

THE TESTING OF BUILDING STONES 409

Conclusions.—The brief account of tests upon building stones in the preceding pages does not by any means exhaust the list of tests that have been, or may be, applied. It remains for us to consider the practical value of building-stone testing. Does the architect employ tests in making his choice of a stone? Rarely indeed. The point of view of the engineer may be gathered from Professor Chatley's ' Stresses in Masonry.' He gives the following simplified results (Table XXVI.) to be used as a basis for calculations:

TABLE XXVI.—ULTIMATE STRESSES (TONS PER SQUARE FOOT).

	Crushing.	Tension.	Shearing.	Bending.
Granite	900	30	50	100
Basalt	800	80	40	—
Slate	800	10	30	50
Sandstone	500	10	30	50
Sandstone (soft)	200	5	10	20
Marble	600	30	50	—
Limestone	500	25	40	60
Limestone (soft)	100	9	35	50
Chalk	10	—	—	—

Low values are given for the ultimate stress, and the stone is assumed to be of good average quality.

' It is useless employing very exact figures, since every specimen, even from the same quarry, will vary somewhat in strength.'

The factor of safety commonly employed is 10.

' Hence the working values for strength should be about one-tenth the ultimate value given [above]. Further, seeing that a compressive strength is accompanied (except when there is a lateral support) by a shearing stress of a maximum intensity equal to half the compressive stress, we may obtain safe wall and pier loads by dividing the ultimate shearing stresses by 5 (*i.e.*, dividing by 10 and multiplying by 2, to convert from shearing to compression).

	Tons per Square Foot.		Tons per Square Foot.
Granite	10	Sandstone	6
Basalt	8	Sandstone (soft)	2
Slate	8	Limestone	8
Marble	10	Limestone (soft)	7

'The strength of the work will depend on the manner of construction and the mortar' (*loc. cit.*, p. 10).

Here we see the question of strength reduced to very simple terms on the assumption that the stone is of *good average quality*. The point is, are we to assume that stones ordinarily supplied are of this quality, or are we to find out by testing whether they are satisfactory, not only as regards strength, but in respect to resistance to decay?

In most cases this assumption is made in reliance upon the merchant and the clerk of works, and in the case of well-tried and familiar stones this is usually sufficient. It is in the case of new and untried stones that the need of some form of test is most felt.

What the architect requires is a series of simple tests for evaluating certain properties of the stone in such a way that a simple formula could be obtained expressing the comparative value of the stone.

The attempt made by Hirschwald to attain this end —the only serious attempt that has been made—shows how laborious and lengthy the process must be.

Whatever tests may be decided upon, the first desideratum is uniformity of method; without this the tests are really of no use for purposes of comparison.

In this country a vast amount of energy has been wasted in repeating very similar experiments in slightly different ways upon a limited number of stones. The results are of little or no use, because the architect cannot have any one of the tests continued upon exactly the same lines, so as to include all the stones employed or new untried stones. What is required is some scheme of *Standard Tests*, and some organization of authority for carrying out the tests regularly and quickly. It would matter little whether the tests were actually carried out by a Government laboratory or by a firm of standing and experience like Kirkaldy's, so long as there was provision for enforcing uniformity and continuity.

There will always be divergence of opinion as to the best tests to be applied, but those which seem to give the best information about the stone are probably Tension, Rate of Absorption, Freezing, and the Microscopic Examination.

MOHS' SCALE OF HARDNESS.

10. Diamond.
9. Corundum.
8. Topaz.
7. Quartz.
6. Orthoclase.

5. Apatite.
4. Fluorite.
3. Calcite.
2. Gypsum or Rock Salt.
1. Talc.

Each mineral in this scale will scratch those below it ; 1 and 2 can be scratched with the finger-nail : 7 to 10 cannot be scratched with a knife.

APPENDICES

APPENDIX A

GRANITE QUARRIES

Cornish Granite.—The principal quarry centre is in the Penryn mass, about Stithians Mabe and Constantine. The usual Penryn type of stone is of medium to fairly coarse texture, with white, irregular-shaped prisms of felspar 1 inch long, with smaller grains, grey quartz and dark mica.

The Cornish granites have been very extensively employed for engineering and architectural work of all kinds. A good example of the Penryn stone is Mr. Verity's new building, 94, Regent Street. See also the Embankment Wall ; New County Hall, L.C.C. ; Vauxhall Bridge ; Keyham Dockyard, etc.

It is quarried in the Maen group of quarries, the Polkannugo group, and Spargo Downs. At Cornsew is a stone of finer grain and cleaner grey colour, similar to the Delank rock of Bodmin Moor.

In the Penzance district the quarries are the Sheffield, Lamorna, New Mill, and Carfury. All the stone from this district is inclined to have a greenish tint, due to the chloritic coloration of the smaller plagioclase crystals ; tourmaline is a common constituent. Large porphyritic felspars up to 3 inches long are very characteristic of Lamorna and Sheffield ; they are less perfectly developed at New Mill. In the Lamorna stone they are arranged less regularly in the rock than they are in the Sheffield Quarry. When used in sufficiently large surfaces, these coarse-grained stones are very effective for polished fronts—*e.g.*, Temple Bar House and Thanet House, Strand.

The stone from the St. Austell district mainly comes from the Luxullian side, not far from Par ; the quarries are the Cottage, Colcerrow, Tregarden, and Carngrey. The Colcerrow and Tregarden Quarries produce a clean grey stone with large porphyritic ortho-

APPENDIX A

clase crystals; the stone is more blue-grey than that from the Penzance district, and it has not the green tinge of Lamorna and Sheffield. The Bodmin granite mass is quarried at St. Breward, Delank, and Cheesewring. The stone resembles on the whole the less porphyritic types of Penryn granite, but it is somewhat richer in quartz and a little darker. The Kit Hill granite is a fairly coarse grey or yellowish grey stone.

On Dartmoor the granite is quarried at Blackenstone, near Moretonhampstead; Swell Tor (Royal Oak), near Princetown; and the Tor Granite Quarry, Merrivale. At all these good grey stone is obtained; pinkish felspars are not uncommon, and the dark mica is rather abundant. In parts it is distinctly porphyritic, but none of the quarried stone is so coarsely porphyritic as Lamorna or Colcerrow. The Blackenstone rock is a dark grey stone, consisting of a ground mass of medium-sized, irregular grains of dark and white felspar, grey quartz, and abundant dark mica in small scales, in which are large clear-cut porphyritic cream-coloured felspars, ranging in size up to 1 inch wide and 4 inches long; they are rather sparsely scattered in the even-textured ground.

Luxullianite is a very striking variety of altered granite; it is apparently coarse-grained, with pink irregular felspars $\frac{1}{4}$ inch to $2\frac{1}{2}$ inches, in a black ground, which consists of quartz and tourmaline. The stone cannot be obtained except in small blocks. It was used for the sarcophagus of the first Duke of Wellington in St. Paul's. The stone was taken from a large loose block; it has never been quarried.

From Lanlivery a very handsome coarse-grained granite has been obtained; the prevailing tone is pale buff. The rock is mainly composed of cream-coloured felspars in small irregular grains and large porphyritic crystals ($\frac{1}{4}$ inch to 3 inches), together with bluish quartz and black tourmaline.

Trowlesworthy Tor, near Shaugh, Dartmoor, was quarried for a short time. The stone is of medium coarse grain, even texture, and not obviously porphyritic; the felspars are pale and white, stained here and there with a strong red, and there is much bluish quartz.

The stone formerly quarried at Heytor (Haytor), on the eastern border of Dartmoor, is of even, medium texture, rather finer than that of Blackenstone, and not porphyritic; the white felspar, grey quartz, and small mica grains are evenly distributed.

Granite from Lee Moor resembles that from the Constantine district in Cornwall, but has a stronger buff tinge.

Gunnislake Quarry produces a grey stone (bluish, pinkish, or buff), that is less porphyritic than the Dartmoor rock.

Aberdeen and Peterhead Granite.—In the Aberdeen district are the following large quarries:

Avochie.

Cairncry: Dark grey. Used mainly for setts.

Clinterty: Grey with reddish tinge.

Corrennie: Pinkish-red, due to the felspars, which are mostly in the shape of irregular grains; very little mica. Used in the Glasgow municipal buildings, Tay Bridge, and extensively used for millstones and rollers for paper and paint.

Dancing Cairns: Bright light bluish-grey to darker grey in the lower part of the quarry. Employed in Trafalgar Square, the Embankment, London Bridge, many buildings in Aberdeen, Broomielaw Bridge, Glasgow, and for setts, crushed rock for macadam, and adamantine slabs.

Dunecht.

Dyce: Dark greyish-blue, fine-grained, with tendency to show laminated arrangement of the minerals.

Fintray (Cothill), Kinaldie.

Kemnay: Medium grained, bright light grey, with dark specks of mica, smaller and thinner than in Creetown. Employed in the Mareschal College, Aberdeen, docks at Hull, Newcastle, Shields, Sunderland, Leith, and piers of Forth Bridge.

Oldtown.

Persley: Light grey.

Rubislaw: Dark greyish-blue, with microcline; darker in the deeper parts of the quarry; fine-grained. Used in Portsmouth and Sheerness Dockyards, Bell Rock Lighthouse, balustrades of Waterloo Bridge, and much used for monumental and polished work.

Sclattie: Like Rubislaw, but a lighter shade of grey; darker in lower part of quarry; mica less abundant. Used for setts.

Tillyfourie: Bluish-grey, sometimes pinkish, medium to coarse grain; felspars up to 1 inch long. The quarry is quite close to Corrennie.

Toms Forest: Like Kemnay, but a trifle darker.

Tyrebagger: Grey.

Both Creetown, Tillyfourie, and Kemnay show slight orientation of mica.

In the Peterhead district the principal quarries are the Admiralty;

APPENDIX A

Stirling Hill, at Boddam; the Blackhill (dark red), at Blackhill and Longhaven; Cruden; and the Cairngall.

A stone somewhat resembling Shap Granite, but paler in the ground mass, has been worked near Aboyne (Birsemore Quarry).

Small quarries are worked in the isolated Pitsligo mass, and in a few other places.

Kirkcudbrightshire Granite.—The larger quarries are:

> Bagbie, Creetown.
> Craignair and Cow Park, Dalbeattie.
> Fell, Creetown (includes the Silver-Grey Quarry).
> Fell Hill, Creetown.
> Glebe, Kirkmabrek.
> Lotus and Kissick Hill, Killywhan.

Craignair: Medium texture, rather coarser than Kemnay; axes pale grey. The grain is very distinct in the polished surface. There are no good crystal outlines, but white and pale pink felspars occur in irregular groups, together with a fair amount of the dark minerals. It is used in the docks at Liverpool, Birkenhead, Newport, and Swansea; in many buildings in Lancashire and other northern towns; Prudential Assurance Offices, Tivoli Music Hall, Thames Embankment, Birkbeck Bank, in London; Glasgow Corporation Waterworks; James Watt Dock, Greenock; Bank of Liverpool Head Offices (red polished Peterhead in lower part); Midland Hotel, Manchester; also heavy rollers for grinding paint, ink, and chocolate, and for macadam, railway ballast, and artificial slabs.

In general the Creetown and Dalbeattie granites are used for architectural work, axed and polished (it has a beautifully clean white appearance when axed); in engineering; also for setts, macadam, and granolithic slabs.

Craigton Granite is a rather darker (browner) red than Stirling Hill, and finer in grain (medium coarse); very little dark mineral is visible.

Irish Granite.—The Moor Quarry employs the greatest number of men, but there are numerous others, including those of Ballynacraigh and Altnaveigh (Bessbrook granite), at the base of Camlough Mountain, whence come the long polished columns in Manchester Town Hall; it is finer-grained and darker than the Ballynacraigh stone, and resembles Cairngall stone. Other quarries are the Sturgeons, Croreagh, and Goraghwood.

The Carlingford Lough granite possesses marked granophyric structure in parts; there is one large quarry in it. The granite of the Mourne Mountains consists of quartz (often with crystal faces), orthoclase, albite, and a green mica; it is quarried at Slieve Donard and Brown Nows quarries.

Other quarries in this district are Castlewellan (stone used in the pedestal and base of the Albert Memorial). (See Finn, Annalong; Pigeon Rock, Tullyfream, etc.)

APPENDIX B

CLASSIFIED LIST OF LARGER SANDSTONE QUARRIES

(M = mine.)

Silurian.

Merioneth.
Wern ddu, Corwen.

Montgomery.
Penstrowed, Newtown.

Radnor.
Llanfawr, Llandrindod Wells.

Old Red Sandstone.

Brecknock.
Penrheol, near Talybont.

Gloucestershire.
"Red Wilderness," Mitcheldean.

Elgin.
Newton, Alves.

Forfar.
Camperdown, Lochee West.
Duntrune, Murrowes.
Leoch, Rosemill, Strathmartins parish.
Pitarlie, Monikie.
Slade, Slade.
Wellbank, Monifieth.
Westhall, near Dundee.

Calciferous Sandstone Series.

Ayrshire.
Monkreddan, Kilwinning.

Dumfriesshire.
Closeburn, Thornhill.

Edinburgh.
Barnton Park, Cramond.
Hailes, Slateford (M).

Fifeshire.
Grange, Grange parish, Burntisland.
Newbigging, Newbigging, Burntisland.

Lanarkshire West.
Robroyston, Bishopsbriggs.

Linlithgowshire.
Pardovan, Philpstoun.

Renfrewshire.
Giffnock, Giffnock.

Buteshire.
Isle of Arran, Corrie.

Carboniferous Limestone Series.

Northumberland.
  Blaxter, Ravensclough, Otterburn.
  Cocklaw, Blackpasture, and Brunton, Chollerford.
  Denwick, near Alnwick.
  Doddington Hill, Doddington.
  Gunnerton, Barrasford.
  Newton Road, High Heaton.
  Prudham, Fourstones.
  Woodburn, West Woodburn.

Yorkshire (N.R.).
  Burtersett (M), Hawes (Yoredale).
  Old Burtersett (M), Hawes (Yoredale).

Lanarkshire.
  Huntershill, Bishopsbriggs.
  Huntershill (M), Bishopsbriggs (Index limestone).

Stirlingshire.
  Craigs, Airth Station.
  Polmaise, Bannockburn.

IRELAND:

Co. Donegal.
  Drumkeelan, Mountcharles.

Millstone Grit.

Cheshire.
  Ball Hill, Kerridge.
  Beeston, No. 1, Bollington.
  Crowden, Moses Hill, Crowden.
  Windyway, Rainow.

Denbigh.
  Bwlchgwyn, near Wrexham.

Derbyshire.
  Bamford Edge.
  Barton Hill, Birchover.
  Birchover, Winster.

Millstone Grit—*continued:*

Derbyshire—continued:
  Bole Hill, Hathersage.
  Brockholes, or Stoke, Grindleford Bridge.
  Crist and Barren Clough, Bugsworth.
  Dukes, Whatstandwell.
  Dukes, Whatstandwell.
  Dungeons, Birchover.
  Hall Dale, Darley Dale.
  Horsley Castle, Coxbench.
  Kinder, No. 1, Hayfield.
  Mouslow, Hadfield.
  New Pilough, Stanton-in-the-Peak.
  Peasenhurst, Ashover.
  Poor Lots, Tansley.
  Stancliffe, Darley Dale.

Lancashire.
  Ashworth Moor, Ashworth Moor.
  Brandwood (M), Nos. 1, 2, 3, Stacksteads.
  Brandwood, Stacksteads.
  Brinscall Sett Rock, Withnell, near Chorley.
  Britannia, near Bacup.
  Broom Hill, Dilworth.
  Butler, Pleasington.
  Chapel, Wine Wall, near Colne.
  Clough Top, Stacksteads.
  Copy, Longridge.
  Cownshore, Whitworth.
  Edgworth, Edgworth.
  Entwistle, Entwistle, near Bolton.
  Fletcher Bank, Ramsbottom, near Manchester.
  Frostholes (M), Lawhead.
  Frostholes, Lawhead, Stacksteads.
  Grane and Heap Clough, Haslingden.
  Greensmoor, Stacksteads.
  Greensmoor (M), Stacksteads.
  Hall Cowm and Ab Top, Britannia, near Bacup.
  Height End, Whitworth, near Rochdale.

Millstone Grit—*continued:*

Lancashire—continued:

Hollow Head, Wilpshire.
Hutch Bank, Haslingden.
Lee (several), Bacup.
Leicester Mill, Anglezarke, near Chorley.
Musbury Heights, Grane Road, Haslingden.
Noggarth, Barrowford.
Nook Fold, Dilworth, Longridge.
Park, Musbury.
Rake Head (M), (2), Stacksteads.
Scout (M), Newchurch.
Shawforth (M), Shawforth.
Scout Moor (Q and M), Shuttleworth, near Bury.
Sheffield Gate (M), Horncliffe.
Spencer, Longridge.
Tong End Pasture (M), Facit.
Tong End, Shawforth.
Withnell Central, Withnell.

Yorkshire.

Arnagill, Masham.
Cat Castle, Lartington.
Close, Salterforth.
Clough, Salterforth.
Dimples, Haworth.
Hallas Rough Park, Cullingworth, near Bradford.
Heights, Salterforth.
Longwoog Edge, near Huddersfield.
Ludgin Hill, Cullingworth.
Middle Tongue, Pateley Bridge.
Moor, Rochdale Road, Greetland.
Moor End (3 Qs), Mount Tabor, Halifax.
Scar, Pateley Bridge.
Scotgate Ash, Pateley Bridge.
Saltaire Road, Shipley.
Stannary, Halifax.
West End, Haworth Moor.
Woodhouse (2), Holmfirth, near Huddersfield.

Millstone Grit—*continued:*

IRELAND.

Co. Clare.

South Lough, Liscannor.
Watson's and Doonagore (Shamrock stone).
Watsonstone, Lahinch.

Coal Measures.

Lanarkshire.

Auchinlea (2), Cleland.
Bothwell Park, Fall Side.
Bredisholm, Uddington.
Clydesdale, Auchenheath.
Earnock, Hamilton.
Overwood (M), Stonehouse.

Linlithgowshire.

Braehead West, Fauldhouse and Braehead East, Union Canal, Linlithgow.
Eastfield, Fauldhouse.
Falahill, No. 2, Fauldhouse.

Stirlingshire.

Brighton, Polmont.

Denbigh.

Cefn, No. 2, near Ruabon.

Durham.

Heworth Burn, Heworth.
High Burn, Heworth.
New, Eighton Banks, Lamesley.
Peareth, Springwell, Gateshead.
Penshaw, Old Penshaw.
Rowley Hill, Eighton Banks.
Windy Nook, Windy Nook.

Flint.

Talacre and Gwespyr, near Mostyn.

Glamorganshire.

Berthgron, Nelson.
Bont Newydd, Trelewis, Treharris.
Cambrian, Clydach Vale.
Colly Isaf, Bedlinog.

Coal Measures—*continued*:

Glamorganshire—continued.

Craig-yr-hesg, Pontypridd.
Craig Daf (M), Treharris.
Cwmcylla, Hengoed.
Cwm Crymlyn, Llansamlet.
Gilfach Main, Trelewis, Treharris.
Gilfach, No. 2, Trelewis, Treharris.

Gloucestershire.

Bixhead, Bixhead.
Bixslade, or Spion Kop, Bixslade.
Coffee Pit (grey), Cannop Valley, near Coleford.
Howlers Hill, Cannop Valley.
Knockley, Parkend.
Oak (blue), Coleford.
Oak, No. 2, Coleford.
Pennant Stone, Nos. 2 and 3, Small Lane, Fishponds.
Pennant Stone, Nos. 4 and 5, Broomhill, Stapleton.

Lancashire.

Blue Delph, Worsthorne, near Burnley.
Catlow, Nelson.
Cox Green, Bromley Cross.
Crag, near Rochdale.
Crisp Delf (M), Dalton.
Cunliffe (2), Rishton.
Dawber Delf, Appley Bridge.
Edge Fold, Middle Hulton.
Enfield, Clayton-le-Moors.
Greengate, Ravenhead, St. Helens.
Hindley Green.
Grimshaw Park, Blackburn.
Houghton's Delph, Houghton Lane, Skelmersdale.
Howitt Hill (2), Parbold.
Mode Wheel, Weaste.
Ousel Nest, Bromley Cross.
Parbold, Parbold.
Pemberton, Highfield, Pemberton.
Pilkington, Horwich.

Coal Measures—*continued*:

Lancashire—continued:

Platts Central, Appley Bridge, Wigan.
Shorrock, Grimshaw Park, Blackburn.
South Field, Nelson.
Victoria, Billinge.
Wharmden, Accrington.
Whitegate, Padiham Road, Burnley.

Monmouthshire.

Trehir, Bedwas.

Northumberland.

Brunton, Gosforth.
Cocklaw, Blackpasture, and Brunton, Chollerford.
Kenton, Kenton.
Tyne, St. Anthony's.

Somersetshire.

Fox's Wood, Keynsham.
Pennant Stone, No. 6, Hallatrow, Clutton.
Temple Cloud, Temple Cloud, Hallatrow.

Yorkshire.

Ackworth Moor Top (2), Pontefract.
Ambler Thorn, Swales Moor, near Halifax.
Apperley Lane, Apperley Lane, Yeadon.
Appleton, Shepley, near Huddersfield.
'Armistone' Stone.
Black Dyke Lane, Thornton, Bradford.
Black Dyke, Thornton, Bradford.
Blue Stone, Park Lane, Handsworth.
Bridge End (M), Garsdale, Sedbergh.
Bolton Woods (6 Qs), Bradford.
Brackenhill, Ackworth Moor Top.

GEOLOGY OF BUILDING STONES

Coal Measures—*continued*:

Yorkshire—continued:

Bracken Hill, Wakefield Road, Ackworth Moor.
Britannia, Stump Cross, Morley, Leeds.
Cardigan, Morley.
College (2 Qs), Ackworth Moor Top.
Constitution Lane, Ackworth.
Crow Trees (2 Ms), Rastrick.
Excelsior, Soothill Wood.
Fagley, Eccleshill, near Bradford.
Finsdale, Morley, near Leeds.
Five Acre (3 Qs), Rastrick.
Fox Hill, Wadsley.
Gaubert Hall (M), Brighouse.
Gelderd Road, Beeston, Leeds.
Gazeby, Wrose Hill, Shipley.
Granney Hall (M), Brighouse.
Hagg Stones, Worrall, Sheffield.
Handsworth, Park Lane, Handsworth.
Hawksworth, Headingley Lane, Leeds.
Hazlehurst, Ambler Thorn, near Halifax.
Hinsell, near Normanton.
Howley Park (3 Qs), Morley, Leeds.
Idle Moor, Bradford.
Klondyke, Handsworth, near Sheffield.
Landermere, Shelf.
Long Close (M), Brighouse.
Long Mill, near Leeds.
Lower Edge, Elland.
Marsh Lane (M), Southowram.
Middle Delph and New Delight, Rastrick.
Midland, Oulton, near Leeds.
Milking Hill (2 Qs), Southowram.
New Farm, Southowram.
Old Park, Crosland Hill.
Oulton, Oulton, near Leeds.
Park (M), Brighouse.

Coal Measures—*continued*:

Yorkshire—continued:

Park, Crosland Hill, Huddersfield.
Park (2 Qs), Lightcliffe, Halifax.
Park Spring (4 small Qs), 2 at Farnley and 2 at Hough Top, Bramley.
Pond (M), Lightcliffe.
Robin Hood, Morley, Leeds.
Robin Hood (2 Qs), Thorpe, near Wakefield.
Ringby, Swales Moor, near Halifax.
Rothwell Haigh (2 Qs), near Leeds.
Rough Heys, Lightcliffe.
Shibden Head, Halifax.
Slack End, Swales Moor, near Halifax.
Snydale, Normanton.
Southages (M), Rastrick.
Sovereign, Shepley.
Speedwell (M), Nab End, Brighouse.
Spinkwell, Bolton Road, Bradford.
Spinkwell, Crosland Hill near Huddersfield.
Spinkwell (M), Bradford.
Stoney Lane, Eccles Hill, near Bradford.
Stork's House, near Bradford.
Swales Moor (2 Qs), Swales Moor, near Halifax.
Thrybergh, Rotherham.
Toller Lane, Heaton, Bradford.
Town Edge Piece, Hazlehead.
Tuck Royds, Hipperholme.
Victoria, Wortley.
Watson, Southowram.
Weatheroyd Wood, Heaton, Bradford.
Westercroft, Shelf, near Halifax.
West Lane, Southowram.
Woodkirk, West Ardsley.
Wood Lane, Bolton, Bradford.
Wood Top, Bolton Woods, Bradford.

APPENDIX B

Permian.

Ayrshire.
  Ballochmyle, Mauchline.
  Barskimming, Mauchline.

Dumfriesshire.
  Corncockle, Lockerbie.
  Corsehill, Annan.
  Gatelawbridge, Thornhill.
  Knowehead, Locharbriggs.
  Locharbriggs, Locharbriggs.

Trias.

Cheshire.
  Rockfield and Weston, Runcorn.
  Weston, Weston, Runcorn.

Cumberland.
  Sandwith, Whitehaven.

Glamorganshire.
  Quarella, Bridgend.

Trias—*continued:*

Lancashire.
  Holt Lane, Whiston.
  Samlesbury, Samlesbury.
  Woolton, Nos. 2 and 4, Woolton.

Shropshire.
  Grinshill, Grinshill.

Elgin.
  Greenbræ, Cummingston.

Younger Formations.

Surrey.
  Cawley's (M). Betchworth.
  Cob Hill and Godstone (M), Godstone.
  Colley Hill (M), Reigate.

Sussex.
  Black Brooks and Perch Hill (M), Brightling, near Battle.

The Carboniferous Limestones of Ireland.—In several districts the Old Red Sandstones and Lower Carboniferous rocks have not yet been clearly differentiated, so that the stones that are here ascribed to one of these systems may eventually be relegated to the other; but for practical purposes this is a trifling matter.

In Counties Cork and Kerry there are numerous small quarries in the Lower Carboniferous, and a few yielding sandstones and flags in the Coal Measures; they are locally known as 'Greystone.'

Stone from Borleigh Quarry, eight miles north of Bandon, was used in Timoleague Abbey. The Calp Sandstone of Kesh (County Fermanagh) was used in the round tower of Devenish, and a finely sculptured cross of the same stone was discovered in 1878; it was in good preservation until it was set up again, and then it peeled badly within a year. Flags and kerbs are obtained from the same beds. In Kilkenny the Lower Sandstones and Coal Measures have been used in Jerpoint Abbey. The mouldings, dressings, and columns retain their tool-marks after 700 years. A red sandstone of this age has been employed for millstones at Drumdowney, and a white stone of good quality comes from the same neighbourhood. Yellowish and light brown Coal Measures sandstones were obtained from Coolcullen for the restoration of St. Canice's Cathedral, Kilkenny, and a hard purple conglomerate of the same age appears as dressing in Inistioge Abbey (1262). A strong free-working stone, obtainable in good lengths, was formerly exported for the Doon Quarry, Co. Limerick; it was used in the staircase in Clarina and Adare mansions. Stone from Morroe was used in Glenstal Castle, and from Knockfierna in St. Oswald's, near Ballingarry; all these stones are from the lower beds. Coal Measure grits, hard and rough when matured, have been used in bridges and for flags in this country.

The lower beds of County Wexford yield coarse gritty stone near Wexford Harbour and elsewhere; it is not promising material, but it seems to wear fairly well. It has been used in the Old Abbey and 'Father Roche's Churches,' Wexford.

In Waterford the basement sandstones, not clearly marked off from the Old Red Sandstone, brown, green, and yellow in colour, have been used locally. The round tower of Ardmore, of local sandstone in coursed work, is in good preservation.

It is of interest to note that in the west of the country several of the quarries—about Lismore, for instance—produce not only sandstone freestone, but flags and roofing slates.

A whitish and greenish sandstone has been used in Clonmel, and a dark red stone is obtained at Bronn House Promontory.

From the lower beds of Drumbane, in Tipperary, a light grey to white stone is obtained in good sizes, and from Dundrum a similar yellow rock is quarried in addition to flagstones. The Coal Measures here are unimportant.

Lower Carboniferous and Coal Measure sandstones are used in Queen's County and King's County; a good example of work in the latter stone may be seen in the carved doorway of Killeshin Church, east of Carlow. The 'Carlow Flags' are obtained from the Coal Measures of Queen's County and Carlow; they are brownish-grey, and well bedded, 10 inches to 2 feet thick.

White, yellow, red, and purple sandstones of the Lower Carboniferous have been worked near Scariff and Mount Shannon, in County Clare; these were used in the Crypt and Cathedral at Killaloe. Flags of fair quality from the Coal Measures have been worked at Ennistimon, Crag, and Money Point, on the Shannon. Some of these resemble the Carlow flags, but are darker and rougher, owing to fossil worm-tracks on the bedding surfaces; they were formerly much used in making chimneypieces at the defunct Killaloe Marble Works.

In County Meath Lower Carboniferous sandstones have been used in the construction of the round tower of Kells, and as quoins in the old church there; also in the round tower of Donaghmore. It has not proved very durable, but has weathered evenly.

Large blocks of fine-grained sandstone are raised at Benmore, County Galway, and in Slieve Dart, north of Dunmore, are coarse grits, used for millstones and laminated flags; some quite thin and slaty. Similar stone is raised in County Mayo.

Lower Carboniferous sandstones are quarried in County Monaghan at Carnmore, Knocknatally, and Donagh—a yellowish-red quartzose friable stone, used in the entrance to Caledon House, County Roscommon. At St. John's Hole is a grey, hard stone, used in Boyle Abbey for all classes of work, including carved mouldings.

In County Cavan, at Latt and Ballyconnell, is a yellowish-grey stone, much used in Cavan (Cullen College, 1871); in County Sligo are flags at Doonbeakin and Ballyglass.

In County Donegal stones of this period were formerly much quarried for millstones. The best known of these is the *Mount Charles Stone*. It is grey or cream-tinted, felspathic, and slightly micaceous, with a siliceous and ferruginous cement; it is hard to work, but wears well. It has been used in the new museum and library, Leinster House, Dublin; dressing in the Town Hall, Sligo. Drumkeelan Quarry produces stone similar to Mount Charles. Lough Eske Castle is built with a dull yellowish stone from Altito, three miles from Donegal; the same stone is used for kerbs.

The Coal Measure sandstone of Cavan was formerly used a good deal, and flags have been raised at Arigna and Keadew, in Roscommon, and in County Sligo.

In Dublin flags of this age were worked and used.

Sandstones of Calp age are largely quarried in County Tyrone, but may have been discontinued through the difficulty of dealing with very thick overburden. Quarries at Tamlaght, Cookstown, Drumquin, and elsewhere, yield cream, yellow, or bluish stone, easy to work and very durable, but not fit for heavy bearing. The produce of these quarries varies a good deal in quality, and care has to be exercised in the selection. Flags are obtained from Carrickmore, Drumquin, and Cookstown.

The Calp sandstones of Dungannon district are quarried at several places. Of these the Ranfurly, or Mullaghana stone, is most used: in Belfast (Post Office, Northern Bank, Clock Tower, St. Patrick's Church, etc.); also in the Royal University, Dublin; Northern Bank, Fintona.

The stone from the small tract of Lower Carboniferous near

Ballycastle, County Antrim, Ballgoly Quarry, may be seen in the viaduct of Glendun and many bridges in the county, including that of Ballycastle, which was rebuilt with this stone in 1852, and has stood very well, even in the trying situation between tide marks; this is a cream-tinted, coarse-grained stone. From the same quarry comes a finer-grained stone of the same colour, which can be dressed to a fine arriss, and bedded any side up. According to Kinahan, it has worn very well in Belfast when other sandstones have had to be redressed or painted. It has been used in the spire of Ballycastle Church (1756); for facing and dressing at Doonhill, County Londonderry (1783-1785), and the fine work and cornices are still fresh; also in the spire of the Charitable Institute (1774), portico of St. George's Church, dressing in the Grain Market, and sometimes for inside dressing and tombstones.

A red stone comes from Fair Head, said to work freely, and to be durable; it is used in Ballycastle Coastguard Station.

In Ballymena, County Antrim, the 'Dungannon Stone' is mostly obtained from Gortnaglink and Carlan, while in Dungannon itself stone from all these and other quarries is used. The colour of the stone ranges from grey to yellow or reddish; it is rather micaceous and somewhat ferruginous, but there is not much cement.

The similar Bloomhill stone (Co. Tyrone) has been much employed in Dublin.

Some fairly good stone—'Dungiven Stone'—comes from Dungiven, in Londonderry; it has been used in Coleraine Church for dressings, and in the Protestant Hall and other buildings, Belfast.

APPENDIX C

CLASSIFIED LIST OF LARGER LIMESTONE QUARRIES

(M=mine.)

Silurian.

Shropshire.
 Bradley, near Much Wenlock.
 Presthope, Much Wenlock.

Staffordshire.
 Coneygre (M), Dudley.
 Phœnix (M), Walsall.
 Rushall (M), Walsall.
 Wren's Nest, West, Dudley.

Devonian.

Devonshire.
 Admiralty, Oreston, Plymouth.
 Bampton.
 Deadman's Bay, Plymouth.
 Hexton, Plymouth.
 Prince Rock, Plymouth.
 Radford, Plymstock.

APPENDIX C

Devonian—*continued*:

Devonshire—continued:
Stony Coombe, Ipplepen, near Newton Abbot.
Westleigh Quarries (Barge and Pitcher, Kiln, Furlong), Burlescombe.
Yalberton, No. 1, Yalberton.

Carboniferous.

Anglesey.

Dinovben, Llangoed.
Flagstaff, four miles from Beaumaris.

Brecon.

Abercriban, Pontsicill.
Clydach, Clydach.
Gilwern, Gilwern.
Llanelly, Clydach.
Llwynon, Penderyn.
Penderyn, Penderyn.
Stuart, Penderyn.
Trevil, No. 1, Trevil.
Trevil, No. 2, Trevil.
Tylerybont, Pontsicill.
Vaynor, Vaynor.

Carmarthen.

Cilvrychen, Llandebie.
Glanwenlais, Llanfihangel-Aberbythych.

Carnarvon.

Little Ormes Head, near Llandudno.
Llysfaen and Pentregwyddel, near Abergele.
Llysfaen, near Abergele.

Cheshire.

Astbury (M), Congleton.

Cumberland.

Alston and Nentforce (Lowbyers), Alston.
Clintz (Lowbyers), Woodend.
Kelton (Lowbyers), Kelton Head, Lamplugh.
Red Hills, Millom.
Rowrah Hall, Rowrah.

Carboniferous—*continued*:

Denbigh.

Chirk Castle, Trevor.
Graig, near Denbigh.
Llanddulas, Llanddulas.
Minera, Wrexham.

Derbyshire.

Ashwood Dale, Ashwood Dale.
Ashwood Dale (Cowdale), Ashwood Dale.
Bold Venture, Peak Dale.
Brierlow, near Buxton.
Buxton Central, Wormhill.
Cawdor, Matlock Bridge.
Crich Cliffe, Crich.
Dale, Wirksworth.
Dove Holes, near Buxton.
Dowlow, Sterndale Moor.
East Buxton, Millers Dale.
Great Rocks, Peak Dale.
Grin, Burbage, Buxton.
Harpur Hill, Harpur Hill Buxton.
Hindlow, near Buxton.
Hoffman, Harpur Hill.
Holderness, Dove Holes, Buxton.
Hopton Wood (2), Middleton, by Wirksworth.
Manystones, Brassington.
Middle Peak, Wirksworth.
Middleton Moor, Wirksworth.
Millers Dale, Millers Dale.
North End, Wirksworth.
Peak Dale, Peak Dale.
Peak Forest, Dove Holes, Dale.
Small Dale, Peak Dale, Buxton.
Station, Matlock.
Victory, Dove Holes, Buxton.

Durham.

Ashes, Stanhope.
Brown's Houses, Frosterley.
Frosterley, Frosterley.
Harehope Gill, Frosterley.
Heights, Stanhope.
Newlandslide, Stanhope.
North Bishoply, Frosterley.
Parson Byers, Stanhope.
Rogerley, Frosterley.

Carboniferous—*continued*:

Flint.
Coed Hendre, near Mold.

Glamorgan.
Coltshill, Norton, Westcross.
Creigiau, Pentyrch.
Morlais Castle, Pant, near Dowlais.
Tynant, Radyr.

Gloucester.
Barn Hill, Yate.
Point, Coleford.
Southmead, Westbury-on-Trym.
Tytherington (2), Tytherington.

Lancashire.
Bank Field, near Clitherhoe.
Bellman Park, Clitherhoe.
Bold Venture, Chatburn, near Clitherhoe.
Coplow, Clitherce.
Crown, Stainton.
Devonshire, Stainton.
Lane Head, Clitherhoe.
Salt Hill, near Clitherhoe.
Scout, Warton.
Trowbarrow, Silverdale near Carnforth.

Leicester.
Breedon Limeworks, Breedon-on-the-Hill.
Cloud Hill, Worthington, near Ashby-de-la-Zouch.

Montgomeryshire.
Llanymynech, near Llanymynech.

Monmouthshire.
Bedwas Colliery.
Ifton, Ifton.
Lady Hill, Liswerry, near Newport.
Limestone (Livingstone), Risca.
Machen, Machen, near Newport.

Carboniferous—*continued*:

Northumberland.
Cocklaw, Blackpasture, and Brunton, Chollerford.
Fourstones, Fourstones.
Haydon Bridge, Haydon Bridge.
Little Mill, Alnwick.
Sandbanks, Berwick.
Whitehouse, Ewesley, near Morpeth.

Pembroke.
Great Western, Tenby.

Radnor.
Old Radnor Lime and Macadam, Old Radnor.

Shropshire.
Nantmawr, Oswestry.
Porthywaen, Porthywaen.

Somerset.
Abbots Leigh (2), Abbots Leigh, near Bristol.
Binegar, Binegar.
Dulcote, St. Cuthbert, Wells.
Emborough Mendip, Emborough.
Moon's Hill, Cranmore.
Sandford Hill (2), Winscombe.
Vallis Quarries (6), Vallis, Frome.
Vobster, Vobster, near Coleford.
Waterlip, Cranmore.
Winsor Hill, Shepton Mallet.

Staffordshire.
Caldon Low, Caldon Low.

Westmorland.
Sandside, Sandside, Beetham.

Yorkshire.
Barton, Barton.
Broughton, Gargrave.
Forcett, Forcett.
Foredale, Horton-in-Ribblesdale.
Hambleton, Draughton, near Skipton.

APPENDIX C

Carboniferous—*continued:*

Yorkshire—continued:
 Horton, Horton-in-Ribblesdale.
 Langcliffe, Settle.
 Meal Bank, Ingleton.
 Raygill, Lothersdale.
 Skipton Rock (Haw Bank), Skipton.
 Swinden, Grassington.
 Thornton Rock, Thornton-in-Craven.
 Threshfield, Threshfield.

SCOTLAND:

Edinburghshire.
 Cousland (M and Q), Dalkeith.
 D'Arcy (M), Newbattle.
 Esperston, Borthwick.

Fifeshire.
 Charlestown (M), Charlestown.

Lanarkshire.
 Auldton (M), Lesmahagow.
 Netherton (M), Auchenneath.
 Thornton (M), Thornton Hall.

Renfrewshire.
 Orchard (M), Cathcart (Index).

IRELAND:

Co. Cork.
 Castlemore, Cork, L. Carb. L.
 Little Island, Caherlag.
 Meelin, Cork, U. Carb. L.

Permian.

Durham.
 Aycliff.
 Bishop Middleham, Ferryhill.
 Fulwell, Sunderland.
 Marsden, South Shields.
 Mainsforth, Ferryhill.
 Raisby Hill, Raisby Hill, Coxhoe.
 Tuthill, Haswell.
 Wingate Grange, Old Wingate.

Permian—*continued:*

Nottinghamshire.
 Pleasley Junction, Mansfield-Woodhouse.
 Steetley and Shire Oaks, Worksop.

Yorkshire.
 Branncliffe, Shireoak.
 Cridling Stubbs, Pontefract.
 Dogkennel, Kiveton Park.
 Kiveton Park.
 Warmsworth Cliffs, Doncaster.
 Womersley, near Pontefract.
 Wood, New Fryston, Castleford.

Yorkshire (West Riding).
 Micklefield, Micklefield, Leeds.
 Newthorpe, Newthorpe.
 South Elmsall, South Elmsall.

Jurassic Lias.

Glamorgan, East.
 Cement, Lower Penarth.
 Cement Works, Rhoose.
 Llandough, Llandough, near Cardiff.
 Porthkerry, Rhoose.

Leicestershire.
 Barrow (M), Sileby.
 Breach Delph, Barrow-on-Soar.

Northamptonshire.
 Blisworth, Blisworth (and iron).
 Burton, Burton Latimer (and iron).
 Carrol Spring, Finedon.
 Finedon, No. 1, Finedon (and iron).
 Thingdon, Finedon (and iron).

Nottinghamshire.
 Barnstone Blue Lias, Barnstone, near Nottingham.

Somersetshire.
 Fourteen Acre, Puriton, Dunball.

Jurassic Lias—*continued*:

Warwickshire.

Calais Blue Lias, Stockton, near Rugby.
Harbury, Bishops Itchington.
New Bilton, New Bilton.
Newbold, Newbold-on-Avon.
Stockton, Stockton, near Rugby.
Warwickshire Hydraulic Blue Lias Limestone, Southam, Rugby.

Jurassic.

Dorsetshire.

Bowers, Portland.
Combefield, Portland.
Croft, Portland.
Fishers' Croft (Straits), Portland.
Independent, Portland.
Inmosthay, West, Portland.
Inmosthay, East, Portland.
Kingbarrow, Portland.
Wakeham, Portland.
Waycroft, Portland.
Weston or Suckthumb, Portland.
Withies, Portland.

Lincolnshire.

Ancaster, Wilsford.
Haydor, Haydor.
Southwitham.
Tunnel, Hibaldstowe.

Northamptonshire.

Weldon.

Somersetshire.

Doulting Stone, Doulting.
Ham Wood, Shepton Mallet.

Yorkshire (North Riding).

Newbridge, Pickering.

Wilts.

Box (Nos. 4 and 6), Clift, and Tyning (M), Box.
Chilmark (M), Chilmark.

Jurassic—*continued*:

Wilts—continued:

Copenacre (M), Corsham.
Dapstone (M), Monkton Farleigh.
Drum and Pit (M), Monkton Farleigh.
Eastern Monk's Park (M), Corsham.
Hartham (2) (M), Box.
Northern Monk's Park (M), Corsham.
Park Lane (M), Corsham.
Pickwick (M), Corsham.
Ridge Park (M), Corsham.
Rockeredge (M), Corsham.
Sands (M), Westwells.
Spring, and No. 6 Corsham (M), Corsham.

Cretaceous (Chalk).

Bedfordshire.

Dunstable, Houghton Regis.
Harlington.
Sundon Lime Works, Sundon, near Dunstable.
Totternhoe Lime Works, No. 2, Totternhoe.

Cambridge.

Norman (Q) and Saxon (Q), Cherry Hinton.

Devon.

Beer, Beer Seaton.

Hampshire.

Lime Works, Buriton, near Petersfield.

Kent.

Barnfield, near Greenhithe (and gravel and sand).
Burham Brick, Burham (and clay).
Crayford, Crayford (also clay, gravel, and sand).
Downsfield, Swanscombe.
East Kent, Gillingham.
Erith Earth Pit, near Crayford (and brick earth).
Histed, Sittingbourne.
Holborough, near Halling.

APPENDIX C

Cretaceous (Chalk)—*continued:*

Kent—continued:
- Huntscliff, Greenhythe.
- Lamb Cliff, near Greenhithe.
- Manor Works, near Greenhithe.
- North Fields, Northfleet.
- Northfleet Chalk, near Northfleet.
- Pilgrims' Road, Swanscombe.
- Red Lion, near Northfleet (and clay and flints).
- Southfleet.
- Stoat's Nest, Coulsdon.
- Stone Castle, Greenhithe.
- Stone Court, near Greenhithe (and sand).
- Town Wharf, Greenhithe (and flints and gravel).
- West Kent Chalk, near Snodland.
- Whorne's Place, Cuxton.
- Wickham, near Cuxton.
- Wouldham Hall, near Folkestone.

Lincolnshire.
- Barton Cliff.
- Hessle Cliff.
- Kirk Ella.
- Melton Ross.
- South Ferriby.
- Vale House, Ulceby.

Surrey.
- Betchworth, Betchworth.
- Merstham Lime Works, near Merstham.
- Oxted Greystone Lime Pit, near Oxted.
- Riddlesdown, near Upper Warlingham, Croydon.
- Star Lane Chalk, Merstham.

Cretaceous (Chalk)—*continued:*

Sussex.
- Amberley (2), Amberley.
- Balcombe, Glynd, near Lewes.
- Beeding Chalk, near Shoreham.
- Meeching, near Newhaven.
- Southerham, near Lewes.

IRELAND:

Co. Antrim.
- Ballyvaddy, Carnlough.
- Criggan, Carnlough.
- Kilcoan, Island Magee.
- Kilwaughter, Kilwaughter, Larne.
- Magheramorne, Magheramorne.
- Parisha, Tickmacrivan (and flints).
- Town, Tickmacrivan (and flints).
- Whitehead, Whitehead (and whinstone).

Kentish Rag.

Kent.
- Allington, Allington, Maidstone.
- Allington, No. 1, Aylesford, Maidstone.
- Allington, etc., Aylesford, Maidstone.
- Castle, Allington.
- Coombe (and Wharf), near Maidstone.
- Postley Fields, near Maidstone.
- Preston Hall, near Aylesford.
- Thong Lane, near Wrotham.
- Tovil (3 Qs), Tovil.

The Lower Lias is quarried at Wilmcote and Binton, near Stratford-on-Avon; Langport, Somerset; Twerton and Weston, near Bath; at Street, near Glastonbury; Saltford, Keynsham (Keynsham Stone, blue lias, used in the railway-station at this place, on the Great Western Railway Works, and in local churches); Stinchcombe and Dursley, in Gloucestershire (rough building stone); Queen Camel, near Sparkford, Dorset; Highbrooks Quarry, between Long Sutton and Kingsdon, between Paulton and Radstock, and near Tainton (Thurlbeer Lias and Knapp Stone).

APPENDIX D

LIST OF CHIEF SLATE QUARRIES

I.—WALES.

Name of Quarry or Mine.	Situation.		Geological Formation.	County.
Dinorwic	Llanberis	—	L. Cambrian	Carnarvonshire
Penrhyn	Bethesda	—	,,	,,
Pant Dreiniog	,,	—	,,	,,
Upper Glynrhonwy	Llanberis	—	,,	,,
Glynrhonwy	,,	—	,,	,,
Cefndu	,,	—	,,	,,
Chwarel Fawr	,,	—	,,	,,
Cook and Ddol	,,	—	,,	,,
Glanrafon	Glanrafon	—	,,	,,
Alexandra	Moel Tryfan, near Bryngwyn	—	,,	,,
Braich	Ditto	—	,,	,,
Vron and Old Braich	Ditto	—	,,	,,
Moel Tryfan	Ditto	—	,,	,,
Cilgwyn	Nantle	—	,,	,,
Cloddfa'r Coed	,,	—	,,	,,
Dorothea	,,	—	,,	,,
South Dorothea	,,	—	,,	,,
Gallt y Fedw	,,	—	,,	,,
Gwernor	,,	—	,,	,,
Llwydcoed	,,	—	,,	,,
New Vronheulog	,,	—	,,	,,
Penybryn	,,	—	,,	,,
Penyrorsedd	,,	—	,,	,,
Tan'y-rallt	Penygroes	—	,,	,,
Talysarn	Nantlle	—	Llandeilo	,,
Welsh Green	,,	—	,,	,,
West Snowdon	Bwlch Cwmllan	—	,,	,,
Penmachno	Cwm Penmachno	M	,,	,,
Rhiwbach	Penmachno	,,	,,	,,
Bugail	Blaenau Festiniog	,,	,,	,,
Craigddu	Ditto	—	,,	Merionethshire
Oakeley	Ditto	M	,,	,,
Llechwedd	Ditto	,,	,,	,,
Maenofferen	Ditto	,,	,,	,,
Votty and Bowydd	Ditto	,,	,,	,,
Rosydd	Ditto	,,	,,	,,
Wrysgan	Ditto	,,	,,	,,
Diphwys	Ditto	,,	,,	,,
Bwlch-y-slater	Ditto	,,	,,	,,

APPENDIX D

I. WALES—continued.

Name of Quarry or Mine.	Situation.		Geological Formation.	County.
Bryn Eglwys	Abergynolwyn	M	Llandeilo	Merionethshire
Cantrybedd	,,	,,	,,	,,
Croesor	Llanfrothen	,,	,,	,,
Park	,,	,,	,,	,,
Aberllefeni	Aberllefeni	,,	,,	,,
Ratgoed	,,	,,	,,	,,
Minllyn	Dinas Mawddy	,,	,,	,,
Aber Coris	Corris	,,	,,	,,
Llanfair	Harlech	,,	,,	,,
Llwyngwern	Llwyngwern, near Machynlleth	—	,,	Montgomeryshire
Rhiwarth	Llangynog	M	,,	Ditto
Gilfach	Gilfach	—	,,	Carmarthenshire
Glogue	Clydey	—	,,	Pembrokeshire
Llandilo	Llandilo, Maenclochog	—	,,	,,
Moelferna	Glyndyfrdwy	M	Wenlock	Merionethshire
Penarth	Corwen	,,	,,	,,
Foel Faen	Llangollen	—	,,	Denbighshire
Cambrian	Glynceiriog	M	,,	,,

II.—LAKE DISTRICT.

Hodge Close	Tilberthwaite	—	Borrowdale Volcanic Series or 'Green Slates and Porphyries'	Lancashire
Moss Rigg	,,	—		,,
Burlington	Kirkby Ireleth	—	—	,,
Parrock	,,	—	—	,,
Saddlestone	Coniston	M	—	,,
Original Ellerwater	Ellerwater	—	—	Westmorland
Lord's	,,	M	—	,,
Steel Rigg	,,	M	—	,,
Dubb	Honister, Keswick	,,	—	Cumberland
Rigg Head	Borrowdale, Keswick	,,	—	,,
Quay Foot	Ditto	,,	—	,,

III.—SCOTLAND.

Name of Quarry or Mine.	Situation.		Geological Formation.	County.
Ballachulish	Ballachulish	—	Metamorphic Series	Argyll
Balvicar	Seil Island	—	Ditto	,,
Cuan	Luing Island	—	Ditto	,,
Cullipool	,, ,,	—	Ditto	,,
Easdale	Easdale Island	---	Ditto	,,
Balnahua	Balnahua Island	—	Ditto	,,
Aberfoyle	Aberfoyle	—	Ditto	Perthshire
Craiglea	Logiealmond	—	Ditto	,,
Birnam	Birnam Hill	—	Ditto	,,
Luss	Luss	—	Ditto	Dumbarton

Blue slates have been quarried in the Ordovician rocks of the Southern Uplands at several points, notably at Traquair and Stobo, in Peeblesshire.

IV.—IRELAND.

Larger Quarries.

Name of Quarry.	Situation.	County.
Garrybeg	Portroe, Nenagh (Killaloe)	Tipperary
Victoria	Clashnasmuth	,,
Madrenna	Leap, near Skibbereen	Cork
Benduff	Rosscarbery	,,

Quarries working on a Small Scale.

Drumaree	Newton Barry	Wexford
Dromosta	Drimoleague	Cork
Knockane	,,	,,

APPENDIX E

SOME USEFUL BOOKS

No complete bibliography of the geology of building stones has ever been published. The following list is intended only as a short introduction to the literature:

GENERAL GEOLOGY.

COLE, G. A. J.: Aids in Practical Geology. 5th edit. 1906. London.
ELSDEN, J. V.: Applied Geology. 2 vols. 1899. London.
GEIKIE, SIR A.: Textbook of Geology. 2 vols., 4th edit. 1903. London.
GEIKIE, SIR A.: Classbook of Geology. 4th edit. 1902. London.
GEIKIE, JAMES: Structural and Field Geology. 1908. Edinburgh.
HARKER, A.: Petrology for Students. 4th edit. 1908. Cambridge.
HATCH, F.: Textbook of Petrology. 5th edit. 1909. London.
JUDD, J. W.: The Students' Lyell. 1896. London.
JUKES-BROWNE, A. J.: The Building of the British Isles. 1888. London.
LAPWORTH, C.: An Intermediate Textbook of Geology. 1898. London.
MARR, J. E.: The Principles of Stratigraphical Geology. 1898. Cambridge.
MARR, J. E.: The Scientific Study of Scenery. 1900. London.
MERRILL, G. P.: A Treatise on Rocks, Rock Weathering, and Soils. 1897. New York.
MIERS, H. A.: Mineralogy. 1902. London.
RUTLEY, F.: Mineralogy. 14th edit. No date. London.
TEALL, J. J. H.: British Petrography. 1888. London.
WATTS, W. W.: Geology for Beginners. 1898. London.
WOODWARD, H. B.: The Geology of England and Wales. 2nd edit. 1887. London.

SPECIAL BOOKS.

BAUSCHINGER, J.: Mitteilungen aus dem Mecanisch-technischen Lab. d k. tech. Hochschüle. Hefte 4, 5, 6, 10, 11, 19. Berlin.
BÖHME: Untersuchungen von natürlichen Gesteinen auf Festigkeit, etc. Mitt. aus dem k.k. Versuchsanstalten, Ergänzungsheft II. 1889. Berlin.
CHATEAU, T.: Technologie du Bâtiment, etc. 2nd edit. 1880. Paris.

- DAVIES, D. C.: A Treatise on Slate and Slate Quarrying. 1878. London.
- GARY, M.: Gesteinsuntersuchungen. Mitt. k. Materialprüfungsamt xxviii., heft 4. Gr. Lichterfelde West. 1910. Berlin.
- GOTTGETREU: Baumaterialien. 2 vols. 1880-81.
- GWILT: Encyclopædia of Architecture.
- HANISCH, A.: Resultate der Untersuch. mit Bausteinen der Österreich-Ungarischen Monarchie. 1896. Vienna.
- HARRIS, G. F.: Granites and the Granite Industries. 1888. London.
- HERMANN, O.: Steinbruchindustrie und Steinbruchgeologie. 1899. Berlin. (A very good book dealing generally with the subject.)
- HIRSCHWALD, J.: Die Prüfung der Natürlichen Bausteine auf ihre Wetterbeständigkeit. 1908. Berlin. (A monumental work on the subject.)
- HULL, E.: A Treatise on the Building and Ornamental Stones of Great Britain and Foreign Countries. 1872. London. (The only general treatise on the geology of British building stones.)
- KARRER, F.: Führer durch die Baumaterial-Sammlung des k.k. Natürhistorischen Hofmuseums. 1892. Vienna.
- KINAHAN, G. H.: Economic Geology of Ireland. Journ. Roy. Geol. Soc., Ireland, vol. viii., N.S., 1889.
- MERRILL, G. P.: Stones for Building and Decoration. 3rd edit. 1903. New York.
- MIDDLETON, G. A. T.: Building Materials. 1905. London.
- MUNBY, A. E.: Introduction to the Chemistry and Physics of Building Materials. 1909. London.
- SCHMIDT, A.: Natürliche Bausteine. 1908. Hanover.
- SEIPP, H.: Die abgekürtze Wetterbestandigkeitsprobe der natürlichen Bausteine. 1905.
- TETMAJER, L.: Methoden und Resultate der Prüfung Künstlicher und Natürlicher Bausteine. 1900. Zurich.

To these must be added the important papers by—

- BEARE, T. H.: Building Stones of Great Britain—Their Crushing Strength and Other Properties. Proc. Inst. Civil Engineers, cvii., part 2, 1892.
- LAPWORTH, H.: The Principles of Engineering Geology. Loc. cit., clxxiii., part 3, 1908.
- BALDWIN-WISEMAN, W. R., and GRIFFITH, O. W.: The Physical Properties of Building Material. Loc. cit., clxxix., part 1, 1910.
- BALDWIN-WISEMAN, W. R.: The Effect of Fire on Building Stones. Trans. Surveyors' Inst., xxxviii., 1906.
- BURROWS, H. W.: Examination of Building Stones. Journ. R.I.B.A., vol. ix., 1893.

APPENDIX E

There are valuable modern bulletins and papers on building stones in the Geological Survey publications of several countries outside Great Britain; notably those of the United States of America and Sweden.

The Report of the Commission on Building Stones for the Houses of Parliament, 1839, is now so out of date that it is practically useless.

Annual lists of all quarries (over 20 feet deep) in Great Britain, Ireland, and the Isle of Man, are issued by the Home Office.

The same Office publishes a list of mines, which includes a number worked for stone.

The following are useful lists of foreign quarries:

Germany.—KOCH, A.: Die Natürlichen Bausteine, Deutschlands. 1892. Berlin.
Germany.—HIRSCHWALD, J.: Die bautechnisch verwertbaren Gesteins - Vorkommnisse des Preussischen Staats. 1910. Berlin.
Austria.—HANISCH, A., and SCHMIDT, H.: Österreichs Steinbrüche. 1901. Vienna.
France.—Répertoire des Carrières de Pierre de Taille Exploitées en 1889. 1890. Paris.
United States.—Mines and Quarries (1902). Bureau of Census. 1905. Washington.

The trade journals—*Quarry*, *Builder*, and *Stone Trade Journal*—frequently contain articles on the geology of building stones, especially the first named monthly.

INDEX

NOTE.—Names in Tables XII., XIV., and XV., and in the tabulated parts of the Appendices are not included.

'Rk.' after a name = Rock.   'St.' after a name = Stone.   'Q.' after a name = Quarry.

Names of *Abbeys, Castl s, Cathedrals, Churches*, and *Colleges* are collected under these words.

ABBEY, Boxby, 266
  Boyle, 423
  Bury St. Edmunds, 220, 222
  Byland, 239
  Croxden, 155
  Dover Priory, 235
  Fonthill, 253
  Glastonbury, 216
  Hayles, 239
  Inistioge, 422
  Jerpoint, 421
  Kirkham, 244
  Lacock, 234
  Malmesbury, 234
  Malton, 244
  Neath, 201
  Newstead, 199
  Peterborough, 222
  Roche, 196
  Romsey, 253
  St. Albans, 223
  St. Pancras Priory, 260
  Timoleague, 421
  Tintern, 136
  Westminster, 168, 253, 263, 269
  Woburn, 262
'Abbey Grey' slate, 300
Abbotsbury, 241, 242
Abdy Rk., 150
Aber, 291
Aberdeen, 72, 74, 138, 414
Aberdovey, 291
Abergavenny, 186
Aberthaw, 191
Aboyne, 415

Abrasion Test, 394
Absorption, 372, 373, 383, 390-392, 405, 406
Abyssal rocks, 36
Acid rocks, 40, 43, 58
Acklington Dyke, 100
Ackworth Rk., 150
Acmite, 25, 28
Actinolite, 25, 26
Acton Rk., 150
Adamellite, 45, 64
Addison, 89
Adie, A. J., 343
Admiralty Q., 414
Adobe, 3
Adularia, 19
Ægirite, 25, 28
Afon Goch, 191
Africa, 266, 331
Agladières, 170
Ahaphaca Valley, 303
Aish, 249
Alabama, 313
Alabaster, 34
Albach, 171
Albite, 19, 20, 21
Alderney, 87
Aldsworth, 323
Alençon, 80
Alet, 170
Alexandra Q., 288
Algæ, 177, 271, 338, 339, 353
Allemagne, 268
Allier, 170
Allington, 321
Allis Mt., 85
Almandine, 30

Alport, 266
Alps, 274
Alsace, 123
Alston, 190
Altberg, 106
Altito, 423
Altnaveigh Q., 415
Alton, 154
Alt-Warthau, 171
Alwalton Marble, 233
Amberley, 260
Amberley Heath, 229
Amberley, Wis., 79
Ambleside, 132
America, 85, 89, 95, 98, 101, 106, 109, 266, 342
Ammanford, 316
Ammonite, 200
Amphibolite, 25
Amstal, 101
Analysis :
  Augite, 29
  Felspar, 63
  Granite, 65, 66, 408
  Hornblende, 29, 63
  Limestone, 183
  Mica, 29, 63
  Sandstone, 127
  Slate, 286, 288, 289
Ancaster, 183, 203, 211
Ancaster St., 213, 222, 374
Andalusite, 31
Andernach, 111, 308
Andesine, 20
Andesite, 27, 43, 103-106
Andryes St., 270
Anglesey, 95, 128, 145, 188

436

INDEX

Anhydrite, 34
Anjou, 305
Annalong, 416
Annan, 156
Anorthite, 19, 20, 21
Anorthoclase, 21
Anston, 183
Anstrudes St., 270
Anthophyllite, 25
Antrim, 101, 109, 135, 168, 424
Aplite, 59
Appleby, 155
Apsley House, 234
Arabia, 123
Aragonite, 14
Ardennes, 170, 269, 288, 290, 305, 306
Ardmore, 422
Ardnamurchan, 89
Ardshiel, 75
Arenig Series, 131
Argyllshire, 75, 301
Arigna, 423
Arkansas, 85, 309, 313
Arklow, 76
Arkose, 113
Armagh Co., 158, 326
Armstead Rk., 141
Arran, 75, 109, 157
Arthur's Seat, 108
Artificial stone, 331
Arzweiler, 172
Asbestos, 29
Aschaffenburg, 172
Ashburton, 186
Ashdown Sand, 162
Ashgill, 299
Ashley, 218
Ashprington, 108
Ashwick, 145
Asia Minor, 266
Aspatria St., 371
Aston Blank, 321
Atford, 323
Atherfield Clay, 162
Atherstone, 100
Atlas St., 331
Atmosphere, 352
Atmosphere, acid in, 336
Attrition Test, 394
Aubigny St., 268
Aude, 170
Augite, 25-28
Augusta Q., 79
Au Sable Granite, 89
Australia, 314
Austria, 80, 170, 271, 308
Auvergne, 106, 316
Avegno, 314

Avening, 217, 323
Avochie Q., 414
Avoncliff, 226, 238
Avon Gorge, 188
Avrigny St., 270
Axbridge, 153
Axe Edge, 146
Axminster, 262
Aylesbury, 166, 255
Aymestry, 184
Ayrshire, 89, 109, 157

Babbacombe, 186
Baccarat, 170, 307
Bachmann, E., 389
Bacon Tier, 249, 259
Bagbie Q., 415
Baggallay, F. T., 327, 328, 329
Baggy and Marwood Beds, 136
Bagshot Beds, 168
Bala Beds, 297
Bala Limestone, 184
Bala or Caradoc beds, 131
Balally, 77
Baldwin-Wiseman, W. R., 377
Ballachulish, 75, 303
Ballard Downs, 249
Ballingarry, 422
Ballochmyle, 157
Ball's Green, 208, 217
Ballybrew, 76
Ballycastle, 424
Ballyconnell, 423
Ballyedmanduff, 77
Ballyglass, 423
Ballygory Q., 424
Ballyknocken, 76
Ballymena, 424
Ballynacraig Q., 415
Balnahua, 301
Banbury, 202
Banchory, 73
Bandon, 421
Banff, 75
Bangor, U.S.A., 309, 310, 313
Banjvea, 106
'Bankers,' 257
Bannisdale Slate, 324
Bantry, 325
Barbadoes Q., 136
Bargagli, 314
Bargate St., 164
Barmouth, 291
Barnack, 211, 212
Barnack Rag, 351
Barnack St., 222

Barnacullia, 77
Barnet, 219
Barnsley, 150
Barnsley Rk., 150
Barnsmore, 95
Barough, 150
Barren Flagstone, 131
Barrow Island, 80
Barrington Park, 231
'Bars,' 54; Bars, 75
Barskimming, 157
Bartlett, 343
Barton, 192
Barton Hill, 261
Barytes, 34, 123
Basalt, 28, 39, 106-109, 365, 385, 409, 410
Base Bed, 246, 247, 249
Basic rocks, 40, 43, 58
Bassignac, 170
'Bastard Freestone,' 204
Bastard St., 228
Bath, 199, 204, 225, 228, 236, 238, 323, 429
Bathampton Down, 238
Bathonian, 268
Bath Oolite, 223, 224, 229, 234
Bathstone, 183, 189
Battlescomb, 321
Bauschinger, J., 344, 345, 370, 396, 402, 403
Bauxite, 331
Bavaria, 123, 272
Baynton St., 238
Bay of Funday, 80
Bayonne, 270
Bayston Hill, 100
Beacon Hill, 96
Beare, T. H., 364, 371, 372, 373, 374, 382
Bedford, 232
'Bedford Oolite,' 272
Bedfordshire, 231, 245, 261, 262
Beds, bedding, 53, 57, 112, 116, 257, 283, 391, 392, 397
Beech Green St., 163
Beer St., 181, 260, 262, 263
Beinn an Dubhaich, 75
Bel Air, 305
Belfast, 424
Belfast Q., 309
Belfort, 169
Belgium, 170, 176, 271, 308
Belper, 146
Belton, 155

438   GEOLOGY OF BUILDING STONES

Benbridge Beds, 264, 265
Ben Cruachan, 75
Bending, stress, 409
Benefield, 233
Bengal, 314
Ben Loyal, 74
Benmore, 423
Ben Rinnes, 74
Benton, 429
Berea Grit, 170
Berkshire, 166, 329
Berry Pomeroy, 186
Bertham Fell, 192
Bessbrook, 415
Bethersden marble, 264
Bettws Gwerfylgoch, 298
Bettws-y-Coed, 297
Beverston, 323
Bewcastle Fells, 141
Bhartpur, 172
Bibury, 321
Bicester, 241
Bidston Hill, 155
Bignor, 167
Binstead, 265
Biotite, 23, 24
Birchover, 146
Birdlip, 204, 218, 231
Birdsall, 244
Birdseye Marble, 176
Birdwell Rk., 149
Birkenhead, 154, 155
Birsemore, 415
Birstal Rk., 149
Bishop's Castle, 131
Bisley Slate, 321
Biszinger, 404
Bitumen, 123, 272
Blackenstone, 413
Black Forest, 81, 95, 97, 109, 172
Black Granite, 44, 89
Blackhill, 415
Black Slate, 301
Blaenau Festiniog, 295
Blaisdon Edge, 185
Blanchard, 312
Blauberg Granite, 81
Blea Wyke Beds, 203
Blekinje, 78
Blenfire Rk., 151
Bletchington, 241
Blisworth Clay, 232
Blockley, 209
Blockley St., 219
Bloomhill St., 424
Blount, B., 288
Blue Lias Lime, 199
Blues, 259

Bluestone, 170, 248, 250
Boddam, 415
Bodmin, 69, 94, 324, 412, 413
Bogny, 306
Bohemia, 81, 109, 170, 271, 308
Böhm-Brod, 170
Bohuslän, 78
Bolsover Moor, 195
Bon Accord Red, 78
Bonawe, 75
Boot, 72
Bootle, 72
Bordeaux, 270
Boring Test, 393, 395
Borleigh Q., 421
Borrowdale, 324
Borrowdale Series, 131, 298
Bothenhampton, 240
Bournemouth, 258
Bourton-on-the-Hill, 209, 218
Bourton, 254
Bourton St., 218
Bourton-on-the-Water, 238
Bowden Marble, 240
Bowel stones, 166
Bowood House, 238
Box, 183, 228, 235
Box Ground St., 234
Boynton, 259
Bozen, 95
Brabant, 272
Brachernagh, 192
Bradford, 148
Bradford Abbas, 318
Bradford Clay, 223, 224, 225
Bradford-on-Avon, 225, 226, 237
Bradford St., 237
Bradgate, 149
Brailes Hill, 318
Brambleditch Q., 216
Bramham Moor, 197
Bramley Fall St., 371
Brancliff, 197
Brandon, 328
Brandsby, 323
Brassington, 191
Brathay, 132
Bratton, 240
Brauvilliers, 269
Bray, 135
Brazil Wood, 87
Breccia, 114, 115
Bredisholm, 151

Bredon Hill, 209
Breitenbrunn St., 271
Brendon Hills, 300
Brent Tor, 108
Breslau, 172
Brest, 80
Bretton, 150
Brickland Green, 167
Bridgend, 201
Bridge Q., 155
Bridlington, 259
Brierly Rk., 150
Brighton, 258
Brill, 255
Brincliffe Edge Rk., 148
Brinscombe, 208
Bristol, 141, 145, 151, 199, 217
British Columbia, 80, 171, 313
Brittany, 80
Brixham, 186
Broadford, 304
Broadway, 209, 218
Brockenham, 101
Brockley Down St., 201
Brockram, 155
Brodrick Wood, 157
Brodsworth, 194, 196
Brognard, L., 266
Brohltal, 110
Bronn House, 422
Bronnville, 312, 313
Bronzite, 27, 28
Brookhampton, 218
Broseley, 145
Brotterode, 81
Broughton, 215
Brown Bed, 247
Brown Jura, 272
Brown Nows Q., 416
Brownstone, 170, 377
Brünn, 170
Brunnlitz St., 271
Brussels, 272
Brympton, 216
Bryozoa limestone, 177, 196
Bubbles in quartz, 17, 26
Bückeberg, 171
Buckley, E. R., 367, 383, 386, 387, 398
Bucks, 166, 169, 231, 237, 245, 260, 261, 262
Budneis, 80
'Building Freestone,' 205
Builth, 132
Bullingdon, 243
Bundenbach, 309
Bunker's Hill obelisk, 361

INDEX

Bunter, 144, 153, 172, 370
Burnard Inlet, 80
Burdie House limestone, 178
Burford, 230, 231, 323
Burgsvik, 172
Burley Down, 317
Burma, 331
Burnley Flags, 150
Burr, 248, 249
Burwell, 261
Buttermere, 95, 287
Bwlch Cwmlan, 297
Byfield, 202
Bytownite, 20

Cadeby, 196, 197
Cader Berwyn, 296
Caen St., 268, 269
Caerfai and Solva Beds, 130
Caerwys, 266
Cailliard, 131
Caithness, 75, 138, 301, 316
Cairncry Q., 414
Cairngall Q., 73, 415
Calcaire de Brie, 268
Calcareous Grit, 162
Calciferous Sandstone, 140, 143, 272, 416
Calcite, 14, 32, 335, 342, 343
Calc-sinter, 267
California, 79, 309, 313, 330
Calkstone, 264
Calne St., 241, 242
Calp, 152, 421, 423
Calton Hill, 109
Calvados, 268
Calverley Q., 163
Cambrian, 128, 129-131, 134, 135, 170, 172, 291, 294, 295, 303, 306, 308, 311, 313
Cambridgeshire, 220, 243, 245, 259, 261
Camlough Mt., 415
Canada, 79, 313, 330
Cap, 248-250, 258
Capel Curig, 297
Capo d'Istria, 271
Carboniferous, 3, 100, 114, 119-121, 126, 134, 137, 139, 170, 176, 182-187, 206, 271, 272, 303, 304, 316, 325, 425-427
Cardiff, 153
Cardiff Q., 311

Carfury Q., 412
Carlen, 424
Carlingford Lough, 76, 416
Carlinglow, 149
Carlow, 76
Carlow Flags, 422
Carlsbad, 81
Carmarthenshire, 297
Carnarvonshire, 77, 84, 95, 97, 100, 188, 295
Carn Boduan, 97
Carngrey Q., 412
Carnmore, 423
Carnsew, 68, 412
Carnsmore of Fleet, 74
Carrick, 135
Carrickmore, 423
Carrock Fell, 89, 97
Carstone, 117, 165, 166
Carth Head, 100
Cassel, 171
Castles:
  Acre, 260
  Arundel, 164, 260
  Auchinleck, 157
  Balmoral, 223
  Belvoir, 223
  Berkeley, 266
  Cary, 202
  Chester, 154
  Chirk, 145
  Corby, 201
  Corfe, 251, 258
  Glenstal, 422
  Gosford, 326
  Harlech, 130
  Inverary, 330
  Kingsbury, 208
  Kylemore, 326
  Lough Eske, 423
  Pfalz, 105
  Pevensey, 168
  Rufus, 246
  St. Angelo, 267
  Sudeley, 239
  Warwick, 154
  Windsor, 169, 238, 253, 260
Castle Hill Q., 166, 213
Castle Howard, 244
Castlewellan Q., 416
Casterton, 203, 211, 213
Casterton St., 219
Castor, 233
Catacluse St., 99
Cathedrals:
  Armagh, 158
  Canterbury, 217, 264, 269

Cathedrals (continued):
  Chester, 155
  Chichester, 199, 253, 260
  Cologne, 105, 171, 270
  Ely, 199, 260
  Exeter, 263
  Iona, 316
  Killaloe, 422
  Kirkwall, 316
  Lincoln, 221, 222, 223
  Llandaff, 157, 201
  Orleans, 270
  Peterborough, 262
  Rochester, 221, 252
  St. Asaph, 155
  St. Canice's, 422
  St. David's, 130
  St. Patrick's, N.J., 101
  St. Paul's, 239, 350, 351
  Southwell, 195
  Truro, 235, 253
  Wells, 216
  Winchester, 263
  Worcester, 134
  York Minster, 220
Caub, 308
Cavan Co., 135, 423
Cave Marble, 215
  Oolite, 215
Caythorpe, 214
Cefn, 145
Cementing material, 119, 123
Cephalopod Bed, 207
Cerithium portlandicum, 246, 252
Ceylon, 331
Chacombe, 318
Chaddington, 231
Chalcedony, 17
Chalk, 3, 163, 183, 259, 327, 409, 428
Challacombe, 186
Chalybite, 32, 335
Chalynch [Chelynch] Q., 216
Chambers Rk., 151
Chamonix, 80
Chancelade, 270
Channel Islands, 84, 86, 89
Chapman Q., 309
Charente-inférieure, 270
Charentenay St., 269
Charlbury St., 239
Charleton, 240

Charlottenburg, 365, 396,
Charlton, 323   [406
Charlwood-by-Box, 323
Charnwood Forest, 128
Chassignelles St., 270
Château-Gaillard St., 270
Châteaulin, 305
Chateay, 268
Chatham, 80
Chatley, Professor, 376, 409
Chatsworth House, 192
  Grit, 146
Chavenage, 323
Chauvigny St., 270
Chazy Beds, 272
Chedworth, 321
Cheesewring, 413
Chelston, 153
Cheltenham, 159, 207, 208, 216
Chemical analysis, 406
Chemnitz, 172
Chepstow, 136
Cher, Dep., 270
Cherty Beds, 247
Cheshire, 3, 123, 151, 155
Chesil Bank, 115
Chesterton, 220
Chevet Rk., 150
Cheviots, 100, 138
Chew Magna, 153
Chiastolite, 31
Chiavari, 314
Chicago, 330
Chickgrove Mill, 254, 259
Chickley, 218
Chili Bar, 313
Chilmark, 183, 252, 253
Chipping Campden, 209,
  Stone, 219
Chipping Norton, 159, 209, 231
  Stone, 218
Chipping Sodbury, 145
China, 330
Chlorite, 24
Christchurch, Yorks, 149
Christiania, 21, 59, 84, 85, 95, 330
Chrudim, 170
Chrysotile, 29
Chrzanow, 271
Chudleigh, 186
Chunar, 172
Church, Sir A. H., 351
Churches:
  All Saints, Dorchester, 259
  Ashridge, 262

Churches (*continued*):
  Awliscombe, 263
  Axmouth, 263
  Ballycastle, 424
  Bentley, 168
  Boston, 222, 223
  Branscombe, 263
  Broad Clist, 263
  Buckerell, 263
  Chard, 263
  Charmouth, 263
  Christchurch Priory, 253
  Christ Church, Richmond, 235
  Clone, 133
  Clyst-Haydon, 263
  Coleraine, 424
  Collyweston, 222
  Colyton, 263
  Combpyre, 263
  Congregational Ch., Caterham, 237
  Dunstable Priory, 262
  Feniton, 263
  Gillingham, 242]
  Grantham, 223
  Henry VII.'s Chapel, Westminster, 237
  Hinton, St. Mary, 242
  Holbeech, 222
  Holy Trinity, Sloane Street, 217
  Honiton, 263
  Isleham, 260
  Kettering, 222
  Ketton, 222
  Killeshin, 422
  Landewednack, 330
  Lateran, Rome, 267
  Lavenham, 329
  Long Melford, 329
  Luton, 262
  Lyme Regis, 263
  Maenturog, 285
  Marham, 261
  Marylebone, 350
  Moulton, 222
  Musbury Seaton, 163
  Ottery St. Mary, 263
  Oulesford, 263
  Our Lady of the Assumption, Cambridge, 237
  Payhembury, 263
  Pulborough, 164
  Roman Catholic, Norwich, 260

Churches (*continued*).
  St. Agatha's, Birmingham, 236
  St. Catherine's Chapel, 242
  St. Dunstan's, 220
  St. George's, Kidderminster, 237
  St. George's, Londonderry, 424
  St. James's, Spanish Place, 236
  St. John's, Cardiff, 217
  St. Lawrence Clyst, 263
  St. Mary's, Cambridge, 217
  St. Mary's, Madras, 351
  St. Mary, Redcliffe, 217
  St. Mary's, Stoke Newington, 269
  St. Mary's, Torquay, 237
  St. Matthew's, Fulham, 235
  St. Matthew's, Hull, 235
  St. Merryn's, 99
  St. Michael-in-Coslany, 329
  St. Pancras, Exeter, 263
  St. Peter's, Rome, 267
  St. Saviour, Moscow, 88
  St. Saviour's, Southwark, 260
  St. Stephen's-in-Brannel, 359
  Shottesbrooke, 329
  Shute, 263
  Spalding, 222
  Sudeley, 239
  Swanage, 258
  Tallaton, 263
  Tisbury, 253
  Tregony, 94
  Uplyme, 263
  Vannes, 80
  Wharram, 244
  Whimple, 263
Cicerchia, 172
Cinder Bed, 249
Cirencester, 225, 230, 241, 323
Clare Co., 133, 304, 422
Claverham St., 153

INDEX

Clay Slate, 277
Clay-with-flints, 169
Cleator, 192
Cleavage, 275, 279-283
    Way, 57
    Bed, 251
Clee Hills, 100, 145
Cleeve, 218
Clegir Q., 298
Cleopatra's Needle, 361
Clermont, 267
Cleve Cloud, 204, 208
Clevedon, 153
Cleveland Dyke, 97, 100
Cliff Field Q. 248
Cliff Hill, 96
Cliff St., 248, 250
Clifton Rk., 149
Clinterty Q., 414
Clipsham, 211, 213
    Stone, 219
Clog, 200
Clonmel, 422
Closeburn, 157
Cloughton, 161
Cloyne, 139
Clumber Park, 199
Clunch St., 261
Clypeus Grit, 205
Coalbrookdale, 145
Coal Measures, 116, 126, 139, 142, 143, 146-151, 303, 316, 418-423
Coblentz, 308
Cobwalls, 3
Cockley Bed, 227, 228, 254
Cogarno, 314
Colcerrow Q., 45, 49, 412
Coldwell, 132
Colleges :
    Balliol, 253
    Brazenose, 217
    Exeter, 322
    Hertford, 217
    King's, 199
    Magdalen, 199
    New, 218
    Peterloo, 222
    Wadham, 243
Collin Station, 101
Colly Farm, 167
Collyweston, 203
Collyweston Slate, 318-320, 321
Colne, 146
Columbier, 270
Colwich, 154
Combe Down, 227, 236
Combe Down St., 237

Combe Martin, 186
Comber, 158
Comblanchien, 270
Combourg, 80
Combrée, 305
Comby Sandstone, 130
Compression, 363
Compton Verney, 202
Condicote, 318, 321
Congleton, 145
Conglomerate, 114, 115
Coniston, 299
    Flags, 133
    Limestone, 184
Connecticut, 79, 170
Constantine, 266, 412
Conway, New Hampshire, 67
Cookstown, 421
Coolcullen, 422
Coomhola Grits, 141
Coral Limestone, 177
Corallian, 241-244
Corby, 154
Coreglio, 314
Corennie Q., 73, 414
Corfe, 324
Cork Co., 139, 304, 325
Cornbrash, 211, 223, 240, 241
Corncockle St., 127, 156
'Corner,' 200
Corn Grit, 228, 235
Cornstone, 186, 200
Cornwall, 3, 4, 50, 51, 53, 54, 60, 68, 69, 89, 93, 94, 99, 108, 120, 135, 136, 274, 284, 288, 299, 324, 359, 412-413
Corrie, 157
Corris, 296
Corrosion Test, 398
Corsehill St., 127, 357
Corsham, 228
    Court, 234
    Down St., 235
Corsica, 86, 314
Corsite, 51, 86
Corwen, 297
    Grit, 132
Coton End, 154
Cotta, 171
Cottage Q., 412
Cotteswold Hills, 203, 204, 206, 209, 318
    Sands, 205, 207
Corrèze, 270
Coryton, 301, 325
Côte-d'Or, 270
Countisbury, 300

Courson St., 269
Cove, 73, 156
Covers, 317
Cowbridge, 157
Cow Park Q., 415
Cox's Q., 232
Cradley, 136
Crag, 422
Craigleith St., 127
Craigmore, 75
Craignair Q., 415
Craig-y-Hesg, 127
Crash, 220, 221
    Bed, 212
Cravant St., 269
Craven Arms, 131
Crawshaw Rk., 147
Crazannes St., 270
Creden Hill, 186
Crediton, 152
Creetown Q., 415
Cresswell Crags, 197
Cretaceous, 117, 142, 162, 170-173, 259, 270-272, 313
Crétaux, 170
Crickhowel, 136
Crickley, 204, 205
Criffel, 74
Crinoidal Limestone, 176-190
Crocusstein, 271
Cromarty, 138
Cromford, 146, 266
Cropper Gate Rk., 149
Croreagh Q., 415
Cross-breaking Test, 377, 378-9
Cross Fell, 100
Crotch Island, 79
Cruden Q., 415
Crushing strength, 363-7, 369, 372-3, 378-9
Crychs, 284
Crypto-perthite structure, 21
Crystallizing force, 340, 346
Crystals, 14
Cuckfield Clay, 162
Cuddesden, 255
Cudowa, 172
Cuellin Hills, 89
Cullington, 154
Culm, 3, 170, 300, 308
Cumberland, 154, 155, 188
Cureton Q., 155
Curf, 246-7, 249
'Curl,' 255

'Curls,' 384
Current-bedding, 341
Cushenden, 135
Cwm Trescol, 297
Cypris Freestone, 259

Dacite, 104
Dagham St., 229
Dalbeattie, 74, 415
Dale, T. N., 49, 275, 279, 286, 287
Dalhousie, 314
Dalkey, 77
Dalradian, 128
Damuda Valley, 172
Dancing Cairns Q., 414
Dancing Ledge Q., 248
Danielsville, 309
Dark Pearl, 84
Darley Dale, 127, 146
Darley Top St., 371
Darlington, 192, 194
Dartington, 186
Dartmoor, 54, 413
Davies, D. C., 293
Dawlish, 152
Dayockwater, 72
Dean and Chapter Pit,
Deanshanger, 232 [244
Decay of stone, 333-361
Deganwy, 192
Deisterzandstein, 171
Dekri, 192
Delabole, 288, 299
'Delabole Butterflies,' 284
Delank, 65, 412, 413
Delesse, A., 65
Delta Q., 311
Denbighshire, 188
Denbighshire Grits, 132, 297
Dene Park, 318
Denner Hill, 169
Density, 372, 373
Dent, 192
Denton St., 239
Derbyshire, 100, 140, 141, 144-146, 188, 191, 192, 194, 195, 316
Derry, 135
Detborough, 160
Devenish, 421
Deville, 306, 307
Devizes, 324
Devonian, 3, 121, 135, 139, 170, 173, 176, 186, 189, 271, 272, 284, 299, 300, 303, 308, 309, 312, 324, 325
Devonport, 187

Devonshire, 3, 93, 99, 100, 108, 135, 136, 152, 153, 167, 186, 262, 300, 324, 325
Devonshire Marble, 177
Dewsbury Bank Rk., 149
Dharmsala, 314
Dhoorn Q., 77
Diabase, 39, 84, 90, 97-101
Diallage, 22, 28
Diamond Jo, 85
Diez, 309
Dilhorne, 145
Dinas Lake, 297
Dingle Promontory, 139
Dinorwic, 287, 288, 290, 293, 294
Diopside, 25
Diorite, 43, 84, 85-87
Dip, 57, 275, 276
Diptford, 300
Dobra, 106
Dogger, 162, 203
Dogger Beds, 272
Dolerite, 27, 28, 39, 43, 90, 97-101
Dolgarth, 297
Dolgelly, 291, 296, 297
Beds, 130
Dolomite, 32, 179, 191, 272, 335, 342, 344, 353, 373, 376
Dolomitic conglomerate, 153
Donagh, 423
Donaghmore, 423
Doncaster, 194, 196
Donegal, 77, 95, 135, 304, 423
Doon Q., 422
Doonagore, 127
Doonbeakin, 423
Dordogne Dep., 270
Dorking, 164
Dornach, 81
Dorsetshire, 115, 158, 199, 203, 207, 224, 241, 245, 248, 249, 254, 268, 323, 429
Dortschen, 308
Doulting, 183, 203, 204, 206
Stone, 216, 374
Dowdeswell St., 216
Dowglas, 94
Down Co., 76, 135, 158
Downside, 201
Downs Vein, 256
Downton Sandstone, 133

Draycot St., 153
Drayes St., 269
Drero, 314
Droitwich, 154
Drumbane, 422
Drumdowney, 421
Drumkeelan Q., 423
Drumquin, 423
Druses, 52
Dryrigg, 132
Drys-lwyn-isaf, 133
Dublin, 76, 133, 192, 423, 424
Dubhaich, 75
Duddington, 318
Dudley, 185
Duluth Granite, 89
Dumbartonshire, 138, 301, 303
Dumfries, 156
Dundalk, 76
Dundonald Q., 158
Dundrum, 422
Dundry, 203, 204
Dundry St., 217
Dunecht Q., 414
Dungannon, 423
Dungannon St., 424
Dungiven St., 424
Dunham, 155
Dunmore, 151, 423
Dunstable, 261
Dunstanburgh Head, 100
Dunston, 214
Durham, 146, 194
Durlston Head, 248, 257
Dursley, 201, 202, 204, 266, 427
Dust, 337
Duston, 160, 211, 318
Duston Slate, 160, 318
Dyce Q., 414
Dykes, 38, 52, 58
Dymock, 133

Easdale, 301
East Blue Hill, 79
East Coker, 241
East Malling, 266
Easton, 318
Ebrington, 209
Ecaussines, 271
Echina, 307
Eckersdorf, 308
Edge Coal Group, 141, 143
Edge Hill, 202
Edinburgh, 109, 143
Eggenstedt, 172
Egremont, 192

INDEX

Egryn, 290
Egypt, 81, 83, 95, 340, 361
Eichstadt, 272
Eifel, 101, 106
Eigg, 109, 117
Eisenbrod, 308
Eisenstadt, 271
Elæolite, 22
Elasticity, 370-377
Elbe Valley, 171
Elfdalen, 95
Elgin, 74, 138
Elidir Mt., 293
Elland Edge, 148
Elland Flagstone, 148
Ellerbeck Bed, 203
Elterwater, 290
Elvan, 60, 92, 94
Ely, 218, 219, 220
Emly Rk., 149
Empire Slate Q., 275
End-grain, 53
Ennerdale, 95
Enniskeen, 325
Ennistimon, 422
Enstatite, 25, 27
Enville St., 176, 269
Eocene, 268, 272, 314
Epidote, 25, 30
Epierre, 80
Epwell, 241
Eriboll, 130
Erzgebirge, 81, 109
Eschenbach, 309
Eskdale, 72
Essex Co., 89
Estuarine Series, 161, 203, 210-213, 232, 233
Etdon, 159
Etnadalen, 313
Euclid Bluestone, 170
Eureka, 310, 313
Exeter, 108, 152, 217
Excelsior St., 331
Exminster, 153
Exton, 218
Extrusive rocks, 36
Eyam, 146
Eycott Hill, 108

Factor of safety, 409
Faen Goch, 293
Fahrenboch, 81
Fairford, 233
Fair Haven, 275
Fair Head, 424
Fairlight Clay, 162
Fairy Castle Q., 77
Falhouse Rk., 149

Falmouth, 325
False cleavage, 282
Fanad, 77
Faringdon, 166, 242
Farleigh Down, 227, 236
Farlei h Down St., 236
Farlow, 136
Farnham. 168
    Stone, 332
Farren Slate, 314
Fault, 191
Feather Bed, 257
Fell Q., 415
Fell Sandstone, 141
Felsite, 92, 95
Felspar, 16, 18-22, 31, 39, 44, 63-65
Felspathoids, 22
Feltsberg, 81
Fermanagh, 421
Ferte-sous-Jouarre, 268
Festining, 291, 295
Festiniog Beds, 130, 131
Fichtelgebirge, 81, 109
Fifeshire, 109, 138
Fingal's Cave, 109
Firestone, 166, 167, 200
Firth of Lorne, 301
Fishguard, 100
Flags, 130-133, 135, 138, 139, 144, 147-151, 158, 159 162, 421-423
Flett, J. S., 397
Flint, 3, 327
    Stone, 248
    Tiers, 249
Flintshire, 188, 191
Flöha, 172
Fluorite, 123
Flysch, 172
Folding, 57, 283
Folkestone, 183
Folkestone Beds, 122, 162, 165
Fontainebleau, 120
Fontenabuona, 314
Fonthill House, 262
Foraminiferal Limestone, 177, 190
Ford's St., 332
Forest Marble, 223, 224, 225, 228, 240, 322, 323
Forest of Dean, 141, 145, 151, 188
Forest St., 151, 165
Forfarshire, 138
Forge Valley, 266
Fort Tourville, 87
Fourche Mountain Granite, 85

Fourneaux St., 269
Foxdale Q., 77
Foxford, 135
Fox Island, 79
Fox-Strangways, C., 244
Foynes, 192
France, 80, 109, 117, 102, 121, 169, 266, 267, 268, 289, 290, 305-308, 316
Franconia, 111, 172
Frankfort, 172
Frankfort-on-Main, 101
Franklin's Pit, 232
Fredeburg, 309
Frederikshald, 330
Fredericksvaern, 84
Freestones, 205, 208, 209, 211, 218, 226-230, 235, 241, 248, 256, 259
Freestone Vein, 256
Freezing Test, 399, 401
Freihermsdorf, 308
Freiwaldau, 271
Fresnois, 305
Fretting Bed, 251, 252
Freudenthal, 308
Friction, 341
Friedberg, 171
Friedeburg, 81
Friedersdorf, 171
Frome, 204
Frome St., 240
Frost, 344
'Frosting,' 322
Frosterley, 192
Frosterley Marble, 177
Fulbeck, 214
Fuller's Earth, 203, 223, 224, 225
Fumay, 281, 306, 307
Furnace, 76
Furness, 188

Gabbro, 27, 28, 43, 87-90
Gablenz, 172
Gäfle, 172
Gaize, 117
Galicia, 170, 271
Galleting, 328
Galloway, 72
Gallymore Hills, 303
Galway, 77, 95, 133, 326, 423
Ganister Measures, 148
Gard, 169, 170
Garnet, 25, 30
Garreting, 328
Garsdale, 192
Garsington, 255
Garth Grits, 131

444 GEOLOGY OF BUILDING STONES

Gary M., 394, 396
Gatelawbridge, 157
Gatton St., 120, 166, 167
Gault, 162, 163, 166, 167
Geaune, 170
Geddington Chase, 240
Geddington Cross, 221
Gembloux, 170
Genoa, 314
Geodes, 17, 52
Geological maps, 7
    Survey, 8
    Museums, 8
Georgia, 79, 313
Germany, 81, 98, 106, 171, 272, 308
Germünden, 309
Gernsbach, 81
Gersdorf, 308
Giant's Causeway, 109
Giffnock St., 127
Gilling, 192
Gillogen, 192
Gimlet Rk., 100
Gingerbread St., 166
Girvan, 131, 138
Glacerville, 313
Glamorganshire, 153, 157, 199, 201
Glan-y-mor, 84
Glasgow, 302, 303
Glasshoughton, 150
Glassy rocks, 43
Glastonbury, 429
Glauconite, 261
Glaucophane, 25
Glebe Q., 415
Glencormick, 133
Glencree, 76
Glencullen, 77
Glendalough, 133
Glendun, 424
Glengariff Grits, 139
Glenpatrick, 304
Glentown, 304
Globe Q., 158
Gloucester, Mass., 79
Gloucestershire, 3, 199, 201, 216, 217, 224, 225, 238, 266, 321, 322, 429
Gmünd, 80
Gneiss, 370
Goat Acre, 242
Godalming, 164
Goddington, 211
Godstone, 166
Gondwana Series, 172
Goraghwood Q., 415
Görlitz, 81
Gortnaglink, 424

Goslar, 309
Gosselet, 307
Göthite, 22, 33
Gothland, 172
Grain, 52, 54, 55, 57, 281
Grain-end, 53
Gramont, 170
Granard, 135
Grande-Combe, 169
Grand Maison, 305
Grands Carreaux, 305
Grange, 158
Granite, 27, 36, 39, 43, 134, 137, 343-345, 365, 367, 370, 372, 373, 375, 377, 379, 380, 385, 390, 392, 405, 408-416
    analysis, 65, 66
    colour, 60
    decay of, 357
    distribution, 69
    minerals, 44
    physical characters, 68
    porphyry, 39
    texture, 48
    varieties, 63
Granitone, 89
Grano-diorite, 45, 64
Granophyre, 92, 95
Granophyric structure, 37, 46, 51
Grantham, 199, 213, 223
Grantown, 74
Granville, N.Y., 80, 275
Graphic Granite, 51
Grasse, 271
Graversfors, 78
Great Hazeley, 255
Great Malvern, 70
Great Milton, 255
Great Oolites, 161, 203, 204, 211
    Clay, 211
    Limestone, 211, 224, 232
    Series, 223
Great Tew, 159
Great Weldon, 211
Great Whin Sill, 100
Green Bed, 251-253
Greenmoor Rk., 148
Greenoside Rk., 148
Greensands, 34, 123, 162-168, 181
Greenstone, 85, 98, 370
Greetwell Road Q., 214
Greisenization, 94
Grenna, 172

Grès de la Couronne, 170
Grey Beds, 216, 256
Grey Limestone, 160, 203, 215
    Series, 323
Grey Royal Granite, 79
Greys, 259
Grey Slates, 317
Greystone, 244, 421
Grignan, 170
Grinding Test, 395, 396
Grinshill, 155
    Stone, 127
Grinstead Clay, 163
Grit, 114, 115
Groby, 96, 331
Grossular, 30
Grosswasser, 308
Ground Bed, 228
Grule, 309
Gryphite Grit, 205, 208, 209
Gudbrandsdalen, 313
Guenfol Q., 306
Guernsey, 87, 89
Guiting Power, 239
Guiting St., 239, 350
Gunnislake, 414
Guy's Cliff, 154
Gwalior, 172
Gwespyr, 151
Gypsum, 34, 335, 351

Habichtswald, 109
Haddingtonshire, 109, 138
Hadley, 154
Hadrian's Wall, 155
Hæmatite, 22, 33
Haighburn Wyke, 161
Hainault, 170, 271
Haldon Hills, Devonshire, 3
Halifax Co., 80, 148
Halkin Mts., 145
Hall, J., 340
Halland, 78
Halleflinta, 110
Hallowell Q., 79
Halsewell Q., 248
Hamdon Hill, 205
Ham Hill, 205, 206, 207
Ham Hill St., 176, 215, 350, 374
Hammam-Mesk-houten, 266
Hampshire, 167, 259, 260, 264, 268
Hampton St., 238
Handöl, 330
Handsworth Rk., 149

INDEX

Hangman and Foreland Grits, 135
Hanisch, A., 392, 403
Hanover, 272
Hanter Hill, 89
Haote Saone, 170
Harbottle Fell, 141
Harbottle Grit, 141
Harbro Rks., 191
Hardness, Mohs's scale, 411
 Tests, 394-397, 411
Hard-way, 57
Hard York St., 332
Haresfield, 204, 218
Harford Co., U.S.A., 311
Harlech, 291, 294
 Grits, 130
Harleston St., 160, 211
Harling, 321
Harnham Bridge, 253
Hartham Park St., 236
Harthill, 100, 130, 150
Hartlepool, 194
Hartwell Park, 255
Harwarden, 266
Harz, 97, 309
Haslingdon Flags, 144
Hastings Beds, 162
 Granite, 163
Hassock, 164, 264
Hatherleigh, 3
Hathersage, 146
Haverfordwest, 184
Hawkestone Hills, 155
Haydor, 203, 211
Haydor St., 223
Hayton, 150
Headbury Q., 248
Headings, 53
Headington St., 243
Heads, 54
Hearthstone, 200
Heathen, 52
Heavitree, 152, 153
Heilbrown, 171
Heimbach Q., 310
Helbron, 272
Helderberg Beds, 272
Helland, 94
Helmedon, 159, 211
Helmsley, 162
Helsby, 155
Helsley Park, 149
Hemingfield, 150
Hendre Ddu, 297
Herbeumont, 308
Hereford, 136, 186
Herefordshire, 89, 133, 185
 Stone, 136

Hermitage, 305
Herrenberg, 309
Hertfordshire, 259, 262
Hertfordshire Pudding St., 169
Hesse, 109, 123
Hestercombe, 87
Herrnskretschen, 171
Hexham, 192
Heythrop, 211
Heythorpe Common, 159
Heytor, 413
Hibaldstow Beds, 214
Hierapolis, 266
Highbrooks Q., 429
High Hazles Rk., 150
High Rk., 75
Highworth, 243
Hildenley St., 244
Hilmarton, 242
Hill-creep, 4
Hillebrand, 287, 289, 290
Hill of Fare Q., 73
Hils-sandstein, 171
Hippodrome, 198
Hirnant Limestone, 184
Hirschwald, Professor J., 83, 108, 335, 345, 368, 369, 383, 384, 388, 393-5, 398-400
Hoar Edge, 131
Hof, 111
Hogback, 275, 284
Hoghton, 146
Hohenelbe, 170
Holderness, 215
Höle, 330
Holland, T., 290
Hollington, 154
Hollybush Sandstone, 130
Holocrystalline, 37
Holywell, 191
Homburg, 108
Honister, 298
Hopton Wood, 183, 192, 193
Hopton Wood St., 347
Horbury Rk., 150
Horethorpe, 240
Hornblende, 25, 26, 28, 29, 342, 343
Hornton St., 202
Hörre-Raumland, 309
'Horses,' 61
Horse-pools, 208, 218
Horsham St., 163, 324
Horton Flags, 132
Houghton, 213
Houghton Common Rk., 150

Houses of Parliament, 195, 201, 217, 354
Howardian Hills, 161, 215
Howley Park, 149
Huddersfield, 148
Huddlestone, 196
Hundsheim St., 271
Hungary, 109, 271
Hunsrück, 308
Hunstanton, 117
Hyatt's Pits, 318
Hypabyssal Rks., 36, 37, 43, 90-101
Hypersthene, 22, 25, 27, 28
Hythe Beds, 162, 164, 263

Ible, 100
Iceland, 106
Ide, 108
Idefjord, 78
Ideford, 186
Idiomorphic crystals, 40
Ightham, 122
Igneous Rks., 35-111, 137
 decay, 357
 map, 71
Iles Chausey, 80
Ilfracombe, 186
Ilkley, 146
Illinois, 272
Ilmenite, 22, 34
Ilminster, 202
Imperial Grey, 79
Imperial Institute, 8, 192
Imperial Pearl, 84
India, 81, 172, 314, 330, 331
Indiana, 272
Inferior Oolite, 123, 159, 161, 203-207, 214, 215, 318
Inghoe Grit, 141
Ingleton Granite, 44
Inoceramus, 261
Intermediate rocks, 40, 43
Intrusive rocks, 36
Inverness, 74, 138
Iowa, 272
Ipplepen, 186
Ireland, 76, 87, 89, 95, 97, 101, 133, 134, 139, 141, 151, 158, 159, 188, 192, 266, 303-304, 316, 325, 415, 421, 432
Iron Bridge, 185
Isle of Man, 77, 101, 188, 325

446 GEOLOGY OF BUILDING STONES

Isle of Wight, 165, 168, 261, 264, 265, 266
Istrian St., 271
Italy, 106

Jackdaw Q., 209
Jämtland, 330
Jasper, 18
Jersey, 77, 87
Joints, 52, 275, 316
Jones, Inigo, 202
Jones, J. C., 389
Jubs, 233
Judd, Professor J. W., 211, 318
Jungfrun Island, 78
Jurassic, 158-162, 173, 268, 269, 271, 272, 313, 317, 428

Kaiserstuhl, 101
Kaiserwerth, 105
Kalberg, 106
Kamel, 109
Kangra district, 314
Kansas, 272
Kaolinite, 31
Kapaonik, 106
Kapfelberg, 171, 272
Karahpur Hills, 314
Kärnten, 271
Keadew, 423
Keesville, 89
Keinton, 176
Keinton Mandefield, 200
Keisley, 184
Kelheim, 171, 272
Kellaways Rk., 123
Kells, 326, 428
Kemble Beds, 229, 230
Kemnay Q., 73, 414
Kendal, 192
Kent, 122, 123, 162, 245, 266
Kentish Rag, 117, 120, 162, 164, 173, 181, 263, 355, 429
Kentallenite, 89
Kents Thick Rk., 150
Kents Thin Rk., 150
Kepwick, 266
Keratophyre, 92
Kerner von Marilaun, 339
Kerridge Rk., 151
Kerry, 136, 139, 304, 421
Kersantite, 80
Kersanton, 80
Kersworth Hill, 261
Kesh, 421
Keswick, 89, 100, 108

Ketton, 183, 203, 210, 212, 213, 220, 222
Ketton St., 348
Keuper, 154, 155
Kexborough Rk., 150
Keynsham St., 429
Kiev, 88
Kilburn, 244
Kilcullen, 325
Kildare Co., 325
Kilkenny, 304, 421, 422
Killaloe Q., 303
Killarney, 326
Killas, 324, 325
Killaways Beds, 211
Killinney Hill, 77
Killywhan, 415
Kiltorcan Beds, 139
Kilverslin, 158
Kimeridge Clay, 245
Kimsbury Castle, 208
Kinahan, 424
Kincardineshire, 72, 73, 138
Kinder Scout Grit, 144, 146
King Island, 314
Kingsbarrow, 246, 247
King's Bridge, 300
King's Co., 422
Kingsdon, 429
Kingskerswell, 152, 153
Kingsthorpe, 160, 232
Kingston, 76, 80, 258
Kinneigh, 325
Kirkaldy and Son, 262, 411
Kirkby, 318
Kirkby Knowle, 161
Kirkby Moor, 299
Kirkby Moor Flags, 133
Kirkby Stephen, 155
Kirkcudbrightshire, 72-74, 415
Kirkdale, 154
Kirkmabrek, 415
Kirtlington, 241
Kirtlebridge, 156
Kirton Beds, 214
Kissock Hill Q., 415
Kit Hill, 413
Kiveton Park, 197
Klein-pflaster, 56
Knapp St., 429
Knaresborough, 194
Knockatally, 423
Knockfierna, 422
Knockroe, 304
Königgrätz, 170
Kreswitz Marble, 271

Kyneton Slate, 318
Kyneton Thorns, 321

Laber, 80
Labrador, Labrador-Granite, 22, 84, 89
Labradorite, 20, 22
La Brie, 268
La Carbonière, 307
Laccoliths, 36, 38
La Corrèze, 170
Ladock, 136
La Forêt, 305
Lairg, 75
Lake Champlain, 311
Lake District, 4, 95, 100, 131, 133, 184, 274, 290, 298, 324, 431
Lake Superior, 377
Lake Superior Brownstone, 171
Lambay Island, 97
Lamb's House Q., 299
Lamorna, 45, 413
Lanarkshire, 159
Lancashire, 141, 143, 146, 148, 150, 154, 188
Lancaster, 146
Lancaster Q., U.S.A., 311
Landon, 268
Land's End, 94
Langecke, 309
Langton, 244
Laning Vein, 256
Lanivet, 94
Lansdown, 321
Larne, 158
Larys-Blanc St., 269
La Saulacé, 305
Laterite, 330
Latschach Marble, 271
Latt, 342
Laurvigite, 84
Lausitz, 101, 109
Lausitz Syenite, 84
Lavagna, 314
'Lavagne,' 314
La Varne, 330
Law Courts, 201
Lazonby, 155
  Stone, 118
Leadenham, 214
Leckhampton, 204, 207, 207, 208
Ledbury, 133
Ledbury Marble, 185
Leeds, 148, 149
Leek, 144, 146
Lee Moor, 413

INDEX 447

Lehigh, Pa., 289, 309, 310
Leicestershire, 70, 84, 87, 96, 188, 199
Leinster, 95
Leitrim Co., 133
Lemunda, 172
Leominster, 136
Lepidolite, 23
Lepidomelane, 23
Lerouville St., 269
Lessines, 97
Leucite, 22
Lexlip, 192
Leysters Pole, 136
Lias, 159, 170, 172, 176, 178, 181, 189, 191, 199-202, 203, 207, 211, 256, 270, 317, 429, 427
Liais de Méreuil, 270
Liais de Morley, 269
Liais de Ravières, 270
Libau, 308
Lichens, 338, 339, 353
Lickey Hills, 100, 130
Liebenerite, 24
Light Pearl, 84
Liholt, 78
Lille, 271
Lillebourne, 266
Lilleshall, 145
Lillington Lovell, 241
Limburg, 309
Limerick, 109, 192, 303, 422
Limestone, 8, 31, 173-272, 345, 347, 365, 370, 373-380, 384, 390, 392, 402, 409, 410, 424-429
    analyses, 183
    colour, 181
    composition, 182
    decay, 347
    distribution, 184
    list of quarries, 190
    map, 189
    organic, 176
    origin of, 175
    principal properties, 182
Limonite, 33
Lincoln, 214
Lincoln Hill, 185
Lincolnshire, 120, 162, 199, 203, 210, 211, 222-224, 231, 232, 245, 259, 266
Lincolnshire Limestone, 203, 210, 211, 212, 213, 214, 321

Lincombe and Warberry Grits, 135
Lindley's Q., 197, 198
Lineover Hills, 204
Lingula Flags, 130, 294
Linlithgowshire, 109
Linton Hill, 241
Linz-on-Rhine, 107
Lion's Haunch, 108
Lipari Islands, 105, 106
Lismore, 422
Little Eaton, 146
Little Rk., 85
Littletown, 149
'Liver,' 200
Liverpool, 154
Lizard, 89, 330
Llanberis, 291, 293
Llandaff, 217
Llandeilo, 131, 297, 298
Llandough, 157
Llandovery, 132, 184
Llanfaglen, 84
Llangollen, 188, 266, 297
Llangristiolns, 145
Llangynog, 297
Llanlivery, 431
Llanllyfni, 291
Lleyn Peninsula, 77, 89, 97
Load Bridge, 202
Locharbriggs, 156, 157
Loch Dee, 74
Loch Etive, 75
Lochfyneside, 76
Loch Linnhe, 89, 301
Loch Lomond, 138
Lochmaben, 156
Lodève, 170
Loire, 170
London, 258
London Basin, 264
Londonderry, 135, 158, 159, 424
London Doors Q., 250
Longborough, 209, 218
Long Burton, 240
Longhaven, 415
Longmynd, 128
Longpost, 429
Longridge, 146
Long Sutton, 429
Lorraine, 271
Lössnitz, 309
Lotus Q., 415
Louth Co., 326
Lovegrove, E. J., 397
Lowenberg, 101
Lower Limestone Shale, 141

Lower Swell, 318
Lower Tier, 247
Lower Trenton, 312
Lower Wych Q., 100
Loxley Edge Rk., 148
Lucas, A., 340
Ludlow, 132, 185
Lugwardine, 136
Luing, 301
Lumachelle, 171
Lumdon, 324
Luss, 303
Lutyens, E. L., 260
Luxembourg, 170, 272, 308
Luxullian, 412
Luxullianite, 31, 48, 94, 413
Lyme, 79
Lyme Regis, 199, 263
Lysekil, 78

Mabe, 412
Macclesfield, 151
Madrenna, 304
Maen, 412
Maenofferam, 288, 296
Maenturog Beds, 130
Main Building St., 251, 252
Maine and Loire, 305
Maine, U.S.A., 79, 89, 95, 309, 312
Magnesian Limestone, 156, 178, 179, 189, 192, 194, 195, 353-354, 376
Magnesite, 29, 33, 335
Magnet Cove, 85
Magnetite, 33, 307
Mähren, 170, 308
Maladrerie, 268
Malhanté, 307
Malmön, 78
Malm Rk., 167, 168
Malvern, 154
Malvern Hills, 100, 131
Manchester, 154, 192
Man, Isle of, 77, 101, 138, 188, 235
Manley, 155
Mannersdorf St., 271
Manor Rk., 149
Mansfield, 180, 181, 194, 195, 197, 198
Mansfield St., 121, 156, 173, 355
Marble, 182, 343, 344, 367, 376, 379, 409, 410
Marble Bed, 221, 212

448 GEOLOGY OF BUILDING STONES

Marcasite, 33
Mariposa, 313
Markfield, 96
Marlham St., 202
Marlstone, 201, 202, 318
Marnhull St., 241, 242
Marseilles, 270
Martigues, 170
Martyrs' Memorial, 195
Maryland, 79, 309, 311
Mascigno, 172
Masey, P. E., 263
Massachusetts, 67, 79, 95, 101, 170
Matlock Bath, 265
Matlock Moor, 146
Mauchline, 157
Maufe, H. B., 303
Mauthausen, 81
Mayen, 308
May Hill, 185
Mayo Co., 135, 423
Mayon, 94
Meadfoot Sandstone, 135
Mealoughmore, 304
Meath Co., 158, 326, 422
Medbourne, 318
Medmenham, 260
Mehlis, 81
Meissen, 81, 101, 109
Melilite, 102
Mells, 145
Melton Mowbray, 218
Mendip Granite, 44
Mendip Hills, 140, 141, 145, 153, 188, 192, 204, 206
Menevian, 131
Mentière, le, 170
Meran, 271
Merionethshire, 95, 100, 295
Merrill, G. P., 79, 267, 272, 398
Merriman, M., 287
Merrivale, 413
Merstham St., 166
Metamorphic rocks, 35
Metamorphic Series, 301, 303
Metz, 272
Meulière, 268
Meuse, 269, 285
Mica, 22-25, 26
analysis, 29, 39, 47
Mica Slate, 277
Micheldean, 127
Michigan, 313
Microbes, 338
Microcline, 16, 20, 21

Micro-granite, 60, 92, 95
Micro-graphic structure, 51
Micro-pegmatitic structure, 51
Micro-perthite, 21
Microscopic examination, 407, 408
Middle Estuarine Series, 203
Middle Rk., 148
Middleton, 192, 193
Middleton Rk., 149
Midford Sand, 203, 205, 206, 207, 208
Midlands, 203
Milford, 135
Milford Granite, Massachusetts, 67
Milford, New Hampshire, 65
Millepore Bed, 203, 215
Mill Grit Rk., 153
Millstone Grit, 139, 141, 142, 143, 144-146, 151, 191, 316, 417
Milton St., 238
Milton - under - Wychwood, 231, 238
Minchinhampton, 225, 229, 238
Minera, 145
Minerals, 13-34
Minerals in granite, 44-48
Minerals in sandstones, 117
Minnesota, 79, 89, 95, 313
Miocene, 170
Mirzepur, 172
Miscoden, 321
Misengrain, 305
Moel Tryfaen, 288
Mögeldorf, 171
Mohs's Scale of Hardness, 411
Moira, 158
Molasse, 121, 170, 370
Mold, 145
Molesmes St., 269
Monaghan Co., 423
Monasterboice, 326
Monestiés, 170
Money Point, 422
Monghyr, 314
Monk's Park St., 235, 236
Monkton Farleigh, 227, 236
Monmouthshire, 136, 186

Monsal Dale, 266
Monson, 312
Montacute, 216
Montauban, 272
Montello, 79
Monte san Giacorno, 314
Montibert, 305
Monthermé, 307
Montmartre, 34
Montpelier, 270
Monzonite, 85
Moon's Hill Q., 96
Moor Grit, 161
Moor Q., 415
Moorstone, 69
Moray Firth, 138
Morbihan, 289, 305, 306
Morcott, 212
Morecambe, 188
Moretonhampstead, 413
Morley Rk., 149
Morroe, 422
Moss, 338, 353
Mottled Sandstone, 154
Moughton Fell, 132
Mouillon, 170
Moulle District, 308
Moulin-Carcé, 305
Moulton Park, 232
Mount Charles St., 423
Mount Desert, 79
Mount Pleasant Q., 215
Mount Shannon, 422
Mount Sorrel, 70, 87
Mount Waldo, 79
Mourne Mts., 76, 416
Much Wenlock, 185
Mudstones, 173
Mugron, 170
Mulatto St., 168
Mull, 75, 109, 316
Mullaghana, St., 423
Müllenbach, 308
Mulroy, 77
Muncaster, 72
Murrel, 227
Muschelkalk, 271, 272, 370
Muscovite, 28
Muxton, 155
Mytton Flags, 131

Nahetal, 172
Nailsworth, 208
Nailsworth St., 217, 218
Nairn 74, 138
Namur, 271
Nantes, 306
Nantlle, 291, 293
Napoleonite, 51, 86

INDEX

Naresborough, 265
Naunton, 321
Nebra, 172
Nebraska, 272
Nepheline, 22
'Nerlys,' 214
Nettetal, 111
Nettlecombe, 321
Neuhaus, 81
Neumarkt, 271
Nevada, 95
Nevin, 97
Newark, 223
New Bailey, 192
New Brunswick, 80, 171
Newfoundland, 314
Newham, 94
New Hampshire, 79, 95
New Jersey, 101
New Mile Q., 412
Newport, 100
Newport, U.S.A., 79
Newquay, 94, 120, 176, 326
New Red Marl, 3
New Red Sandstone, 152-158
New Rockland Q., 313
Newry, 76
Newton Abbot, 153, 186, 187
Newton Dale, 266
Newtonwards, 158
New Vein, 257
New Westminster, 313
New York, 89, 170, 272, 289, 309, 311, 330
New Zealand, 105, 314
Nicaragua, 13
Niagara Beds, 272
Niedermendig Schwamstein, 105
Niederposta, 171
Nigg Q., 73
Nill's Hill, 131
Nist Bed, 248
Nithsdale, 157
Nordmarkite, 84
Norfolk, 166, 259-261, 329
Norite, 89
Normanby, 233
Normandoux, 270
Normandy, 80, 267
Northampton Sand, 123, 203, 210-212, 214, 318
Northamptonshire, 159, 160, 199, 210, 214, 221, 222, 224, 231, 239, 318
North Anston, 196

North Carolina, 89, 330
North Cheriton, 240
Northfield, 312
North Grimston, 244
North Hill, 70
Northleach, 230
North Perrott, 206
Northumberand, 100, 140, 141, 146, 194
Norton Bridge, 240
Norrtälje, 78
Norway, 21, 30, 78, 84, 95, 313, 330
Norwegian Gabbro, 88
Norwich, 328
Notgrove, 209
Nottinghamshire, 123, 146, 194, 195
Nova Scotia, 80, 313
Nova Scotia St., 171
Nummulites, 177
Nuneaton, 100
Nuremberg, 171, 357
Nutfield, 164
Nuttlar, 309

Oakenshaw Rk., 149
Oakham, 218
Oakley, 289
Oakley Mine, 289, 296
Oakridge Common, 321
Oaks Rk., 150
Oberhausen, 101
Oberhesse, 101
Oberkirch, 81
Oberkirchleithen, 171
Oberlausitz, 81
Obernkirchen, 171
Oberstreit, 81
Oberwesel, 368
Obsidian, 39, 43, 102
Odd Down, 228, 236
Odenwald, 81, 95, 109
Odenwald Syenite, 84
Offa, 330
Offenstetten, 272
Ogwell, 186
Ohio, 170
Okehampton, 325
Oldford, 204
Old Fold, 161
Old Franklin Q., 289
Oldham Edge, 151
Old Lawrence Rk., 150
Old Q., 213
Old Red Marl, 3
Old Red Sandstone, 114, 119, 120, 126, 129, 134-139, 301, 303, 304, 316-421, 422, 416

Old Red Sandstone, limestone in, 185, 189
Oldtown Q., 73, 414
Oligocene, 264, 268
Oligoclase, 20
Olivine, 28, 39
Olmütz, 308
Olney, 232
Ombersley, 154
Onesacre, 148
Onesmoor, 148
Ontario, 80, 171
Oolites, Great, 223-241, 317
 Inferior, 203-223, 317
 limestone, 178
 map, 189
Opal, 18
Ophitic structure, 98
Oppenheim, 123
Ordnance Dept., U.S. Army, 343, 364, 367, 375, 380
Ord of Caithness, 74
Ordovician, 126, 129, 131, 133, 134, 135, 137, 184, 291, 295-298, 303-306, 309, 311-313, 326
Orival St., 268
Orkney Island, 138, 139
Ormskirk, 154
Orthoceras Limestone, 272
Orthoclase, 16, 18, 342
Osbournby, 266
Ose Fjord, 313
Osmington Oolite, 241
Ossett, 150
Østre Slidre, 313
Oswestry, 145
Otter Creek, 79
Oughterlinn, 135
Oulton Rk., 149
Oundle, 233, 240
Ourthe Valley, 271
Oven St., 159
Övedskloster, 172
Owmby, 233
Owram, North and South, 148
Oxford Clay, 211
Oxfordian, 269
Oxfordshire, 159, 224, 225, 238, 239, 243, 245, 255, 321, 322
Oystermouth, 191

Pachelbronn, 123
Padstow, 99
Paignton, 153

450 GEOLOGY OF BUILDING STONES

Painswick, 203, 204, 208
Painswick St., 217
Palotte Q , 269
Paludina, 258, 264
Paludina Limestone, 176
Par, 412
Paragonite, 23
Parenzo, 271
Paris Basin, 268, 270
Parkgate Rk., 149
Park Nook, 196
Parnell's Q., 76
Parrock, 298
Parsley Q., 414
Partabpur, 172
Parys Mts., 137
Passage Beds, 244
Pateley Bridge, 146
Paulet, 275
Paulton, 429
Paving, 241, 272
Paving-stone, 198, 199, 200, 201, 233, 318
Peach Bottom Slate, 311
Pea Grit, 159, 205, 207, 208, 209, 218
Peak, 161
Peak Forest, 100
Pebble Beds, 154
Peckforton, 155
Peel, 138
Peel Fells, 141
Pégauds, 170
Pegmatitic structure, 52, 58
Pembrokeshire, 89, 108, 132, 145, 191, 297
Pen Argyl, 309, 310
Pendle, 132, 160, 232, 233
Pendle Grit, 144
Pendleside Beds, 144, 151
Pendock St., 154
Penistone Flags, 148
Penmaenmawr, 97
Pennant Series, 126, 151
Pennant St., 127
Pennsylvania, 170, 272, 287, 289, 309, 311, 330
Penrith, 155
Penrhyn, 288, 293
Penryn, 45, 68, 69, 94, 412
Penshurst, 163
Pentivy, 80
Pentuan St., 94, 331
Penyghent, 146
Pen-y-glog, 132
Penzance, 413
Peridotite, 28
Perlitic structure, 39

Perlonjour, 271
Permian, 121, 126, 150, 152, 155, 156, 158, 170, 172, 189, 192, 195, 272, 421, 427
Perthite, 16, 21
Perthshire, 138, 301, 303
Peterborough, 218, 220, 222
Peterhead, 72, 73, 414
Petersburg-Jechnitz, 81
Petit Granite, 44, 176, 271
Petits-Carreaux, 305
Petit Tor, 186
Petworth, 164
Pewsdown, 321
Pfalz, 172
Pflaff, F., 404
Phenocrysts, 40
Philadelphia, 330
Phlogopite, 23
Phonolite, 28
Phyllade aimantifère, 103, 104-106, 290, 316
Picking Bed, 227, 228
Pickwell Down Beds, 136
Piedmont, 330
Pier Limestone, 161
Pierre de Màne, 170
Pier St., 215
Pietra di Maschine, 89
Pigeon Rock Q., 416
Pilsener Granite, 81
Pinney Bed, 251, 252, 253
Pirna, 171
Pisolite, 179, 203, 208
Pitchstone, 92
Pitsligo, 415
Plagioclase, 16, 19-22
Planking, 230, 241
Planterie St., 270
Plaster of Paris, 34
Pleasley, 197
Plutonic rocks, 36, 43, 82-90
Plymouth, 186, 187, 325
Pockets, 117
Poliĉka, 271
Polkannugo, 412
Pollard, W., 384
Polyphant St., 99
Pond Freestone, 248, 249, 250
Ponk Hill, 100
Pontefract, 194
Pontefract Rk., 150
Pontesbury, 131
Ponton, 213

Porcellanite, 110
Porchères, 170
Pore cement, 356
Porfido verde antico, 97
Porosity, 378-379, 383, 384
Porphyrite, 27, 90, 96-97
Porphyritic structure, 40
Porphyry, 36, 90-96, 133, 365, 385
Portisham, 254, 323
Portland, 183, 192, 204, 244, 245, 249, 258, 268, 323
Portlandian, 244-255
Portland Sand, 245, 249
   screw, 246, 254
   Stone, 170, 176, 244-255, 347, 349
Portmadoc, 291
Portroe, 303
Portsmouth, 258
Portsoy, 75
Portugal, 85, 313
Postlewitz, 171
'Posts,' 200
'Pot-lids,' 319, 322
Potsdam Sandstone, 170, 171
Potstone, 329
Potten, 330
Pouéze, 305
Poulton Slates, 323
Pre-Cambrian, 184, 128
Precelly, 297
Prehan, 135
Pressure. See Crushing
Pressure strength, 392, 402, 409
Prestatyn, 266
Preston, 212
Pretoria, 314
Princetown, 341
Prudham, 192
Prussia, 84
Przibram, 81
Pseudo-bedding, 57
Puddlecote St., 239
Pullborough St., 164
Pulteney Town, 301
Pumice, 105, 332
Punjab, 314
Purbeck Beds, 176, 178, 245, 246, 248, 249, 251, 255-259, 268, 323
Purbeck, Isle of, 245, 248-250
Purbeck Marble, 258, 347
Purbeck - Portland St., 250

INDEX

Purton, 242
Puzzolan, 332
Pwlgwyn, 266
Pwllheli, 100
Pyle, 153, 157
Pyrenees, 305
Pyrites, 33, 301, 353, 356
Pyrope, 30
Pyroxene, 25

Quader Sandstein, 171
Quantock Hills, 186
Quar Hill, 208, 218
Quarella St., 127, 157
Quarry water, 122
Quartz, 15, 16, 17, 39, 46, 342, 343
Quartz-felsite, 60
Quartzite, 121, 125, 130, 131, 133, 135, 344
Quartz-Porphyry, 43, 60, 90, 92
Quebec, 171, 180, 313
Queen Camel, 317, 422,
Queen's Co., 422   [429
Quenast, 97
Quernstone, 166
Quilly St., 268
Quincy Granite, 67, 79
Quoin St., 165

Raasey, 109
Radipole, 241
Radnorshire, 132
Radstock, 202, 429
Radyr St., 153
Rag, 12, 148, 160, 204-213, 220-233, 251, 254, 256, 263, 264, 322, 348
Raichur, 81
Rain, 405, 406
Rajpatana, 314
Ramsay, 77
Randwick, 204
Randwick Hill, 205
Ranfurly St., 423
Rannoch Moor, 75
Ransome's St., 332
Rastrick, 148
Rathmichael, 133
Rattler Test. See Rumbler
Reade, T. M., and Holland, 285, 287, 288, 290
Reading Beds, 168, 169
Recco Valley, 314
Red Barn, 158
Red Beach, 79
Redcar, 199]
Red-free St., 158

'Red Granite,' 135
Red Hills, 75
Redlinch, 240
Red Rk., 150
Red Sandstone Group, 143
Redstone Granite, 67
Red Swedish Granite, 78
Red Wilderness, 127
Reginald's Tower, 133
Reigate St., 166, 168
Remiremont, 80
Renazé, 305
Rencontre St., 270
Rendcombe, 321
Renscombe, 250
Repallo, 314
Retyn, 94
Revin, 306, 307
Rewari, 314
Rhætic, 157, 178, 202
Rhaunen, 309
Rhewarth, 297
Rhine, 95, 109, 272, 308
Rhiwbach, 296
Rhiwlas Limestone, 184
Rhode Island, 79
Rhodonite, 25
Rhomb-porphyry, 21
Rhone Valley, 106, 109
Rhune, 170
Rhyolite, 36, 43, 102, 104-105
Rhynconella Cuvieri, 263
Riccal Dale, 161
Richmond Co., Quebec, 313
Richmond, Yorks, 146
Ribblesdale, 132
Riddle Scout Rk., 150
Rideal, 352
Riebeckite, 25, 27, 67, 68
Riendolte, 170
Riesengebirge, 97
Rift, 52, 54, 57
Rimogne, 281, 307
Ringstead Bay, 241
Roach, 204, 246, 249, 254-256
Roach Rocks, 146
Roads, U.S. Office of Public, 396, 397
Robeston Walthen Limestone, 184
Robin Hood's Well, 196
Robinhood Rk., 149
Rochdale Flags, 148, 150
Rochefort-en-Terre, 289, 306
Rocher, 151
Rochlitz, 111

Rock Bed, 201
Rockeries, 59
Rockingham, 160
Rockport, 79
Rockport Grey Granite,
Rock Valley, 198   [67
Rodborough, 204, 207
Roe-stone, 178
Rome, 267
Romerstadt, 308
Romsey, 260
Roncourt, 272
Roscommon Co., 423
Rosecraw, 95
Rosiwal, 407
Roslin Sandstone, 141
Ross of Mull, 76, 316
Rothbury Grit, 141
Rotherham, 150
Rough Rk., 144
Rovigno, 271
Rowley Rag, 100
Rowley Regis, 100
Royal   Institute   of British Architects, 6
Royal Oak Q., 413
Roydon, 261
Rubislaw, 50, 73, 414
Rudesheim, 308
Rugby, 199
Ruhla, 81
Rum, 75, 109
Rumbler Test, 395, 397
Runcorn, 155
Ruscombe, 205
Rushnacora, 325
Russia, 88
Rutlandshire, 159, 201, 210, 218, 219
Rutley, F., 65
Ryde, 265

Salcombe, 325
Salisbury, 258
Salperton, 321
Saltergate, 266
Saltford, 429
Salzburg, 170, 271
'Sand,' 54
Sandbars, 54
Sand-blast Test, 395, 396
Sandgate Beds, 162, 164
Sandringham Hall, 220
Sandsfoot, 241
Sandstone, 114, 115, 343-345, 365, 367, 370-373, 376-378, 380, 384, 390, 391, 401, 402, 409, 410, 416-421

Sandstone, analyses of, 127
  cement of, 119
  colour of, 124
  composition of, 126
  decay, 354-357
  distribution of, 128-172
  minerals of, 117
  types, 125
Saint :
  Alban's Head, 248
  Aldhelm St., 234
  Aubyn, 87
  Austell, 69, 94, 412
  Barnabé, 307
  Bees, 154
  Brelades, 77
  Budeaux, 325
  Clement's Bay, 87
  Columb Minor, 94
  David's Head, 89, 97, 100, 108
  George, New Brunswick, 80
  Germain, 169
  Germains, 170
  Germans, 325
  Goar, 308
  Hilary, 157
  John's Hole, 423
  Juste, 170
  Kevin's Kitchen, 133
  Kilda, 75
  Mary Church, 186
  Neots, 94
  Nicholas Q., 153
  Oswald's, 422
  Pancras, 198
  Pierre Canivet, 269
  Robert, 270
  Savinien St., 270
  Teath, 299
  Theresa Marble, 271
San Francisco, 313
Sanidine, 19, 39
Santa Catalina, 330
Sarsden, 231
Sarsen St., 122, 169
Saturation, 384
Saturation - coefficient, 384, 389, 391
Savoie, 305
Savonnières St., 269
Saxby, 233
Saxony, 81, 85, 95, 97, 101, 111, 172, 309
Scallet, 228, 235
Scandinavia, 22, 59, 78, 272, 330

Scania, 172
Scarla, 301
Scarborough Limestone, 160, 161, 215
Scariff, 422
Sceauteaux, 170
Schalstein, 325
Schiller structure, 84
Schillerization, 27
Schists, 128, 315, 316, 325
Schultz, R. W., 6
Schwäbisch-Hall, 171
Sclattie Q., 73, 414
Sclerometer Test, 394, 395
Scotland, 72, 84, 87, 89, 95, 97, 101, 108, 128, 130, 132, 137, 138, 140, 141, 151, 156, 158, 162, 188, 266, 301-303, 316, 325, 432
Scrabby, 135
Scrabo, 158
Sea air, 337
Seacombe, 248
Seacombe St., 250
Sea Green Slate, 275, 311, 312
Sedgley, 185
Sedimentary rocks, 35, 112
See Finn Q., 416
Seil, 301
Seizincote, 218
Selbornian, 166
Selsby Hill, 204
Sericite, 24
Serpentine, 29, 330
Serravezza, 314
Servia, 106
Settle, 146
Setzdorf, 81
Seven Churches, 133
Sevenhampton Common, 321
Shale, 114
Shalstone, 240
Shannon, 422
Shantallow, 77
Shamrock St., 127
Shap, 45, 50, 51, 67, 70
Shaugh, 413
Shearing, 367
Shear-zones, 284
Sheets, 53
Sheffield, 148, 149, 150
Sheffield Rk., 149
Sheffield Q., 412
Shell Limestone, 176

Shelve District, 131
Shepton Mallet, 96, 201, 216
Shetland, 330
Shingle, 256, 315
Shipton - under - Wychwood, 230
Shireoaks, 197
Sholesbrook Limestone, 184
Shotover Hill, 255
Shotover Limestone, 243
Shotover Sands, 166
Shrewsbury, 100, 155
Shrimp St., 248, 250
Shropshire, 3, 108, 128, 130, 131, 136, 145, 155, 185, 188, 318
Siebel's St., 332
Siebengebirge, 101, 106, 109
Silesia, 81, 95, 109, 271, 308
Silica, 15-18
Siliceous Limestone, 180
Silkstone Rk., 149
Sills, 38
Silsley Hill, 218
Silurian, 126, 131-137, 169, 171-173, 272, 297, 298, 303, 316, 424
Silurian Limestone, 184
Silurian (map), 189
Silver Bed, 214
Silver Grey Q., 415
Simonside Fell, 141
Sink St., 214
Sisana, 271
Skerries, 192
Skertchly, S. B. J., 328
Skiddaw Slates, 131, 298, 324
Skillereen, 325
Skye, 75, 89, 109
Slack Bank Rk., 149
Slant Bed, 251
Slate, 114, 133, 273-326, 343, 376, 385, 390, 394, 395, 409, 410, 422, 430, 432
Slatedale, 309
Slatington, Penn., 289, 309, 310
Slatt, 241, 249, 259
Slatt Beds, 323
Sleaford, 222
Slickenside, 53
Slieve Donard Q., 416
Sligo Co., 423
Slip-cleavage, 282-284

INDEX

Sloley, 136
Småland, 78
Smawse, 197
Smith, Angus, 336
Smith Sound, 314
Snettisham, 166
Snow, 337
Snowdon, 291
Snowshill, 318
Soaking test pieces, 386-388, 390
Soapstone, 30, 329
Soft Bed Flags, 147
Soft Burr, 258, 259
Soignies, 271
Solenhofen Limestone, 335, 404
Somersetshire, 96, 100, 151, 153, 186, 188, 200-202, 216, 224, 300, 317, 429
Soudley Sandstone, 131
Souppes, 268
South Ranceby, 213
South Shields, 194
Spargo Downs, 412
Sparkford, 202, 317, 429
Sp. grav., 378-379, 380-382
Sphero-siderite, 32
Spilsby, 120
Spremberg, 84
Spy Wood Grit, 131
Stack of Scarlet, 109
Staining Test, 398, 400
Staffa, 109
Staffordshire, 141, 144, 145, 154, 185
Stamford, 211, 212, 218, 219, 222, 318, 321
Stamford Marble, 212, 214, 219
Stamford St., 213
Stancliffe, 146
Stancliffe St., 127
Standard Tests, 411
Stanley Hill, 209
Stanstead, 80
Stanton, 146
Stanway, 209, 218
Staple Ashton, 242
Stavanger, 330
Staynerville, 80
Steatite, 30, (329)
Steeple Ashton, 159
Steetley, 195, 196
Steiermark, 80, 106
Sternberg, 170, 308
Stepper Q., 99
Steward's Q., 247

Stinchcombe, 201, 429
Stipstone Q., 200
Stirlingshire, 138
Stirling Hill Q., 415
Stithian, 412
'Stocked,' 319
Stockbridge, 260
Stockholm, 78
Stockholm, Pembrokeshire, 316
Stockport, 154
Stoke Ferry, 261
Stoke Ground St., 237
Stonesfield, 225
Stonesfield Slate, 229, 230, 321-323
Stone planks, 240
Stone tiles, 315-324
Stoney Houghton, 197
Storeton, 155
Stowe-Nine-Churches, 232
Strain-slip-cleavage, 280, 283
Strata, Table of, 11
Stratford-on-Avon, 199, 429
Strath Halladale, 75
Street, 429
Strehlen, 81
Stress-strain curves, 371, 374
Stresses, ultimate, 409
Striegau, 81
Strike, 57, 276
Stripe, 275, 282, 295, 301, 309
Strong Blue St., 150
Stroud, 204, 217, 218, 229
Stroud Hill, 207, 208
Studdfield, 152
Sturgeon's Q., 415
Stuttgart, 172
Stylolite, 197
Sudbrook, 211
Suffolk, 259, 327-329
Sunderland, 194
Surrey, 162, 167
Surveyor's Institute, 223
Susquehanna Co., U.S.A., 311
Sussex, 162, 167, 245, 260, 324, 329
Sussex Marble, 264
Sutherland, 75, 138, 158
Sutton, 201
Sutton-in-Ashfield, 197
Sutton Mallet, 153
Sutton St., 263

Sutton Waldron, 242
Swales Moor, 148
Swanage, 249, 257, 258, 323
Swanage St., 255
Sweden, 78, 95, 172, 330
Swedish Rose, 78
Swedish Syenite, 84
Swill Tor, 413
Swinbrook, 231, 238
Swindon St., 254
Swinford, 135
Switzerland, 109, 330
Syenite, 28, 43, 81-85

Tabert, 326
Tabular Hills, 266
Tachylyte, 43, 107
Taconic Range, 311
Tadcaster, 194
Tadmaston, 240
Tainton (Taynton), 231, 429
Stone, 238
Talacre, 145, 151
Talc, 29, 30
Tamlaght, 423
Tasmania, 314
Taunton, 87
Tavistock, 108
Taynuilt, 75
Tealby Beds, 166
Teffont Q., 253, 259
Teignmouth, 152
Telemark, 30
Temperature, change of, 342
Temple Combe, 241
Temple Guiting, 218, 239
Tenant's Harbour, 89
Tensile Strength, 367-369
Tension Test, 393, 401, 409
Tercé St., 270
Terne, 307
Tertiary, 142-148, 172, 175, 264, 270, 272
Testing of stones, 8, 362-411
Tetbury, 229, 323
Tetmajer, L., 382
Teutoburger Wald, 171
Tewkesbury, 154, 199
Texas, 272
Thame, 255
Thanet Sand, 117, 123
Third Grit, 144, 146
Thistleton, 213
Thom, T. M., 332
Thomm, 308

454 GEOLOGY OF BUILDING STONES

Thornback, 256
Thornhill, 157
Thornhill Rk., 149
Thornton, 148
Thorpe Mandeville, 160
Three Elms St., 136
Threlkeld, 95
Througham, 321
Thulite, 30
Thuringia, 81, 95, 97, 109
Thurlbeer Lias, 429
Tideswell Dale, 100
Tilberthwaite, 240
Tilestones, 148, 317, 318
Tilgate St., 163
Tillyfourie Q., 73, 414
Tilly-Whim, 233, 248, 250, 251
Tinsley, 150
Tintagel, 299
Tipperary, 303, 422
Tisbury, 252, 254
Titanite, 34
Tiverton, 429
Tivoli, 267
Toadstone, 100
Todbere St., 241, 242
Torridonian, 119, 128
Toms Forest Q., 73, 414
Tonalite, 87
Tor, 413
Torbay, 153
Torcross, 300
Torquay, 186, 187, 300, 325
Totland Bay, 266
Totnes, 108, 186, 300
Totten, Col., 343
Totternhoe St., 176, 261
Toughness Test, 397
Tough-way, 57
Toulon, 271
Tourmaline, 25, 31
Tournay, 271
Trachyte, 103, 104-106, 370
Transvaal, 314
Transverse strength, 377
Transylvania, 85
Trass, 111, 332
Travertine, 178, 265
Travertin moyen, 268
Treborough, 300
Treeton Rk., 150
Trefdraeth, 145
Tregarden Q., 412
Trélazé, 305
Tremadoc Beds, 294
Trembowla, 170
Tremolite, 25, 26

Tremore, 94
Tronton Beds, 171
Tretzendorf, 171
Trevailes, 94
Trevose Head, 99
Trias, 118, 124, 126, 129, 134, 152-154, 158, 170, 171, 271, 421
Triberg, 81
Tribogna, 314
Trier (Treeves), 308
Trigonia gibbosa, 243
Trigonia Grit, 205, 208
Tring, 255
Trondhjem, 330
Troppau, 170, 308
Trough Bed, 251, 252, 253
Troutbeck, 132, 299
Troutdale, 266
Trowlesworthy, 413
Trowlesworthy Tor, 69
Truro, 94
Trusham Granite, 100
Tudeils, 170
Tuedian Grits, 141
Tufa, 178, 201, 267, 272
Tuff, 38, 109-111, 345, 385
Tullamore, 192
Tully Beds, 272
Tullyfream, 416
Tunbridge Wells Sand, 162
Tuscany, 172, 314
Tutzendorf, 171
Tyrebagger Q., 414
Tyrol, 80, 85, 95, 271
Tyrone Co., 423
Tytherington, 145

Udelfangen, 170
Uley Bury, 218
Unfading Green Slate, 311, 312
Union Q., 306
United States, 79, 170, 272, 390
Untersberg Marble, 271
Upholland Flags, 148, 150
Upper Building St., 251, 252
Upper Calcareous Grit, 244
Upper Limestone Shales, 144
Upper Natley, 169
Upper Pickwick, 236
Upper Red Measures, 151

Upper Westwood, 226
Uppingham, 166, 212, 218
Upton Noble, 224, 241
Upware, 243
Upwey, 254, 259
Uralite, 28
Uralitization, 28
Ural Mts., 85
Usica, 314
Ussher, W. A. E., 152, 214
Utah Territory, 79
Uthammar, 78

Vacuoles in quartz, 17, 26, 47, 55
Valders, 313
Valentia Island, 304
Vale of Pickering, 162
Vale of Wardour, 251
Valongo, 313
Val-Rot, 270
Vancouver, 313
Vanelles, 307
Vånevik, 78
Var, 330
Varberg Granite, 78
Vaze's Q., 229
Veins, 38, 52, 58, 191, 256, 305
Velenhelli, 288, 290
Vergennes, 89
Vermiculite, 24
Vermont, 89, 272, 275, 285, 289, 290, 309, 311, 312
Verne Fort, 245
Via Gellia, 266
Vianshill, 153
Victoria Q., 303
Stone, 303
Vienna, 170, 308
Vienne Dept., 270
Villach, 271
Villars, 270
Vinalhaven, 79, 89
Vindhyan Series, 172
Virbo, 78
Virginia, 79, 101, 309, 313
Virgo Granite, 78
Vitré St., 169
Vogelsberg, 109
Volcanic rocks, 36, 38, 43, 101-111
Vorarlberg, 170
Vosges, 80, 95, 97, 106
Vossebygden, 313
Vughs, 52

INDEX

Wadhurst Clay, 162
Wakeham Q., 247
Wakesley, 212
Waldershuter, 81
Wales, 4, 77, 89, 95, 97, 100, 126, 128-132, 136, 140, 145, 151, 188, 192, 201, 266, 274, 284, 298, 316, 324, 430-431
Walheim, 272
Walling Bed, 214
Walsall, 100
Walter's Ash, 169
Waltersdorf, 308
Waltham, 213
Walton, 154
Wansford, 211
Wanstrow, 240
Wardour Q., 252
Ware, 263
Warmefontaine, 308
Warsaw, Wis., 79
Warwick Sandstone, 154
Warwickshire, 154, 200, 202
Warrington Q., 232
Washenborough, 214
Washington, 79, 330
Washington Co., U.S.A., 311
Wastdale, 72
Wastwater, 95
Waterford, 101, 133, 304, 422
Waterstones, 154
Waupaca, 79
Waushara, 79
Wave action, 346
Weakening coefficient, 346, 392-395, 402
Weald Clay, 162
Weald District, 122, 162, 163, 166, 168
Wealden Beds, 171, 176, 324
Weathering Test, 403-405
Weaver's Down, 165
Weight per cubic foot, 372, 373, 378-380
Weissenberg, 171
Weldon, 183, 211, 212
    Marble, 212, 221
    Rag, 212
    Stone, 221
Wenlock Beds, 185, 297, 299
Wensley Dale, 140, 190
Wentreath, 80

Wenvoe, 192
Wépion, 170
West Africa, 342
West Cranmore, 206
West Derby, 154
Westerham, 166
Western Isles, 72
Westerwald, 106
Westington Hill, 209, 219
Westmorland, 70, 192
Weston, 155, 246, 247, 429
Weston Hill, 325
Westphalia, 272, 309
West Virginia, 313
Westwood Ground St., 237
Wexford, 133, 422
Weymouth, 241, 259, 323
Whatstandwell, 146
Wheatly, St. 243
Whernside, 146
Whit Bed, 246-249, 251
Whitby, 162
Whitchurch, 255
White, Gilbert, 165, 167
White Hill, 208
'White Lias,' 202
'White Lime,' 185
White Limestone, 229-231
White Pendle, 318
Whitwell Oolite, 215
Whitwood, 150
Wicken, 243
Wicklow, 76, 101
Widsworthy, 263
Wieburn, 111
Wight, Isle of, 165, 168, 261, 264-266
Wilmcote, 429
Wilmscote Q., 201
Wilsford, 211
    Stone, 213, 222
Wiltshire, 165, 224, 241, 242, 245, 318, 322, 323
Wincanton, 240
Winchcombe, 239
Winchester, 217, 258
Wind action, 341
Windrush, 231
    Stone, 238
Winsley, 277
Winsley Ground St., 237
Winspit Q., 248, 250
Wintrich, 308
Wirksworth, 191-193
Wisconsin, 79, 95, 171, 272

Wisconsin Tests, 377, 380
Withiel, 94
Wittering Pendle, 321
Wockley, 254
Wöllersdorf St., 271
Wolmer Forest, 165
Wood, silicified, 13, 18
Woodhead Hill Rk., 150
Woodhouse Rk., 149
Woodstock, 321
Woodward, H. B., 201, 202, 253
Woolhope Limestone, 185
Woolley Edge Rk., 150
Worcestershire, 3, 70, 130, 136, 200
Worksop, 194
Worsborough Dale, 150
Worthing, 163
Wrekin Q., 130
Wrexham, 266
Wrington, 201
Wünschelburg, 171
Würtemberg, 172
Würzburg, 172
Wych, 70
Wyke, 259
Wyoming Valley St., 170

Xenoliths, 52

Yutton, 153
Yellowstone Park, 267
Yeovil, 206, 207, 216
Yonne, 269
Yoredale, 140, 141, 190, 192
York Co., Pa., 101
Yorkshire, 100, 140, 141, 144, 146-148, 158-162, 188, 192-196, 199, 203, 214, 215, 224, 233, 243, 245, 259, 266, 316, 317, 323
Yr Eifl, 97
Yvoir, 170

Zeil, 171
Zell, 308
Zircon, 31
Zittau, 171
Zogersdorf St., 271
Zoisite, 30
Zone of drainage, 389
    of immersion, 389
    of soakage, 389

Mr. Edward Arnold's List of
Technical & Scientific Publications

Extract from the LIVERPOOL POST of Dec. 4, 1907:

"During recent years Mr. Edward Arnold has placed in the hands of engineers and others interested in applied science a large number of volumes which, independently altogether of their intrinsic merits as scientific works, are very fine examples of the printers' and engravers' art, and from their appearance alone would be an ornament to any scientific student's library. Fortunately for the purchaser, the publisher has shown a wise discrimination in the technical books he has added to his list, with the result that the contents of the volumes are almost without exception as worthy of perusal and study as their appearance is attractive."

The Dynamical Theory of Sound. By HORACE
LAMB, D.Sc., LL.D., F.R.S., Professor of Mathematics in the Victoria University of Manchester. viii + 304 pages, 86 Illustrations. Demy 8vo., 12s. 6d. net (inland postage 5d.).

An Introduction to the Theory of Optics. By
ARTHUR SCHUSTER, Ph.D., Sc.D., F.R.S., Honorary Professor of Physics at the University of Manchester. Second Edition (Revised). xvi + 352 pages. Demy 8vo., 15s. net (inland postage 5d.).

The Becquerel Rays and the Properties of
Radium. By the Hon. R. J. STRUTT, F.R.S., Fellow of Trinity College, Cambridge; Professor of Physics at the Imperial College of Science and Technology. Second Edition (Revised and Enlarged). vi + 215 pages. Demy 8vo., 8s. 6d. net (inland postage 5d.).

Physical Determinations. Laboratory Instruc-
tions for the Determination of Physical Quantities. By W. R. KELSEY, B.Sc., Principal of Taunton Technical Institute. Second Edition. xii + 329 pages. Crown 8vo., 4s. 6d.

Advanced Examples in Physics. By A. O.
ALLEN, M.A., B.Sc., Assistant Lecturer in Physics at Leeds University. With Answers. Second Edition (Revised and Enlarged), with additional examples. Crown 8vo., 2s. (inland postage 5d.).

Notes on Practical Physics. By A. H. FISON,
D.Sc., Lecturer in Physics at the Medical Schools of Guy's Hospital and London Hospital. Crown 8vo., 3s. 6d.

Five-Figure Tables of Mathematical Functions.
By J. B. DALE, M.A., Assistant Professor of Mathematics, King's College, London. Demy 8vo., 3s. 6d. net.

LONDON: EDWARD ARNOLD, 41 & 43 MADDOX STREET, W.

Logarithmic and Trigonometric Tables (To Five Places of Decimals). By J. B. DALE, M.A. 2s. net.

Mathematical Drawing. Including the Graphic Solution of Equations. By G. M. MINCHIN, M.A., F.R.S., Formerly Professor of Applied Mathematics at the Royal Indian Engineering College, Cooper's Hill; and J. B. DALE, M.A. 7s. 6d. net (inland postage 4d.).

Graphs and Imaginaries By J. G. HAMILTON, B.A., and F. KETTLE, B.A. Crown 8vo., 1s. 6d.

Homogeneous Co-ordinates. By W. P. MILNE, M.A , D.Sc., Mathematical Master, Clifton College. Crown 8vo., 5s. net.

An Introduction to Projective Geometry. By L. N. G. FILON, M.A., F.R.S., Assistant Professor of Mathematics, University College, London. Crown 8vo., 7s. 6d.

Vectors and Rotors (with Applications). By O. HENRICI, Ph.D., F.R.S., LL.D., and G. C. TURNER, B.Sc. 4s. 6d.

The Strength and Elasticity of Structural Members. By R. J. WOODS, M.E., M.Inst.C.E., Fellow and formerly Assistant Professor of Engineering, Royal Indian Engineering College, Cooper's Hill. Second Edition. xii + 310 pages. Demy 8vo., cloth, 10s. 6d. net (inland postage 4d.).

BY THE SAME AUTHOR.

The Theory of Structures. xii + 276 pages. Demy 8vo., 10s. 6d. net (inland postage 4d.).

The Calculus for Engineers. By JOHN PERRY, M.E., D.Sc., F.R.S., Professor of Mechanics and Mathematics in the Royal College of Science. Tenth Impression. Crown 8vo., 7s. 6d.

Oblique and Isometric Projection. By JOHN WATSON, Lecturer on Mechanical Engineering and Instructor of Manual Training Classes for Teachers for Ayrshire County Committee. 3s. 6d.

The Balancing of Engines. By W. E. DALBY, M.A., B.Sc., M.Inst.C.E., M.I.M.E., Professor of Engineering, City and Guilds (Engineering) College. Second Edition. xii+ 283 pages. Demy 8vo., 10s. 6d. net (inland postage 4d.).

Valves and Valve Gear Mechanisms. By W. E. DALBY, M.A., B.Sc., M.Inst.C.E., M.I.M.E. xviii + 366 pages. Royal 8vo., 21s. net (inland postage 5d.).

Machine Sketches and Designs for Engineering Students. By A. CRUICKSHANK, A.M.I.Mech.E., and R. F. McKAY, M.Sc. Demy 4to., 1s. 6d.

Steam Turbine Design. With especial reference
to the Reaction type, and including chapters on Condensers and Propeller Design. By JOHN MORROW, M.Sc., D.Eng., Lecturer on Engineering at Armstrong College, Newcastle-on-Tyne. viii + 472 pages. Demy 8vo. [*Now Ready.*

Hydraulics. For Engineers and Engineering
Students. By F. C. LEA, M.Sc., A.M.Inst.C.E., Senior Whitworth Scholar, A.R.C.S.; Lecturer in Applied Mechanics and Engineering Design, City and Guilds (Engineering) College, London. Second Edition. xii + 536 pages. 15s. net (inland postage 5d.).

Hydraulics. By RAYMOND BUSQUET, Professeur
à l'École Industrielle de Lyon. Translated by A. H. PEAKE, M.A viii + 312 pages. Demy 8vo., 7s. 6d. net (inland postage 5d.).

The Practical Design of Motor-Cars. By JAMES
GUNN, lately Lecturer on Motor-Car Engineering at the Glasgow and West of Scotland Technical College. viii + 256 pages. Demy 8vo. 10s. 6d. net.

Power Gas Producers: their Design and
Application. By PHILIP W. ROBSON, sometime Vice-Principal of the Municipal School of Technology, Manchester. iv + 247 pages. Demy 8vo., 10s. 6d. net (inland postage 4d.)

The Foundations of Alternate Current Theory.
By C. V. DRYSDALE, D.Sc. (Lond.), M.I.E.E. xii + 300 pages. Demy 8vo., 8s. 6d. net (inland postage 4d.).

Electrical Traction. By ERNEST WILSON, Whit.
Sch., M.I.E.E., Professor of Electrical Engineering in the Siemens Laboratory, King's College, London; and FRANCIS LYDALL, B.A., B.Sc. Two volumes, sold separately. Vol. I., Direct Current; Vol. II., Alternating Current. 15s. net each (inland postage 5d. each).

A Text-Book of Electrical Engineering. By
Dr. A. THOMÄLEN. Translated by G. W. O. HOWE, M.Sc. Second Edition. viii + 464 pages. Royal 8vo., 15s. net (inland postage 6d.).

Alternating Currents. A Text-Book for
Students of Engineering. By C. G. LAMB, M.A., B.Sc., A.M.I.E.E., Clare College, Cambridge; Associate of the City and Guilds of London Institute. 333 pages. 10s. 6d. net (inland postage 5d.).

Electric and Magnetic Circuits. By ELLIS H.
CRAPPER, M.I.E.E., Head of the Electrical Engineering Department in the University College, Sheffield. viii + 380 pages. Demy 8vo., 10s. 6d. net (inland postage 5d.).

Applied Electricity. A Text-Book of Electrical
Engineering for "Second Year" Students. By J. PALEY YORKE, Head of the Physics and Electrical Engineering Department at the London County Council School of Engineering and Navigation, Poplar. Second Edition. xii + 420 pages. Cloth, 7s. 6d. (inland postage 4d.).

Exercises in Electrical Engineering. By
T. MATHER, F.R.S., M.I.E.E., Professor of Electrical Engineering; and G. W. O. HOWE, M.Sc., M.I.E.E., Assistant Professor of Electrical Engineering, City and Guilds (Engineering) College, South Kensington. viii + 72 pages. 1s. 6d. net.

Physical Chemistry: its Bearing on Biology
and Medicine. By J. C. PHILIP, M.A., Ph.D., B.Sc., Assistant Professor of Chemistry in the Imperial College of Science and Technology. Illustrated. 7s. 6d. net.

Lectures on Theoretical and Physical Chemistry.
By Dr. J. H. VAN 'T HOFF, Professor of Chemistry at the University of Berlin. Translated by R. A. LEHFELDT, D.Sc.
Part I. CHEMICAL DYNAMICS. 12s. net.
Part II. CHEMICAL STATICS. 8s. 6d. net.
Part III. RELATIONS BETWEEN PROPERTIES AND COMPOSITION. 7s. 9d. net.

A Text-Book of Physical Chemistry. By R. A.
LEHFELDT, D.Sc., Professor of Physics at the Transvaal University College, Johannesburg. xii + 308 pages. Crown 8vo., 7s. 6d. (inland postage 4d.).

Organic Chemistry for Advanced Students.
By JULIUS B. COHEN, Ph.D., B.Sc., Professor of Organic Chemistry in the University of Leeds, and Associate of Owens College, Manchester. viii + 632 pages. Demy 8vo., 21s. net (inland postage 6d.).

The Chemistry of the Diazo-Compounds. By
JOHN CANNELL CAIN, D.Sc. (Manchester and Tübingen), Editor of the Publications of the Chemical Society. 176 pages. Demy 8vo., 10s. 6d. net (inland postage 4d.).

The Chemical Synthesis of Vital Products and
the Inter-relations between Organic Compounds. By RAPHAEL MELDOLA, F.R.S., V.P.C.S., F.I.C., etc.; Professor of Chemistry in the City and Guilds of London Technical College, Finsbury. Vol. I., xvi + 338 pages. Super royal 8vo., 21s. net (inland postage 5d.).

Organic Analysis: Qualitative and Quantitative.
By H. T. CLARKE, B.Sc., A.I.C., Lecturer in Stereo-Chemistry in University College, London. With Introduction by Professor J. NORMAN COLLIE, Ph.D., LL.D., F.R.S. viii + 264 pages. Crown 8vo., 5s. net.

Elements of Inorganic Chemistry. By the late
W. A. SHENSTONE, F.R.S., Lecturer on Chemistry at Clifton College. New Edition (Enlarged and Revised). xii + 554 pages. Crown 8vo., 4s. 6d.

A Course of Practical Chemistry. Being a
Revised Edition of "A Laboratory Companion for Use with Shenstone's 'Inorganic Chemistry.'" By the late W. A. SHENSTONE, F.R.S. xii + 136 pages. Crown 8vo., cloth, 1s. 6d.

Inorganic Chemistry. Covering the Syllabus
of the London Matriculation Examination. By W. M. HOOTON, M.A., M.Sc., Chief Chemistry Master at Repton School. Crown 8vo., 3s. 6d.

Outlines of Inorganic Chemistry. With special reference to its Historical Development. By E. B. LUDLAM, D.Sc., Head of Chemical Department, Clifton College. With Introductory Note by Professor Sir W. RAMSAY, K.C.B., F.R.S. Crown 8vo., 4s. 6d.

Outlines of Experimental Chemistry. By E. B. LUDLAM, D.Sc., and H. PRESTON. Demy 8vo., 2s.

A History of Chemistry. By Dr. HUGO BAUER, Royal Technical Institute, Stuttgart. Translated by R. V. STANFORD, B.Sc. (Lond.). Crown 8vo., 3s. 6d. net (inland postage 4d.).

Physical Chemistry for Beginners. By Dr. CH. M. VAN DEVENTER. With a Preface by Dr. VAN 'T HOFF. Translated by R. A. LEHFELDT, D.Sc. xvi+146 pages, with Diagrams and Tables. Crown 8vo., cloth, 2s. 6d.

Experimental Researches with the Electric Furnace. By HENRI MOISSAN. Translated by A. T. DE MOUILPIED, M.Sc., Ph.D. xii+307 pages. Demy 8vo., 10s. 6d. net (inland postage 4d.).

Electrolytic Preparations. Exercises for use in the Laboratory by Chemists and Electro-Chemists. By Dr. KARL ELBS, Professor of Organic and Physical Chemistry at the University of Giessen. Translated by R. S. HUTTON, M.Sc. xii+100 pages. Demy 8vo., 4s. 6d. net (inland postage 4d.).

Introduction to Metallurgical Chemistry for Technical Students. By J. H. STANSBIE, B.Sc. (Lond.), F.I.C., Associate of Mason University College, and Lecturer in the Birmingham University Technical School. Second Edition. xii+252 pages. Crown 8vo., 4s. 6d. (inland postage 4d.).

On the Calculation of Thermo-Chemical Constants. By H. STANLEY REDGROVE, B.Sc. (Lond.), F.C.S. iv+102 pages. Demy 8vo., 6s. net (inland postage 4d.).

First Steps in Quantitative Analysis. By J. C. GREGORY, B.Sc., A.I.C. viii+136 pages. Crown 8vo., 2s. 6d.

Manual of Alcoholic Fermentation and the Allied Industries. By CHARLES G. MATTHEWS, F.I.C., F.C.S., etc. xvi+295 pages. Crown 8vo., 7s. 6d. net (inland postage 4d.).

An Introduction to Bacteriological and Enzyme Chemistry. By GILBERT J. FOWLER, D.Sc., Lecturer in Bacteriological Chemistry in the Victoria University of Manchester. Illustrated. Crown 8vo., 7s. 6d. net.

ARNOLD'S GEOLOGICAL SERIES.

General Editor: DR. J. E. MARR, F.R.S.

THE economic aspect of geology is yearly receiving more attention, and the books of this series are designed in the first place for students of economic geology. They will, however, also be found of great use to all who are concerned with the practical applications of the science, whether as surveyor, mining expert, or engineer.

The Geology of Coal and Coal-Mining. By WALCOT GIBSON, D.Sc., F.G.S. 352 pages. With Illustrations. 7s. 6d. net (inland postage 4d.).

The Geology of Ore Deposits. By H. H. THOMAS and D. A. MACALISTER, of the Geological Survey of Great Britain. Illustrated. 7s. 6d. net (inland postage 4d.).

The Geology of Building Stones. By J. ALLEN HOWE, B.Sc., Curator of the Museum of Practical Geology. Illustrated. 7s. 6d. net (inland postage 4d.).

The Geology of Water Supply. By H. B. WOODWARD, F.R.S. Illustrated. Crown 8vo., 7s. 6d. net (inland postage 4d.).

A Text-Book of Geology. By P. LAKE, M.A., Royal Geographical Society Lecturer in Regional and Physical Geography at the University of Cambridge; and R. H. RASTALL, M.A., F.G.S., Demonstrator in Geology in the University of Cambridge. Illustrated. Demy 8vo., 16s. net.

The Dressing of Minerals. By HENRY LOUIS, M.A., Professor of Mining and Lecturer on Surveying, Armstrong College, Newcastle-on-Tyne. x + 544 pages. With 416 Illustrations. Royal 8vo., 30s. net.

Traverse Tables. With an Introductory Chapter on Co-ordinate Surveying. By HENRY LOUIS, M.A., and G. W. CAUNT, M.A. Demy 8vo., flexible cloth, rounded corners, 4s. 6d. net (inland postage 3d.).

Mines and Minerals of the British Empire. Being a Description of the Historical, Physical, and Industrial Features of the Principal Centres of Mineral Production in the British Dominions beyond the Seas. By RALPH S. G. STOKES, late Mining Editor, *Rand Daily Mail*, Johannesburg, S.A. xx + 403 pages, 70 Illustrations. Demy 8vo., 15s. net (inland postage 5d.).

Geological and Topographical Maps: their Uses for the Geologist and Civil Engineer. By A. R. DWERRYHOUSE, D.Sc., F.G.S., Lecturer in Geology at the Queen's University, Belfast.
[*In the Press.*

Modern Methods of Water Purification. By
JOHN DON, A.M.Inst.Mech.E., and JOHN CHISHOLM, A.M.Inst.Mech.E. xvi + 368 pages. 96 Illustrations. Demy 8vo., 15s. net.

Practical Photo-micrography. By J. EDWIN
BARNARD, F.R.M.S., Lecturer in Microscopy, King's College, London. Illustrated. Demy 8vo., 15s. net.

The Chemistry and Testing of Cement. By
C. H. DESCH, D.Sc., Ph.D., Lecturer in Metallurgical Chemistry in the University of Glasgow. Illustrated. 276 pages. Demy 8vo., 10s. 6d. net.

The Chemistry of Breadmaking. By J. GRANT,
M.Sc., Head of the Fermentation Industries Department at the School of Technology, Manchester. [*In the Press.*

Wood. A Manual of the Natural History and
Industrial Applications of the Timbers of Commerce. By G. S. BOULGER, F.G.S., A.S.I., Professor of Botany and Lecturer on Forestry in the City of London College. Second Edition. xi + 348 pages, with 48 Plates and other Illustrations. Demy 8vo., 12s. 6d. net (inland postage 5d.).

A Class Book of Botany. By G. P. MUDGE,
A.R.C.Sc., and A. J. MASLEN, F.L.S. With over 200 Illustrations. Crown 8vo., 7s. 6d.

Elementary Botany. By E. DRABBLE, D.Sc.,
Lecturer on Botany at the Northern Polytechnic Institute. 234 pages, with 76 Illustrations. Crown 8vo., cloth, 2s. 6d.

An Experimental Course of Chemistry for
Agricultural Students. By T. S. DYMOND, F.I.C., lately Principal Lecturer in the Agricultural Department, County Technical Laboratories, Chelmsford. 192 pages. Crown 8vo., 2s. 6d.

The Development of British Forestry. By
A. C. FORBES, F.H.A.S., Chief Forestry Inspector to the Department of Agriculture for Ireland. Author of "English Estate Forestry," etc. Illustrated. Demy 8vo., cloth, 10s. 6d. net.

English Estate Forestry. By A. C. FORBES,
F.H.A.S. x + 332 pages, Illustrated. Demy 8vo., 12s. 6d. net (inland postage 5d.).

Astronomical Discovery. By HERBERT HALL
TURNER, D.Sc., F.R.S., Savilian Professor of Astronomy in the University of Oxford. xii + 225 pages, with 15 Plates. Demy 8vo., cloth, 10s. 6d. net (inland postage 5d.).

The Evolution Theory. By Dr. AUGUST WEISMANN, Professor of Zoology in the University of Freiburg in Breisgau. Translated, with the Author's co-operation, by J. ARTHUR THOMSON, Regius Professor of Natural History in the University of Aberdeen; and MARGARET THOMSON. Two vols., xvi+416 and viii+396 pages, with over 130 Illustrations. Royal 8vo., cloth, 32s. net.

The Chances of Death and Other Studies in Evolution. By KARL PEARSON, M.A., F.R.S., Professor of Applied Mathematics in University College, London. 2 vols., xii+388 and 460 pages, with Illustrations. Demy 8vo., 25s. net (inland postage 6d.).

Hereditary Characters. By CHARLES WALKER, M.Sc., M.R.C.S., Director of Research in the Glasgow Cancer Hospital. Demy 8vo., 8s. 6d. net.

The Life of the Salmon. With reference more especially to the Fish in Scotland. By W. L. CALDERWOOD, F.R.S.E., Inspector of Salmon Fisheries for Scotland. Illustrated. 7s. 6d. net.

A Text-Book of Zoology. By G. P. MUDGE, A.R.C.Sc. (Lond.), Lecturer on Botany and Zoology at the London School of Medicine for Women, and Demonstrator on Biology at the London Hospital Medical College. Illustrated. Crown 8vo., 7s. 6d.

House, Garden, and Field. A Collection of Short Nature Studies By L. C. MIALL, F.R.S., late Professor of Biology in the University of Leeds. viii+316 pages. Crown 8vo., 6s. (inland postage 4d.).

Animal Behaviour. By C. LLOYD MORGAN, LL.D., F.R.S., Professor of Psychology in the University of Bristol. viii+344 pages. Second Edition. 7s. 6d. net (inland postage 5d.).

BY THE SAME AUTHOR.

Psychology for Teachers. New Edition, entirely rewritten. xii+308 pages. Crown 8vo., cloth, 4s. 6d.

An Introduction to Child-Study. By W. B. DRUMMOND, M.B., C.M., F.R.C.P.E., Medical Officer and Lecturer on Hygiene to the Edinburgh Provincial Committee for the Training of Teachers. 348 pages. Crown 8vo., 6s. net (inland postage 4d.).

BY THE SAME AUTHOR.

Elementary Physiology for Teachers and Others. 206 pages. Crown 8vo., 2s. 6d.

The Child's Mind: its Growth and Training. By W. E. URWICK, M.A. Crown 8vo., cloth, 4s. 6d. net.

LONDON: EDWARD ARNOLD, 41 & 43 MADDOX STREET, W.